Urban Surveying
and Mapping

Teodor J. Blachut
Adam Chrzanowski
Jouko H. Saastamoinen

Urban Surveying and Mapping

Springer-Verlag

New York   Heidelberg   Berlin

Teodor J. Blachut
National Research Council of Canada
Ottawa, Canada K1A 0R6

Adam Chrzanowski
University of New Brunswick
Fredericton, New Brunswick, Canada E3B 5A3

Jouko H. Saastamoinen
National Research Council of Canada
Ottawa, Canada K1A 0R6

With 153 figures and 2 color plates.

Library of Congress Cataloging in Publication Data

Blachut, T. J.
  Urban surveying and mapping.

  Bibliography: p.
  Includes index.
  1. Surveying.   2. Cartography.   I. Chrzanowski, Adam.
II. Saastamoinen, Jouko H.   III. Title.
TA549.B57   526.9′09173′2   78-31768

Printed in the United States of America.

9 8 7 6 5 4 3 2 1

ISBN 0-387-90344-5 Springer-Verlag   New York
ISBN 3-540-90344-5 Springer-Verlag   Berlin   Heidelberg

Preface

The idea of writing a textbook on urban surveying and mapping originated with the Commission on Cartography of the Pan American Institute of Geography and History (PAIGH) because of the urgent need for planned and integrated surveying and mapping in urban communities of the American Hemisphere. It is obvious, however, that, with the exception of some European countries, the same situation exists in most cities of the world. The undersigned was asked to undertake the task.

The task was not simple. The only available comprehensive text in the field is *Geodezja Miejska*[1], which was published recently in Poland and reached the authors only after most of the present text was written. It is tailored to a very specific market and different requirements. Although it is an impressive book, it differs vastly from our own approach. Other reference texts are fragmentary or obsolete.

During the last two decades, revolutionary changes have occurred in surveying and mapping technology which have had a profound effect on actual procedures. In addition, the traditional concepts of urban surveying and mapping are undergoing rapid evolution. It is recognized that administration and planning require a great variety of continuously updated information which must be correlated with the actual physical fabric of the community, as determined by surveying and mapping. Modern urban surveying and mapping is therefore the foundation of the broad and dynamic information system that is indispensable in any rational municipal effort.

It was decided that the text should provide, above all, the general guidelines needed to establish uniform standards in this grossly neglected field. This

---

[1] Bramorski, K., Gomoliszewski, J., and Lipinski, M., *Geodezja Miejska*. Państwowe Przedsiębiorstwo Wydawnictw Kartograficznych, Warsaw, 1973.

approach was supported by the fact that, in most countries, the complex task of urban surveys is in the hands of personnel who lack formal education in geodetic disciplines.

This manual is not a conventional textbook on surveying. Therefore, no attempt has been made to explain basic theories, and descriptions of construction and operating techniques of instruments are limited to procedures and instrument features particularly suitable for urban work. Examples of various operations and maps have also been included.

Another important decision concerned the selection and recommendation of techniques and procedures. A manual of this scope and character cannot be simply a collection of formulas. Since it is designed to assist surveying engineers throughout the world who are not sufficiently familiar with city work, it was necessary for the authors to give preference to certain procedures. Because of the lack of more comprehensive studies in the field of urban surveying and mapping, however, the authors would be hard-pressed to demonstrate convincingly the overall superiority of certain methods. Also, some techniques were developed too recently to have been fully evaluated. Finally, the ultimate choice of a procedure or technique cannot be made solely on the strength of its superior technical characteristics, since other important factors imposed by local circumstances, such as financial restrictions, must be taken into account. These factors might even preclude use of the most attractive technique.

Preparation of the book was possible thanks to modest but essential financial support given by PAIGH and by the federal Ministry of State for Urban Affairs in Canada. The Canadian support was used for studying, by an initially much larger group of authors, some problems and solutions in the field of urban surveying and mapping in various Canadian, American, and European cities. A part of the PAIGH funds was used for the actual preparation of the text (illustrations and technical editing), but most of the funds were employed for the Spanish translation and the Spanish publication of the book.

Most helpful in reviewing the manuscript were Dr. V. Kratky and Professor R. Sanchez. Mr. G. H. Schut contributed an essential part of the text on analytical aerial triangulation. The chapter on ground surveying is, in significant part, an adaptation of a much larger manuscript on the same subject written by Dr. H. Ziemann. Mr. R. A. Smith provided the text of Chapter 10 on surveying and mapping data banks. Very tedious linguistic and technical editing was taken care of by Mr. A. Richens and Mrs. M. Giroux. Mr. D. Honegger drew the figures. To these persons the authors wish to express their sincere appreciation and thanks.

Although the manual will not, by itself, solve the difficulties that beset the surveying and mapping of our cities, it may mark the beginning of a more coordinated effort, and as the ancient Greek proverb says, "The beginning is half of the whole."

Ottawa, Canada   T. J. BLACHUT
March, 1979

Contents

Chapter 1

Organizational Schemes in Urban Surveying and Mapping   1

Background Considerations   1
A Brief Review of Existing Organizational Schemes in Urban
    Surveying and Mapping   2
Further Comments on Structural and Functional
    Characteristics of City Survey Work   5
The Functional Relationship Between the City Survey Office and
    Cadastral Operations   8
Suggested Organizational Models of City Survey Offices   9
References   11

Chapter 2

Map Projection Systems for Urban Areas   12

Geodetic Map Projections   12
Formulas for the Reference Ellipsoid   16
Transverse Mercator Grid Coordinates   22
Grid Convergence and Scale Factor   27
Transverse Mercator Projection Corrections   32
Establishment of Local Plane Coordinate Grid   38
References   41
Additional Readings   41

Chapter 3

Horizontal Control Surveys   42

Introduction   42
Density and Accuracy Requirements of Horizontal Control
    Networks   47

Network Design   50
Monumentation of Control Points   69
Electromagnetic Distance Measurements (EDM)   77
Measurement of Horizontal Angles   93
Reduction of Observations   110
Net Adjustment and Computation of Coordinates   121
Maintenance and Record-Keeping   134
New Developments in Control Surveys   135
References   136
Additional Readings   137

Chapter 4

Vertical Control   139

General Characteristics of Vertical Control Nets in Urban Areas   139
Accuracy Specifications   140
Design of Vertical Control   142
Monumentation   144
Instruments and Field Procedures   146
Adjustment and Computations of Heights   149
Trigonometric Determination of Heights   157
Record-Keeping and Maintenance   159
References   159

Chapter 5

Ground Surveying   161

Introductory Remarks   161
Brief Review of Surveying Methods   162
Measurement of Distances   174
Measurement of Angles   185
Surveying Procedures   191
References   219
Additional Readings   219

Chapter 6

Utility Surveys   221

Introduction   221
Surveying of Utilities   222
Cadastre of Utilities   226
References   233
Additional Readings   233

Chapter 7

Remarks on Cadastral Surveys   235

Introduction   235
Basic Functions and Characteristics of an Urban Cadastre   236
Field Work   238
Graphical, Numerical, and Computational Cadastre   239
Accuracy Considerations   240
Cadastral Maps   243
References   245
Additional Readings   245

Chapter 8

Use of Photogrammetry in Urban Areas   246

Introduction   246
Notions Concerning Aerial Photography   247
Planning of Photographic Missions   258
Photogrammetric Determination of Supplementary Control
    Points: Aerial Triangulation   266
Pictorial, Graphical, and Numerical Presentation of
    Terrain Contents   283
Photogrammetric Cadastre in Urban Areas   306
Special Applications of Photogrammetric Methods in Urban
    Areas   314
References   327
Additional Readings   329

Chapter 9

City Maps   330

Introduction   330
Base Map   331
Derived City Maps   345
1 : 25000, 1 : 50000, 1 : 100000 and Smaller-Scale
    Topographical Maps   350
Maintenance of City Maps   350
Thematic Maps   351
Computer-Supported Mapping Systems   353
References   358

Chapter 10

Principles of Surveying and Mapping Data Banks
in Cities by R. A. Smith   359

Index   365

Chapter 1

Organizational Schemes in Urban Surveying and Mapping

BACKGROUND CONSIDERATIONS

The need is urgent for an organized, systematic surveying and mapping program of all land and water bodies, and particularly cities and towns, with their ever-growing complexity, heavy concentration of population, and resultant problems. Most of a country's population (in some countries, over 80 % of the total population) live in cities and towns, where decisions on the social, economic, cultural, and political future of the countries are made. Most children spend the formative years of childhood and adolescence in cities. With the rapidly increasing world population, which apparently will double within the next 30 years, the situation in many cities will deteriorate further. Worst of all, if the present trend of negligence continues, yet unbuilt cities, especially those of the emerging countries, will develop into monstrous slums.

To correct this situation, we urgently need comprehensive planning, supported by legislation that will ensure the rudimentary requirements for sound development. However, planning, its implementation, and the monitoring of the resultant physical changes require maps and other information provided by surveying. Without reliable knowledge of the physical structure, terrain, and ambient conditions of the city, as supplied by surveying, planning is impossible. Moreover, if engineering projects are not supported by reliable surveying data, losses due to such factors as delays and errors quickly amount to a significant percentage of the total cost of the project. On the other hand, the cost of surveying and mapping usually represents a negligible fraction of the overall cost of a project. Therefore, administrative authorities should secure complete surveying data as the first step in meeting their civic responsibility. In reality, however, some of the largest metropolises do not have adequate maps, at useful scales, with information on general topography, individual properties, buildings, other structures, and extremely vital underground, surface, and overhead utilities.

In spite of that, the average citizen of an urban community is paying handsomely, for the most part indirectly, for frequent, uncorrelated surveying operations. Every engineering project involves surveying, but if this surveying is done for a single purpose and without adequate geodetic support and adherence to generally valid specifications, the results are not recoverable and cannot be integrated into a single uniform system with multipurpose characteristics, and this is also true of legal surveys, cadastral-type operations, and evaluation of properties for taxation purposes. There are hundreds of examples illustrating thoughtless squandering of energy and resources due to a lack of integrated survey systems in urban communities. For example, a study conducted recently for the city of Toronto (population, 2 million) proved that an integration of surveys might save the community about 1 million (American) dollars yearly (1).

Financial loss is only one consequence of the lack of proper organization of surveying. A more serious problem is the confusion and chaos resulting from trying to manage a highly complex city without reliable survey information. An example of this confusion was seen in some large North American cities where explosions of gas pipes occurred; the city authorities responsible were not able to take proper emergency measures because of their ignorance of the location, condition, and other pertinent features of the gas installation.

In spite of the obvious arguments for an organized, technically and economically sound, systematic survey of urban areas, only a few countries can point to real accomplishments in this area. This situation results from general ignorance of the social and economic significance of surveying and mapping. Many authorities, particularly elected public representatives who make decisions in these matters, prefer to support more spectacular projects that may be recognized and acclaimed by the public rather than the development of a valuable surveying system not immediately visible and understood by taxpayers. Lack of competent professional surveying cadres in many countries does not help to improve the situation.

A BRIEF REVIEW OF EXISTING ORGANIZATIONAL SCHEMES IN URBAN SURVEYING AND MAPPING

No effort is made to list all the organizational schemes encountered in various countries and cities; rather we will point out, in Schemes A through E, the main differences between the most common solutions.

Scheme A. The municipality does not have a survey office, but has a public works office that oversees the essential technical services of the city, as for example, water supply and sewers, and may retain some rudimentary planning functions. There are no uniform city maps at the scale of 1:2000 or larger, and fragmented maps at smaller scales are of a very general nature and mostly obsolete. Land properties are shown on diagram-type maps that indicate only the relative location of individual parcels. If a major engineering project is considered, such as the construction of an important transportation artery, higher-level authorities (the regional or central government, for example) usually take care of the project and the respective surveying and mapping requirements.

Scheme B. The municipality assumes a more active role in the planning and development of the city. There is still no survey office, but a public-works office or city planning office may either purchase uniform city maps, usually photogrammetric, at the scale of 1:1000 or 1:2000 or be prepared to provide maps of sections of the city as the situation requires. The "cadastral" operation is completely separated from city authority, as in Scheme A, but the municipality may have more precise data on individual real property, including buildings.

Schemes A and B are most frequently encountered in cities of up to a few hundred thousand inhabitants.

Scheme C. Scheme C can be encountered in the larger cities of the United States and Canada. There is a city survey office; however, it does not produce and maintain the maps needed by the city but is the custodian of maps occasionally ordered. It provides surveying services requested by other technical departments of the municipal administration. Also, the survey office may try to maintain and extend horizontal and vertical control nets over the municipal territory for various surveying and mapping projects.

The existence of this city "survey" office does not preclude parallel, and even extensive and technically sound, surveying activity by other departments, usually at a higher administration level—the regional, provincial, or federal government. Provincial or state highway departments frequently carry out their own surveying and mapping in urban areas for the planning and development of highways that cross cities or municipal territory. Any effort to coordinate the various surveying activities within a municipal area depends very much on the professional competence of those in charge of city survey and of the external survey departments involved.

Cadastral survey is separate from the city's survey office, but for administrative purposes, this office may use legal surveying information to produce uniform, large-scale maps showing boundaries of individual parcels. In recent years, orthophoto maps have been used for this purpose in some cities. Orthophoto maps, a relatively inexpensive product, contain complete planimetric information of the terrain, although this information remains uninterpreted. These maps can also be produced very quickly in an automatic mode by using the vertical information of the terrain.

Scheme D. The city survey office is responsible for carrying out the surveying and mapping required by all branches of city administration, with the exception of the cadastral operation. The city survey office therefore initiates the necessary surveying and mapping work. In addition, it may produce numerous thematic maps using its own information and data from other information files. On the other hand, it may provide only the basic geometric framework for the production of thematic maps required by other departments, such as the planning office or the bureau of statistics.

Municipalities operating under this scheme have their own professional surveying and support personnel. They also have their own field survey equipment and occasionally even photogrammetric plotters. In some European countries, e.g., West Germany (2), cities of more than 300 000 inhabitants often have their own map-printing facilities.

The size of the surveying office and the number of persons employed depend primarily upon the scope of the surveying and mapping for which the office is responsible and the extent to which private surveyors and mapping companies are used.

In countries with more highly developed surveying standards, the city survey office and the cadastral office usually maintain a close working relationship. Survey operations of both institutions are based on a common control net, and the surveying data produced by both offices are exchanged. In this scheme, the city survey office is usually functionally connected with several city departments, such as planning, public works, building control, taxation, police, and fire protection. For instance, if a building permit is necessary and the building control authority requires that the edifice be constructed strictly according to the approved plan, the city survey office must verify that specific regulations have been met at critical construction phases; city building-control authorities then act accordingly.

Municipalities that adopt this scheme usually have well-integrated surveying and mapping programs and are in possession of complete map series at such scales as 1:1000, 1:5000 and 1:10 000.

Scheme E. This scheme is characterized by city survey offices that also have the prerogatives of cadastral offices. This very logical and efficient solution exists in Switzerland but is seldom encountered elsewhere. The city survey office operates very much as in Scheme D but, in addition is responsible for cadastral work and therefore generates and is the custodian of cadastral documents within the city area. The office does not have to undertake all the field work, but it exercises complete control, guided by state law, over the form and quality of the cadastral work and is responsible only to the higher cadastral authority of the country. This responsibility for cadastral work demands of the head of the survey office and the engineers the appropriate professional license in addition to technical training and knowledge.

By placing cadastral work under the responsibility of the city survey office, a complete integration of the surveying and mapping effort throughout urban areas is attained. Also, duplication caused by two independent offices exercising separate authority over similar surveying operations is eliminated.

FURTHER COMMENTS ON STRUCTURAL AND FUNCTIONAL CHARACTERISTICS OF CITY SURVEY WORK

In the study of structural and functional characteristics of city survey work, special attention was given to North American, South American, and European countries. Certain observations were made, and conclusions that should be of interest to those involved in city survey work were drawn.

A rather disconcerting general impression is that only a few countries make a serious and coordinated effort to make surveying and mapping facilities accessible to city survey offices or to provide survey offices with their own facilities. There is also a conspicuous lack of reference publications in this area.

Some countries, particularly Eastern European countries with a nationwide program backed by sufficient and expert technical cadres, may not always have access to the most advanced technology, and one may be suspicious of the general efficiency claimed in the field, because of cumbersome bureaucratic practices. These countries, however, do have well-thought-out, very detailed technical specifications covering all phases of surveying and mapping operations, including interaction with other departments and services; however, the massive specifications and rules leave the impression of being too detailed and unwieldy. Also, the obsessive secrecy in the field of surveying and mapping found in these countries makes their survey effort most difficult to assess properly. There is, however, little doubt about their high technical competence in this area.

In Central European countries, which probably have the longest tradition of integrated city surveys, a different situation is encountered. Because of the administrative independence of municipalities, or the provinces and states to which the municipalities belong, there is a certain fragmentation and variety in the organizational form and function and scope of the surveying work carried out by cities. Nevertheless, many similarities and uniform tendencies, particularly in technical aspects, are found, which is particularly interesting since these similarities occur in spite of the lack of rigid central regimentation. These tendencies are evident in Table 1-1, which was drawn from similar tables of West German cities in the study of Schriever (2). The table, arranged according to the size of the population of the cities, primarily contains data on the technical characteristics of the surveying and mapping work carried out by the cities as well as the technical characteristics of facilities provided by municipalities for the use of city survey offices. As Schriever observed, it is most interesting to find that, in West Germany, cities with a population of 300 000 usually require complete reproduction and printing facilities to respond adequately to demands. West Germany, which has a population density of about 230 persons/km^2 as well as advanced industrialization and high living standards, has requirements in the field of city surveying and mapping that are not directly applicable to cities in other countries. Nevertheless, the scope of city surveying and mapping that is encountered in West Germany and other Central European

Table 1-1. Basic Technical Characteristics of City Survey Offices in West Germany

City	Population (in thousands)	Area (km²)	Base maps 1:500	Base maps 1:1000	Cadastre	General planimetry	Marked floor number on buildings	Elevation contours or spot elevations
Berlin	2142	481		×	×		×	×
Hamburg	1862	747		×	×			×
Munich	1261	311		×	×			
Köln	854	251	×	×	×		×	
Essen	702	189	×	×	×	×	×	×
Düsseldorf	686	153		×	×	×	×	×
Frankfurt	661	194	×	×	×	×		×
Dortmund	646	278	×	×	×	×	part	part
Stuttgart	615	207	×		×	×		×
Bremen	605	324	Cad. map		1:250–1:2000			
Hannover	524	135		×	×	×	×	×
Duisburg	465	143	×	×	×		×	part
Nürnberg	447	137	×	×	×	×	×	×
Wuppertal	412	149	×	×	×	×	×	×
Gelsenkirchen	355	104		×	×	×	×	
Bochum	347	121	×	×	×	×		×
Mannheim	324	145	×	×	×			
Kiel	269	82	×	×	×	×	×	×
Wiesbaden	259	164	×		×	planned		
Oberhausen	251	77		×	×			
Karlsruhe	254	117	×	×	×			
Lübeck	242	300		×	×			
Braunschweig	227	77	×	×	×	×		×
Krefeld	225	113	×		×	planned		
Kassel	212	105		×	×	×		×
Augsburg	211	86	×	×	×	×	×	×
Hagen	200	88	×	×	×			
Münster	203	74	×	×	×	×	×	part
Mühlheim	190	88		×	×	×	×	
Aachen	176	59	×		×	×		
Bielfield	169	48	×	×	×			
Bonn	138	31	×	×	×	×		
Darmstadt	140	117	×		×	×	×	×
Recklinghausen	126	66	×	×	×	×	×	
Herne	102	30	×	×	×	×		

Table 1-1. (*Continued*)

Use of photogrammetry		Derived maps			Use of photogrammetry		Reproduction facilities				Duplication technique				Number of colors used	
											Base map		Derived		Maps	
Original mapping	Updating	1:2000	1:2500	1:5000	Original mapping	Updating	Reproduction camera	Printing Plate	Offset print	Screen print	Printing	Copying	Printing	Copying	Base	Derived
part	part			×	part	part	×	×			×		×		1	1
		×	×			part	×	×	×		×	×	×		1	4
	part		×	×		part	×	×			×		×		1	1
	part		×	×		part	×	×	×.			×		×	1	1
×	×		×	×	part	part	×	×	×		×		×		2	5
			×	×			×	×				×		×	1	1
	part	×		×		part	×	×	×		×		×		2	4
	×			×	part	part	×	×	×		×		×		1	2
part	part		×	×			×	×	×			×	×		1	5
			×	×		part	×	×	×			×		×	1	1
×	×			×		part	×	×	×		×		×		1	2
	part	×		×			×	×				×	×		1	1
				×		part	×	×	×			×	×		1	4
		×		×		part	×	×	×			×	×		1	2
part	part	×		×		part	×	×	×		×		×		2	1
part	part	×		×		part	×	×	×		×		×		2	2
	part		×	×			×	×	×			×	×		1	3
part	part		×	×			×	×	×			×	×		1	4
			×	×		part						×	×	×	1	1
		×		×			×	×			×		×		1	2
		×		×			×					×		×	1	1
	part		×	×		part	×					×		×	1	1
	part			×	×	part	×	×	×		×		×		1	1
			×	×		part	×	×		×		×		×	1	1
×	×		×	×		part	×					×		×	1	1
			×									×		×	1	1
		×		×	×	×	×					×	×	×	1	1
		×		×			×	×		×		×	×	×	1	1
			×	×			×	×	×			×	×	×	1	1
		×		×		part						×		×	1	1
				×								×		×	1	1
	part		×	×		part	×				×		×	×	1	1
	part	×	×	×		part	×	×			×		×		1	2
			×	×			×	×	×		×		×	×	1	1
			×	×		part	×	×		×		×		×	1	1

countries is an example that should be carefully studied by less advanced countries.

The need for an adequately equipped city survey office is further supported by rapid technological changes and developments in the surveying and mapping field. Attention should be drawn to the fact that the growing use of photogrammetric and electromagnetic surveying and mapping methods, together with subsequent numerical processing and automatic drafting or display of data, is part of an operational system that requires a certain organizational integration under one roof to achieve optimum efficiency (3).

The fact that most of the world's cities do not have properly organized survey offices must not be taken as proof that the cities can operate adequately without these offices. Rather, it is another example of lack of logic in the management of important human affairs, caused in part by the ignorance of authorities and in part by circumstance. Most urban municipalities have serious financial problems. Lack of facilities such as sewers, water, paved roads, hospitals, and schools is evident and painfully felt by the citizens. The pressure for financial support for these necessities is therefore understandable. If delays occur in providing basic coordinated surveying and mapping data, however, the community will have to pay dearly for blunders and chaos in administration and planning.

THE FUNCTIONAL RELATIONSHIP BETWEEN THE CITY SURVEY OFFICE AND CADASTRAL OPERATIONS

Whatever the characteristics of the cadastral system used in the country, its basic purpose is to provide reliable information on land ownership. In most systems the *usage* and *value* of real estate (land and buildings) and data necessary for general management, including taxation purposes, must also be recorded. The only two reliable ways of defining boundaries of land parcels are: (1) by monuments or marks set directly along the boundaries or (2) by measurement of the property boundaries in reference to monumented terrain points (control points), which as a rule are not boundary points.

From the point of view of individual owners, the first method, barring exceptional situations, would suffice if there were no exterior requirements and obligations, such as payment of taxes; landowners usually do not need to know the precise size or exact geographical positions of their properties. From the point of view of an organized urban community, however, the first method of defining land parcels is not satisfactory. The precise geographical location of individual properties is essential for the planning and execution of various engineering projects. This requires a second approach: surveying based on a general control net. This is particularly important in urban areas because of the complex content, specific technical requirements, and high value of real estate. It follows that, when referring to "cadastral surveying," we have in mind a relatively comprehensive and precise operation resulting in, as its final product, a map.

Compared to cadastral surveying, "technical" city surveying and mapping must meet similar requirements, but also have a wider scope. In addition to the knowledge of property boundaries and buildings that is required in a cadastre, a technical survey requires knowledge of the topography of the terrain and the location and dimensions of other details, such as street curbs, manholes, catch basins, vegetation, and surface, underground, and overhead installations. Since the cadastral map, which is continuously kept up to date, provides the best foundation for a technical city map, the need for the integration of both types of surveying activities for reasons of general economy and efficiency is evident.

In practice, the operational integration of cadastral and technical surveying in city areas seldom occurs. The main difficulties are traditional and imaginary, not technical. Since cadastre was initially established to facilitate an equitable tax collection by recording the size and value of individual properties, the responsibility for cadastre was usually assigned to either the department of finance or justice, or both, whereas general surveying and mapping were the responsibility of military or technical departments. Far too often this prevented the cadastre from being properly integrated into the general survey of the country, particularly in municipalities where cadastral offices continued to operate within their narrow confines.

As is proven in countries such as Switzerland, however, the function of the cadastre, or at least its surveying part, can be successfully integrated within the work of the city survey office. Obviously, in matters concerning cadastre, the city office must follow the procedures and specifications laid down by the national cadastre system. This does not present difficulties, since the methods of surveying are identical, although the application differs. On the contrary, full coordination and meaningful integration of both operations can be achieved simply by making the same technical personnel responsible for technical and cadastral surveys.

Another important argument for the integration of city survey offices with cadastral offices is the basic requirement of cadastre, i.e., that any changes in real properties, such as sale, subdivision, or erection of a building, must be approved and recorded immediately since only then can the alteration be permitted or acquire the power of a legal act. This information is most essential to, and also part of, city administration. There is, therefore, no rationale for separating cadastre from technical surveying in cities.

SUGGESTED ORGANIZATIONAL MODELS OF CITY SURVEY OFFICES

It is impossible to suggest a universally acceptable organizational model of a city survey office. Number of personnel, type of equipment, range of responsibility, and other features depend not only upon the size and character of the city but also upon the organizational characteristics of the country and the current practices in the surveying and mapping field. In some countries, the

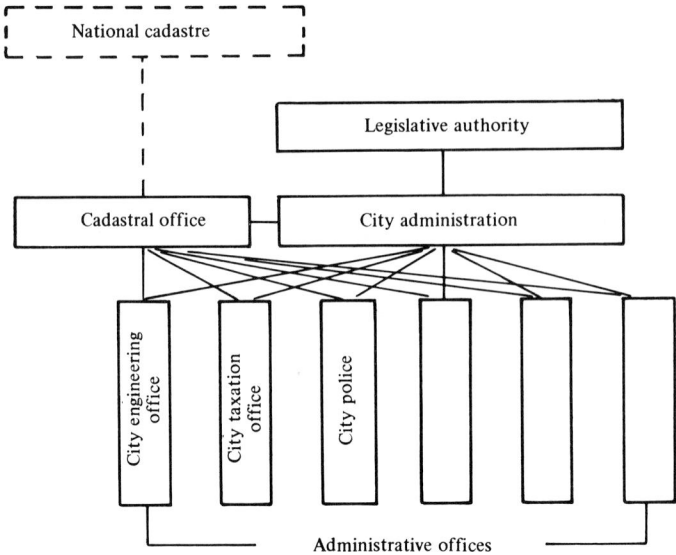

Figure 1-1. Organizational scheme for communities without a survey office proper.

city survey office may contract the bulk of the surveying work, whereas in others, everything may be accomplished by the city survey office itself.

Cadastral surveying is a countrywide operation, with offices spread over all national territory, primarily in cities and towns. In small communities, therefore, the responsibility for integrated cadastral and technical surveying operations can best be assumed by the existing cadastral office, until the establishment of a fully fledged city survey office with a more complete and sophisticated surveying and mapping program can be justified. For these relatively small municipalities, the organizational scheme presented in Figure 1-1 is envisaged.

In this organizational scheme, the city technical office (general planning, building control, water and sewer, etc.) relies on the services of the local cadastral office for surveying and mapping. Since this constitutes a functional integration of two offices belonging to two independent authorities (cadastral service of the country and the municipal authority), a special arrangement must be made, and the mode of cooperation must be well defined.

In larger municipalities, existence of a technically competent survey office is indispensable. A large part of city planning, development, and management is directly based on and supported by the work of the city survey office. This office must have the authority to *initiate projects* in the field of its responsibility. Independent of the mode of execution of various surveying projects (in-house or through contracts), the survey office is the depositer—and custodian of survey data and maps. In addition, the surveying component of the cadastre operation is part of the city survey office. Consequently, the head of the office and the survey engineers must also hold licenses for the execution of cadastral work.

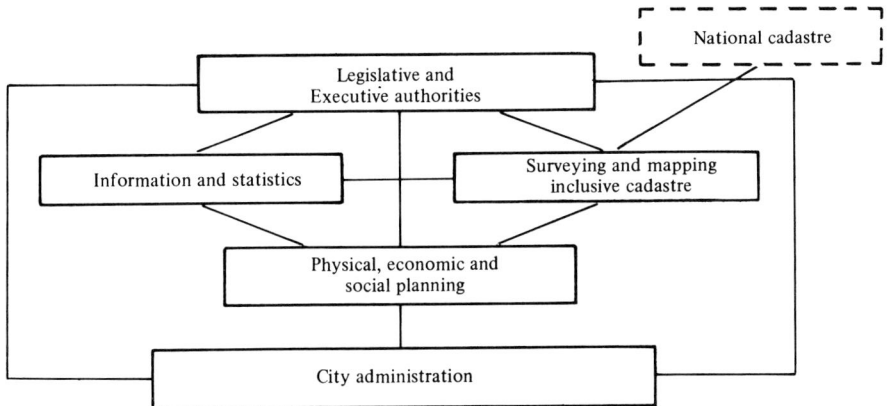

Figure 1-2. Organizational scheme for cities with a survey office proper.

In conclusion, in accordance with the basic function and responsibility of urban surveying and mapping, the city survey office must occupy a primary place in the structural scheme of city administration, as shown in Fig. 1-2.

References

1. Bleaney, J. H. *A Study for Metropolitan Toronto, Borough of North York, Ont.*, Public Works Department, 1973.
2. Schriever, H. Vom Plan zur Karte—Kartographische Aufgaben der Kommunalen Vermessungsämter, *Deutsche Kartographie der Gegenwart*, 1970.
3. Blachut, T. J. The Role of Urban Surveying and Mapping, *Plan Canada*, Town Planning Institute of Canada, 80 King St. W., Toronto 1, Ontario, May 1971.

Chapter 2

Map Projection Systems
for Urban Areas

GEODETIC MAP PROJECTIONS

Introduction

The purpose of surveying and mapping is to determine relative positions of points at or near the surface of the earth, which requires the establishment of a system of reference in three-dimensional space. A natural choice of elevation, or height Z, as one spatial coordinate, leaves a set of two horizontal coordinates X and Y subject to definition. In urban surveying and mapping, the horizontal coordinates are most conveniently referred to a rectangular plane coordinate system, or grid, which is oriented so that the positive directions of the coordinate axes point toward the north and the east. Unfortunately, there is no general agreement as to whether X or Y should be taken in the north direction; *Figure 2-1 shows the convention accepted throughout in this text.*

The curved surface of the earth cannot be presented on a plane surface without distortion, and this distortion increases with the size of the area involved. It is obvious that, although a city lot can be surveyed accurately with complete ignorance of the earth's curvature, a map of a whole continent will require the use of a suitable map projection. Less obvious is the need of a map projection for an urban plane coordinate grid. Usually such surveys are so limited in area that the internal discrepancy caused by ignoring the curvature is unimportant compared to errors in the measurements themselves (Fig. 2-1).

However, there are other compelling reasons why the concept of a "flat earth" is unworkable in urban surveying. A city has a tendency to grow in size, often in an unpredictable manner, and as a result, its coordinate grid is likely to need

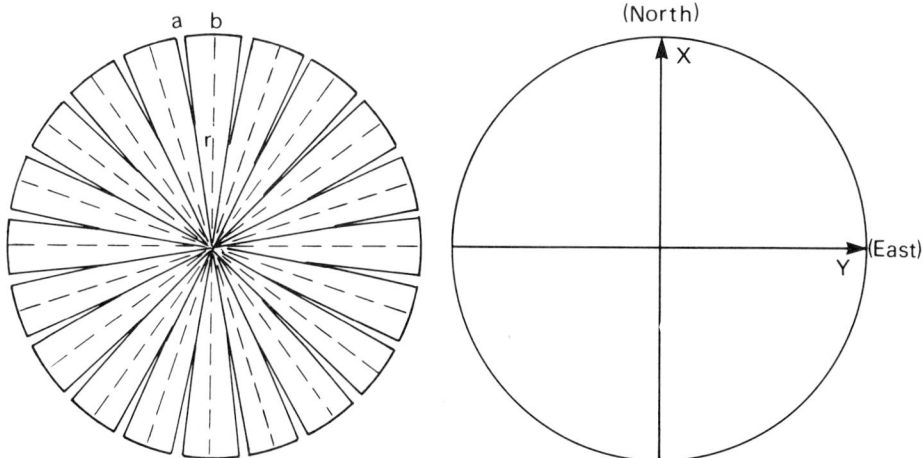

Figure 2-1. Diagram showing a spherical segment of the earth's surface cut open and flattened onto a plane. Gaps (*a*) will cause distortion in plane coordinates *X* and *Y*.

Distance *r* along the earth's surface (km)	Ratio *a/b*
10	1 : 2 400 000
20	1 : 600 000
30	1 : 300 000
50	1 : 100 000
100	1 : 24 000

extension from time to time, beyond the originally intended limits. Random additions to the grid area would result in an uncontrolled accumulation of internal discrepancies. Establishment of the grid upon a map projection is the only satisfactory way to provide for future growth.

The most important argument for a plane coordinate grid built upon a map projection is the *universality of its coordinate values*: Such plane coordinates can be transformed readily into any similar system that may cover the same physical area. This is a decisive advantage since coordinates of local control points are frequently required in a regional system for such purposes as general mapping and highway planning. A large expenditure for urban field surveys can hardly by justified if provision for such coordinate transformation does not exist.

General Criteria for an Urban Map Projection

There are an infinite number of ways to establish a map projection, i.e., a one-to-one correspondence between points on the surface of the earth and points on a plane surface. Clearly, one solution is given by two families of intersecting lines drawn on the plane, one representing the meridians, and the other the parallels of latitude. As in Mercator's projection, such lines can be straight lines intersecting at right angles, concentric circles and straight lines radiating from a common center (as in the ordinary conical and polar stereographic projections), and so on. Of a multitude of various kinds of map projections, however, only a few conformal projections meet the requirements of geodetic surveying.

A map projection is called orthomorphic, or *conformal*, if it retains a similarity in detail to the original. It follows from this definition that the angles between intersecting lines do not change in a conformal projection. Furthermore, if dS denotes the length of a line-element on the projected surface and ds its counterpart on the projection plane, the ratio

$$m = \frac{ds}{dS} \tag{2-1}$$

which is called the *scale factor*, though varying from point to point, is at every point independent of the direction of dS.

Similarity in detail can be achieved only at the expense of larger areas; a familiar example is the exaggerated size of Greenland on a conformal map of North America. Choice of a conformal projection for a particular application implies that the preservation of the angular relationships is considered more important than that of the areas. Geodetic map projections dealt with in this chapter belong to this category.

A geodetic map projection for urban surveys should possess the following general characteristics:

1. The one-to-one correspondence between the projected surface and the projection plane should be expressible in terms of mathematical formulas that permit numerical computation to any predetermined precision.
2. The distortion of angles and distances caused by the projection should be reasonably small and easy to calculate.
3. The projected surface representing the earth should be an ellipsoid of revolution rather than a reference sphere.

With these conditions met, the map projection becomes a powerful mathematical tool for performing rigorous geodetic computations on a plane surface, at the same time retaining its original function in the graphical presentation of maps.

The fundamental theory of geodetic map projections is due to the works of classical mathematicians, notably Lambert, Lagrange, and Gauss; little if anything has been added to it during the past 50 years. However, the traditional ways of handling numerical computations, which until recently required the use of precomputed *grid tables*, have been greatly simplified through electronic computing. For the contemporary user, a geodetic map projection system is best presented in the form of a self-contained collection of formulas necessary for the practical application of the system. Grid tables are about to become obsolete; program libraries tailored to the needs of the individual user will take their place.

The Transverse Mercator Projection and its Variants

Except for surveys of polar regions, which are best treated in stereographic projection, the *Transverse Mercator* projection of the reference ellipsoid onto a plane (also known as the Gauss–Krüger projection) has superseded all other conformal projections as a standard map projection for geodetic purposes. The Transverse Mercator plane is a conformal representation of the reference ellipsoid such that one selected *central meridian* is rectified into a straight-line segment (*NS* in Fig. 2-2). The equator is represented by a straight line perpendicular to the central meridian; the other meridians and parallels of latitude are curved lines of a more complex nature. Rectangular plane coordinates are referred to the central meridian as the *x*-axis and to the equator as the *y*-axis. The scale factor increases both eastward and westward from the central meridian (where it is unity), rapidly becoming excessive and eventually infinite. The effective projection area is therefore restricted to a narrow zone embracing the central meridian. Larger areas can be divided by equally spaced central meridians into the number of zones required. In such zonal grid systems based on a set of fixed central meridians, each zone is usually extended by 30′ beyond its ordinary longitude limits. This provides an overlap of 1° between adjacent zones, within which plane coordinates are expressible in both zones, and provision can be made for the conversion of coordinates from one zone to the other.

Outside the central meridian, the scale factor of the Transverse Mercator projection is always greater than unity, which means that all the grid distances

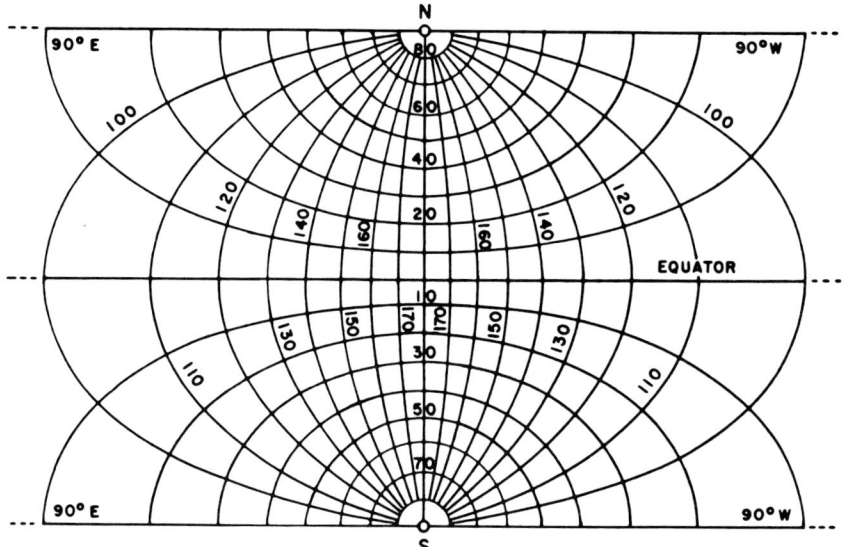

Figure 2-2. Graticule of a hemisphere on the Transverse Mercator projection. Note that the horizontal lines extend to infinity and that only a narrow central zone is relatively free from exaggeration (1).

will be longer than the corresponding true distances on the projected surface. If so desired, the *maximum* distortion of lengths can be made smaller if all the grid lengths are reduced by a constant factor, the size of which depends on the width of the projection zone. Such a reduction has been adopted on the *Universal Transverse Mercator*, or UTM, grid, as well as on other commonly used Transverse Mercator variants (pp. 25–26). It is debatable, however, whether a scale reduction should always be considered a definite improvement over the original Transverse Mercator grid, since it brings about a widespread systematic distortion near the central meridian.

A programmable set of mathematical formulas for the Transverse Mercator projection, which is necessary in its practical application to urban surveying, will be presented here. The various formulas in the text have been designed with the view of maintaining a high formal precision (1 mm) in computed grid coordinates, including UTM coordinates. A comparison of the formulation with that required by any other conformal projection leaves little doubt that the Transverse Mercator, or one of its variants, is the standard projection system best suited for urban surveys. In particular, this projection allows the easiest calculation of scale factors and other projection corrections that are so important in the daily work on the plane coordinate grid.

FORMULAS FOR THE REFERENCE ELLIPSOID

Dimensions

The mathematical figure of the earth used in geodetic computations is an ellipsoid of revolution, called the *reference ellipsoid*, or spheroid, which is generated by a slightly flattened ellipse rotating about its minor axis (Fig. 2-3). Such a surface is fully defined by two parameters that can be selected in several different ways. Some of the most common parametric constants of a reference ellipsoid are as follows.

	International Ellipsoid (Hayford, 1909)	Clarke 1866 Spheroid
Equatorial semiaxis (m)	$a = 6\ 378\ 388.000\ 00^*$	$6\ 378\ 206.400\ 00^*$
Polar semiaxis (m)	$b = a(1 - f) = 6\ 356\ 911.946\ 13$	$6\ 356\ 583.800\ 00^*$
Flattening	$f = (a - b)/a = 1/297.000\ 000^*$	$1/294.978\ 698$
Polar radius of curvature (m)	$c = a^2/b = 6\ 399\ 936.608\ 11$	$6\ 399\ 902.551\ 59$
Minor eccentricity squared	$e'^2 = (a^2 - b^2)/b^2 = 0.006\ 768\ 170\ 197$	$0.006\ 814\ 784\ 946$

* Exact value.

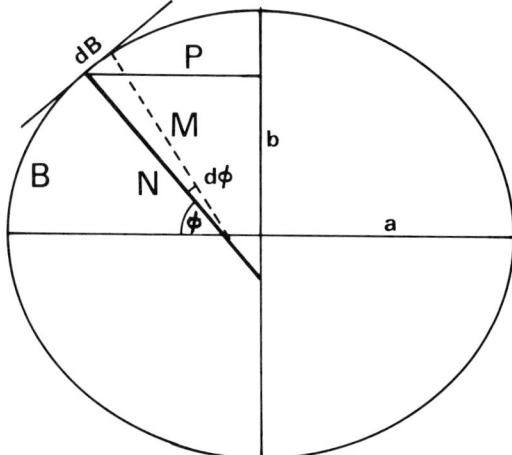

Figure 2-3. The meridian ellipse.

The dimensions of reference ellipsoid to be used in urban surveying are those specified in the official **geodetic datum** for the primary geodetic networks of the country. Currently, the 1927 *North American Datum* specifies the above values of the Clarke 1866 Spheroid in Canada, Mexico, and the United States, whereas the International Ellipsoid is being used in many countries of Europe and South America.

Latitude and Longitude

Positions on the surface of the reference ellipsoid are given in geographical coordinates, which are *geodetic latitude* ϕ and *geodetic longitude* λ. At a given point, geodetic latitude is the angle between the normal to the ellipsoid and the equatorial plane. Longitude is the angle between the meridian plane passing through the point and the plane of the Greenwich standard meridian. Latitudes north of the equator and longitudes east of Greenwich are considered in computations as *positive* quantities, latitudes south of the equator and longitudes west of Greenwich as *negative* quantities.

Some fundamental quantities of the reference ellipsoid, their notations, and basic relationships are listed on page 18. They will be used without further reference.

Length of the Meridional Arc

Rectification of the Transverse Mercator central meridian involves numerical evaluation of the *elliptic integral* (Eq. 2-8) and the *elliptic function* (Eq. 2-9) in terms of series expansions since neither has a solution in closed form.

Functions of geodetic latitude ϕ		Eq. no.
Auxiliary quantity V squared	$V^2 = 1 + e'^2 \cos^2 \phi$	(2-2)
Radius of curvature in the meridian	$M = c/V^3 = c \cdot V^{-3}$	(2-3)
Radius of curvature in the normal section perpendicular to the meridian (length of normal)	$N = c/V = c \cdot V^{-1}$	(2-4)
Radius of curvature in azimuth α	$R_\alpha = c[V + (V^3 - V)\cos^2 \alpha]^{-1}$	(2-5)
Mean radius of curvature	$R = \sqrt{MN} = c \cdot V^{-2}$	(2-6)
Radius of parallel circle	$P = N \cos \phi$	(2-7)
Length of meridional arc from equator to latitude ϕ	$B = \int_0^\phi M \, d\phi$	(2-8)
Latitude corresponding to given length of meridional arc $B = x$	$\phi_1 = \int_0^x M^{-1} \, dB$	(2-9)

Note that the sign convention for latitude implies a negative value for B south of the equator. Auxiliary quantity V is always taken as positive.

Either of the following two methods, the conventional series or the mid-latitude series, may be chosen for programing these important subroutines for electronic computing.

Conventional formulas. A general formula (up to e'^{10}) for computing the length of the meridional arc from equator to any latitude is

$$B = A_0 c\phi - A_1 c \sin \phi \cos \phi (1 + A_2 \sin^2 \phi + A_4 \sin^4 \phi$$
$$+ A_6 \sin^6 \phi + A_8 \sin^8 \phi), \qquad (2\text{-}10a)$$

where

$$A_0 = 1 - \frac{3}{4}e'^2 \left\{ 1 - \frac{15}{16}e'^2 \left[1 - \frac{35}{36}e'^2 \left(1 - \frac{63}{64}e'^2 \left(1 - \frac{99}{100}e'^2 \right) \right) \right] \right\}$$

$$A_1 = \frac{3}{4}e'^2 \left\{ 1 - \frac{25}{16}e'^2 \left[1 - \frac{77}{60}e'^2 \left(1 - \frac{837}{704}e'^2 \left(1 - \frac{2123}{1860}e'^2 \right) \right) \right] \right\}$$

$$A_2 = \frac{5}{8}e'^2 \left[1 - \frac{139}{144}e'^2 \left(1 - \frac{1087}{1112}e'^2 \left(1 - \frac{513\,427}{521\,760}e'^2 \right) \right) \right]$$

$$A_4 = \frac{35}{72}e'^4 \left(1 - \frac{125}{64}e'^2 \left(1 - \frac{221\,069}{150\,000}e'^2 \right) \right)$$

$$A_6 = \frac{105}{256}e'^6 \left(1 - \frac{1179}{400}e'^2 \right)$$

$$A_8 = \frac{231}{640}e'^8.$$

Numerically, for the *Clarke 1866 Spheroid*, the formula is

$$B = 6\,367\,399.689\,17\,\phi$$
$$- 32\,365.186\,93\,\sin\phi\,\cos\phi(1 + 0.004\,231\,4080\,\sin^2\phi$$
$$+ 0.000\,022\,2782\,\sin^4\phi + 0.000\,000\,1272\,\sin^6\phi \qquad (2\text{-}10b)$$
$$+ 0.000\,000\,0008\,\sin^8\phi),$$

and for the *International Ellipsoid*,

$$B = 6\,367\,654.500\,06\,\phi$$
$$- 32\,146.297\,86\,\sin\phi\,\cos\phi\,(1 + 0.004\,202\,6520\,\sin^2\phi$$
$$+ 0.000\,021\,9764\,\sin^4\phi + 0.000\,000\,1246\,\sin^6\phi \qquad (2\text{-}10c)$$
$$+ 0.000\,000\,0008\,\sin^8\phi),$$

where ϕ is expressed in radians and B in meters.

The same formula can be used for computing latitude ϕ_1 that corresponds to given length of meridional arc $B = x$. The procedure is based on successive approximations $\phi_{(1)}, \phi_2, \ldots, \phi_n$,

$$\phi_{(1)} = \frac{x}{A_0 c} \qquad \text{with } B_{(1)} \text{ computed as above,}$$

$$\phi_2 = \phi_{(1)} + \frac{x - B_{(1)}}{A_0 c} \qquad \text{with } B_2 \text{ computed as above,}$$

$$\vdots$$

$$\phi_1 = \phi_n, \text{ when } B_n = x.$$

The trigonometric functions may be computed from exponential function e^ϕ by using the formulas

$$\sin\phi = \frac{e^\phi - e^{-\phi}}{2} - \frac{\phi^3}{3}\left[1 + \frac{\phi^4}{840}\left(1 + \frac{\phi^4}{7920}\left(1 + \frac{\phi^4}{32\,760}\right)\right)\right] \quad (2\text{-}11)$$

$$\cos\phi = \frac{e^\phi + e^{-\phi}}{2} - \phi^2\left\{1 + \frac{\phi^4}{360}\left[1 + \frac{\phi^4}{5040}\left(1 + \frac{\phi^4}{24\,024}\left(1 + \frac{\phi^4}{73\,440}\right)\right)\right]\right\}$$
$$(2\text{-}12)$$

which, in 12-digit calculation, will give at least 10 correct decimals for $-90° \le \phi \le +90°$.

Conversion into radians. One radian is, in seconds of arc,

$$\rho'' = 206\,264''.806\,247.$$

Example 2-1. On the Clarke 1866 Spheroid, find the latitude halfway between the north pole and the equator.

One half of the meridian quadrant is, from Eq. (2-10b),

$$x = \tfrac{1}{2} \times 6\,367\,399.689\,17 \times \frac{90 \times 3600''}{206\,264''.806\,247} = 5\,000\,944.021.$$

First approximation:

$$\phi_{(1)} = \frac{5\,000\,944.021}{6\,367\,399.689} = 0.785\,398\,1634 \qquad (=45°)$$

$$x = 5\,000\,944.021$$

$$\sin \phi_{(1)} = 0.707\,106\,7812 \qquad B_{(1)} = 4\,984\,727.100$$

$$\cos \phi_{(1)} = 0.707\,106\,7812 \qquad x - B_{(1)} = \quad 16\,216.921.$$

Second approximation:

$$\phi_2 = \phi_{(1)} + \frac{16\,216.921}{6\,367\,399.7} = 0.787\,945\,0305$$

$$x = 5\,000\,944.021$$

$$\sin \phi_2 = 0.708\,905\,3930 \qquad B_2 = 5\,000\,944.056$$

$$\cos \phi_2 = 0.705\,303\,5828 \qquad x - B_2 = \quad -0.035.$$

Third (and final) approximation:

$$\phi_3 = \phi_2 + \frac{-0.035}{6\,367\,400} = 0.787\,945\,0250 = 45°\,08'\,45''.3279 \text{ N.}$$

The same computation on the International Ellipsoid would give latitude 45° 08' 41".7467 N.

Midlatitude formulas. Equation (2-10a) is universal in application but, since it involves iteration in the computation of ϕ_1, the following expansions into Taylor's series offer numerical formulas that are more convenient to use in urban surveying.

The length of the meridional arc from equator to latitude ϕ is given by the series

$$B = B_m + B_1(\phi - \phi_m) + B_2(\phi - \phi_m)^2 + B_3(\phi - \phi_m)^3 + B_4(\phi - \phi_m)^4 + \cdots,$$

$$(2\text{-}13a)$$

where B_m denotes the length of the meridional arc to selected midlatitude ϕ_m and the coefficients are

$$B_1 = c(1 + e'^2 \cos^2 \phi_m)^{-3/2} = cV_m^{-3}$$

$$B_2 = \tfrac{3}{2}ce'^2 \sin \phi_m \cos \phi_m V_m^{-5}$$

$$B_3 = \tfrac{1}{2}ce'^2(2 \cos^2 \phi_m - 1)V_m^{-5} + \tfrac{5}{2}ce'^4 \sin^2 \phi_m \cos^2 \phi_m V_m^{-7}$$

$$B_4 = -\tfrac{1}{2}ce'^2 \sin \phi_m \cos \phi_m V_m^{-5} + \tfrac{15}{8}ce'^4 \sin \phi_m \cos \phi_m(2 \cos^2 \phi_m - 1)V_m^{-7}$$

$$+ \tfrac{35}{8}ce'^6 \sin^3 \phi_m \cos^3 \phi_m V_m^{-9}$$

$$\vdots$$

Provided that latitude ϕ is within $\phi_m \pm 2°$, five terms of the series are sufficient for computing B with an error <0.3 mm.

Example 2-2. For the Clarke 1866 Spheroid, $\phi_m = 44°$, and from Eq. (2-10b), $\sin \phi_m = 0.694\,658\,3705$, $\cos \phi_m = 0.719\,339\,8003$, and $B_m = 4\,873\,606.0900$. Substitute into Eq. (2-13a)

$$\phi - \phi_m = \frac{7200''}{\rho''}(\phi)$$

and compute coefficients for the powers of (ϕ).

Clarke 1866 Spheroid, latitudes $42°00'-46°00'N$,

$$B = 4\,873\,606.0900 + 222\,222.2705(\phi) + 39.4834(\phi)^2$$
$$+ 0.0399(\phi)^3 - 0.0160(\phi)^4, \tag{2-13b}$$

where

$$(\phi) = \frac{\phi° - 44°}{2°} = \frac{\phi'' - 158\,400''}{7200''}.$$

Latitude ϕ_1, which corresponds to given length of meridional arc $B = x$, can be similarly expressed by the series

$$\phi_1 = \phi_m + C_1(x - x_m) + C_2(x - x_m)^2 + C_3(x - x_m)^3 + C_4(x - x_m)^4 + \cdots, \tag{2-14a}$$

where ϕ_m is the latitude corresponding to selected midvalue of meridional arc $B = x_m$, and the coefficients are

$$C_1 = c^{-1}(1 + e'^2 \cos^2 \phi_m)^{3/2} = c^{-1}V_m^3$$
$$C_2 = -\tfrac{3}{2}c^{-2}e'^2 \sin \phi_m \cos \phi_m V_m^4$$
$$C_3 = -\tfrac{1}{2}c^{-3}e'^2(2\cos^2 \phi_m - 1)V_m^7 + 2c^{-3}e'^4 \sin^2 \phi_m \cos^2 \phi_m V_m^5$$
$$C_4 = \tfrac{1}{2}c^{-4}e'^2 \sin \phi_m \cos \phi_m V_m^{10} + \tfrac{15}{8}c^{-4}e'^4 \sin \phi_m \cos \phi_m(2\cos^2 \phi_m - 1)V_m^8$$
$$\quad - \tfrac{5}{2}c^{-4}e'^6 \sin^3 \phi_m \cos^3 \phi_m V_m^6$$
$$\vdots$$

The series in Eq. (2-14a) is subject to the same limitations in application as Eq. (2-13a).

Example 2-3. Clarke 1866 Spheroid, $x_m = 4\,900\,000$. From Eq. (2-10b) by iteration,

$$\phi_m = 0.772\,090\,728\,15, \quad \sin \phi_m = 0.697\,634\,6724, \quad \cos \phi_m = 0.716\,453\,6723.$$

Substitute into Eq. (2-14a)

$$x - x_m = 200\,000(x)$$

and compute coefficients for the powers of (x).

Clarke 1866 Spheroid, x: 4 700 000–5 100 000;

$$\phi_1 = 0.772\,090\,728\,15 + 0.031\,414\,593\,66(x) - 0.000\,005\,024\,65(x)^2$$
$$- 0.000\,000\,002\,09(x)^3 + 0.000\,000\,001\,65(x)^4, \qquad (2\text{-}14b)$$

where $(x) = (x - 4\,900\,000)/200\,000$.

Formulas (2-13b) and (2-14b) can be checked against each other by setting $(\phi) = 1$ and computing ϕ_1 ($=46°$).

Precision of Geographical Coordinates

The latitude and longitude of primary geodetic stations are usually listed in the sexagesimal system (degrees, minutes, and seconds of arc) with three or four decimal digits in the seconds. On the surface of the ellipsoid,

$1''$ in latitude equals approximately 31 m
$1''$ in longitude equals approximately 31 cos ϕ m.

This means that latitude and longitude rounded to three decimal digits correspond to a positional precision of ± 15 and ± 15 cos ϕ mm, respectively, on the ellipsoid.

Reference (2) gives a detailed description of the geometry of the ellipsoid; formulas for computing the length of the meridional arc to any desired accuracy are derived in Ref. (3).

TRANSVERSE MERCATOR GRID COORDINATES

Conversion of Geographical Coordinates into Transverse Mercator Coordinates

Geographical coordinates (ϕ, λ) of a geodetic station are transformed into Transverse Mercator coordinates (x, y) by applying the general formulas

$$x - B = a_2 l^2 + a_4 l^4 + a_6 l^6 + \cdots$$
$$y = a_1 l + a_3 l^3 + a_5 l^5 + \cdots, \qquad (2\text{-}15a)$$

where $l = \lambda - \lambda_0$ is the difference in longitude from central meridian λ_0 in radian measure, B is the length of the meridional arc from equator to latitude ϕ, and the coefficients

$$a_1 = P = N \cos \phi = c \left[\left(\frac{1}{\cos \phi} \right)^2 + e'^2 \right]^{-1/2}$$

$$a_2 = \tfrac{1}{2} a_1 \sin \phi$$

$$a_3 = \tfrac{1}{6} a_1 (-1 + 2 \cos^2 \phi + e'^2 \cos^4 \phi)$$

$$a_4 = \tfrac{1}{12} a_2 (-1 + 6 \cos^2 \phi + 9 e'^2 \cos^4 \phi + 4 e'^4 \cos^6 \phi)$$

$$a_5 = \tfrac{1}{120} a_1 [1 - 20 \cos^2 \phi + (24 - 58 e'^2) \cos^4 \phi + 72 e'^2 \cos^6 \phi + \cdots]$$

$$a_6 = \tfrac{1}{360} a_2 (1 - 60 \cos^2 \phi + 120 \cos^4 \phi + \cdots)$$

$$\vdots$$

are functions of latitude ϕ. Provided that longitude λ is within $\lambda_0 \pm 3°30'$, three terms of the series are sufficient for computing x and y accurately from geographical coordinates rounded to four decimal digits in the seconds. The following example illustrates the arrangement of computation.

Example 2-4. Clarke 1866 Spheroid; Central Meridian 75°W

Given: Lat. 45° 53′ 38″.3864 N   Compute: x, y (in meters).

Long. 77° 55′ 03″.8473 W

$l = -2° 55' 03''.8473$

$= -0.050\,924\,0888.$

From (2-10b) or (2-13b), $\sin \phi = 0.718\,053\,3721$, $\cos \phi = 0.695\,988\,0422$, and $B = 5\,084\,085.5903$. Then

$$P = \frac{6\,399\,902.5516}{\sqrt{(1/\cos \phi)^2 + 0.006\,814\,784\,95}} = 4\,446\,921.878,$$

and

$$x = B + \frac{Pl^2}{2} \sin \phi \left[1 + \frac{l^2}{12} (-1 + 6 \cos^2 \phi + 0.061\,33 \cos^4 \phi \right.$$

$$\left. + 0.000\,19 \cos^6 \phi) + \frac{l^4}{360} (1 - 60 \cos^2 \phi + 120 \cos^4 \phi) \right],$$

$$\qquad (2\text{-}15\text{b})$$

$$y = Pl \left[1 + \frac{l^2}{6} (-1 + 2 \cos^2 \phi + 0.006\,8148 \cos^4 \phi) \right.$$

$$\left. + \frac{l^4}{120} (1 - 20 \cos^2 \phi + 23.6047 \cos^4 \phi + 0.4907 \cos^6 \phi) \right].$$

Result: $x = 5\,088\,227.618$ m, $y = -226\,452.508$ m. The obtained result can be checked by inverse computation (Example 2-5).

Conversion of Transverse Mercator Coordinates into Geographical Coordinates

Transverse Mercator coordinates (x, y) of a geodetic station are transformed into geographical coordinates (ϕ, λ) by applying the general formulas

$$\phi = \phi_1 + b_2 y^2 + b_4 y^4 + b_6 y^6 + \cdots$$
$$\lambda = \lambda_0 + b_1 y + b_3 y^3 + b_5 y^5 + \cdots, \tag{2-16a}$$

where λ_0 is the longitude of the central meridian, ϕ_1 is the latitude corresponding to length of meridional arc $B = x$, both in radian measure, and the coefficients

$$b_1 = P_1^{-1} = N_1^{-1} \sec \phi_1 = c^{-1} \left[\left(\frac{1}{\cos \phi_1} \right)^2 + e'^2 \right]^{1/2}$$

$$b_2 = -\tfrac{1}{2}b_1^2 \sin \phi_1 \cos \phi_1 (1 + e'^2 \cos^2 \phi_1)$$

$$b_3 = -\tfrac{1}{6}b_1^3 (2 - \cos^2 \phi_1 + e'^2 \cos^4 \phi_1)$$

$$b_4 = -\tfrac{1}{12}b_1^2 b_2 [3 + (2 - 9e'^2)\cos^2 \phi_1 + 10e'^2 \cos^4 \phi_1 - 4e'^4 \cos^6 \phi_1]$$

$$b_5 = \tfrac{1}{120}b_1^5 [24 - 20 \cos^2 \phi_1 + (1 + 8e'^2)\cos^4 \phi_1 - 2e'^2 \cos^6 \phi_1 + \cdots]$$

$$b_6 = \tfrac{1}{360}b_1^4 b_2 (45 + 16 \cos^4 \phi_1 + \cdots)$$

$$\vdots$$

are functions of latitude ϕ_1. Equation (2-16a) serves as the inverse function of Eq. (2-15a) and has the same limits of application.

Example 2-5. Clarke 1866 Spheroid; Central Meridian 75°W.

$$\text{Given: } x = 5\,088\,227.618 \qquad \text{Compute: } \phi, \lambda.$$
$$y = -226\,452.508$$
$$\lambda_0 = -1.308\,996\,9390.$$

From Eq. (2-10b) by iteration or from Eq. (2-14b),

$$\phi_1 = 0.801\,651\,747\,84$$
$$\sin \phi_1 = 0.718\,505\,8956$$
$$\cos \phi_1 = 0.695\,520\,8682$$

$$b_1 y = \frac{y\sqrt{(1/\cos \phi_1)^2 + 0.006\,814\,784\,95}}{6\,399\,902.5516} = -0.050\,957\,5207$$

$$\phi = \phi_1 - \frac{(b_1 y)^2}{2} \sin \phi_1 \cos \phi_1 (1 + 0.006\ 814\ 785 \cos^2 \phi_1) \left[1 - \frac{(b_1 y)^2}{12} \right.$$

$$\times\ (3 + 1.938\ 67 \cos^2 \phi_1 + 0.068\ 15 \cos^4 \phi_1 - 0.000\ 19 \cos^6 \phi_1)$$

$$\left. + \frac{(b_1 y)^4}{360} (45 + 16 \cos^4 \phi_1) \right],$$

(2-16b)

$$\lambda = \lambda_0 + b_1 y \left[1 - \frac{(b_1 y)^2}{6} (2 - \cos^2 \phi_1 + 0.006\ 8148 \cos^4 \phi_1) \right.$$

$$\left. + \frac{(b_1 y)^4}{120} (24 - 20 \cos^2 \phi_1 + 1.0545 \cos^4 \phi_1 - 0.0136 \cos^6 \phi_1) \right].$$

Result:

$$\phi = 0.801\ 001\ 3410 = 45° 53' 38''.3864 \text{ N}$$

$$\lambda = -1.359\ 921\ 0279 = 77° 55' 03''.8473 \text{ W}.$$

This problem is the inverse of Example 2-4.

Translation into Grid Coordinates

Transverse Mercator coordinates (x, y) form a system that contains positive as well as negative coordinate values; the y-coordinates west of the central meridian and the x-coordinates south of the equator are always negative. A system of *grid coordinates* (X, Y) in which all the values are positive is obtained from the Transverse Mercator coordinates by translation of the origin:

Northern Hemisphere	Southern Hemisphere	
$X = x$	$X = X' + x$	(2-17)
$Y = Y' + y$	$Y = Y' + y.$	

"False easting" (e.g., $Y' = 500\ 000$ m) must be added to the y-coordinates. South of the equator, "false northing" (e.g., $X' = 10\ 000\ 000$ m) is added to the x-coordinates.

Example 2-6. Grid coordinates of the point in Example 2-4 may be written

$$X = 5\ 088\ 227.618$$
$$Y = 273\ 547.492,$$

false easting $Y' = 500\ 000$ being added to the y-coordinate.

UTM Coordinates and Modified TM Coordinates

Universal Transverse Mercator (UTM) is a worldwide (up to 80° lat.) grid system in 6°-wide zones based on the transverse Mercator projection; the central meridians are at longitudes 3°, 9°, etc., east and west of Greenwich. UTM coordinates N, E (northing, easting) are related to corresponding Transverse Mercator coordinates x, y through equations

Northern Hemisphere (meters)	Southern Hemisphere (meters)	
$N = 0.9996x$	$N = 10\,000\,000 + 0.9996x$	
$E = 500\,000 + 0.9996y$	$E = 500\,000 + 0.9996y$	(2-18)

or

$x = N/0.9996$	$x = (N - 10\,000\,000)/0.9996$	
$y = (E - 500\,000)/0.9996$	$y = (E - 500\,000)/0.9996.$	(2-19)

Constant factor $m_0 = 0.9996$ in the above equations is called the *central scale factor*, its purpose is to reduce the maximum distortion of scale in the projection zones.

Transverse Mercator grid systems applying a central scale factor are often referred to as "modified systems." One of the most popular is the "3° Modified TM System," applying $m_0 = 0.9999$ in projection zones 3° wide in longitude.

Example 2-7. UTM coordinates of the point in Example 2-4 are, according to Eq. (2-18),

$$N = 5\,086\,192.327 \qquad E = 75°273\,638.073.$$

The central meridian has been indicated in front of the E-coordinate.

Transformation of Grid Coordinates into Another Plane Coordinate System

Conversion formulas introduced on pp. 22–25 express algebraically the one-to-one correspondence that holds between Transverse Mercator coordinates x, y and geographical coordinates ϕ, λ. Computation of the latter as an intermediate step allows the transformation of plane coordinates from one grid to another with a different central meridian.

Example 2-8. Transform UTM coordinates in Example 2-7 into the adjacent zone (81° W) given $N = 5\,086\,192.327$, $E = 75°273\,638.073$.

From Eq. (2-19), $x = 5\,088\,227.618$, $y = -226\,452.508$; whence $\phi = 45°\,53'\,38''.3864$ N, $\lambda = 77°\,55'\,03''.8473$ W, as in Example 2-5. Conversion of ϕ, λ, as in Example 2-4 except that $l = 3°\,04'\,56''.1527$, gives $x = 5\,088\,708.144$, $y = 239\,221.659$. From Eq. (2-18), $N = 5\,086\,672.661$, $E = 81°739\,125.970$.

If a great number of points within a limited area are to be transformed from one conformal plane coordinate system into another (not necessarily Transverse Mercator), the transformation is best carried out through ϕ, λ for selected points at intervals; the rest of the points are interpolated by using the method of divided differences (4).

GRID CONVERGENCE AND SCALE FACTOR

The two related quantities grid convergence and scale factor are both of central importance in the theory and practice of Transverse Mercator projection corrections. Unlike the grid convergence, which is seldom computed directly, numerical values of the scale factor are frequently required in practical work. Midlatitude formulas for the scale factor, given on p. 31, have been designed to facilitate electronic computing.

Convergence

The positive direction of the x-axis defines, on the Transverse Mercator plane, the direction of *grid north*. Outside the central meridian and the equator, the direction of grid north differs from geographical north, i.e., from the direction of the meridian, by an angle C, called *grid convergence*, the value of which varies from point to point.

On the reference ellipsoid, *geodetic azimuths* are counted from geographical north, clockwise from $0°$ to $360°$, whereas on the Transverse Mercator plane, *projected grid azimuths* are similarly counted from grid north. As shown in Figure 2-4, grid convergence C must be subtracted from geodetic azimuth α_{12} to obtain corresponding projected grid azimuth T_{12}

$$T_{12} = \alpha_{12} - C. \tag{2-20}$$

At a given point, the difference between two projected grid azimuths equals the corresponding horizontal angle (difference of geodetic azimuths) on the ellipsoid

$$T_{13} - T_{12} = \alpha_{13} - \alpha_{12}, \tag{2-21}$$

convergence C being cancelled out.

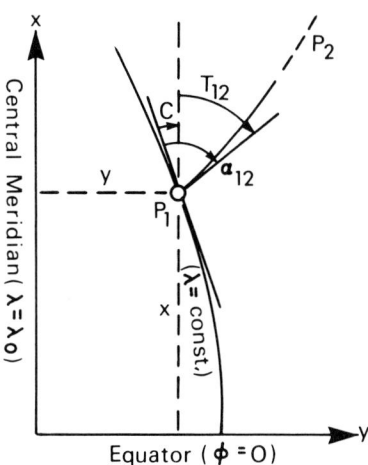

Figure 2-4. Geodetic azimuth α_{12} of line $P_1 P_2$ on conformal projection plane. Grid convergence C is counted clockwise from geographical north to grid north.

The following formulas for the accurate calculation of grid convergence serve for the occasional conversion of geodetic azimuths into projected grid azimuths, and vice versa. Such conversions are not routinely made in practice; they would be superfluous on lines whose end points have already been projected onto the plane.

Grid convergence C at a geodetic station is computed from geographical coordinates (ϕ, λ) by applying the general formula

$$C = a_7 l + a_9 l^3 + a_{11} l^5 + \cdots, \tag{2-22a}$$

where $l = \lambda - \lambda_0$ is the difference in longitude from central meridian λ_0 in radian measure, and the coefficients

$$a_7 = \sin \phi$$
$$a_9 = \tfrac{1}{3} \sin \phi \cos^2 \phi (1 + 3e'^2 \cos^2 \phi + 2e'^4 \cos^4 \phi)$$
$$a_{11} = \tfrac{1}{15} \sin \phi \cos^2 \phi (-1 + 3 \cos^2 \phi + \cdots)$$
$$\vdots$$

are functions of latitude ϕ. The convergence can also be computed from Transverse Mercator coordinates (x, y) by the formula

$$C = b_7 y + b_9 y^3 + b_{11} y^5 + \cdots, \tag{2-23a}$$

where the coefficients

$$b_7 = P_1^{-1} \sin \phi_1 = c^{-1} \sin \phi_1 \left[\left(\frac{1}{\cos \phi_1} \right)^2 + e'^2 \right]^{1/2}$$
$$b_9 = -\tfrac{1}{3} P_1^{-3} \sin \phi_1 (1 - e'^2 \cos^4 \phi_1 - 2e'^4 \cos^6 \phi_1)$$
$$b_{11} = \tfrac{1}{15} P_1^{-5} \sin \phi_1 (3 - \cos^2 \phi_1 + \cdots)$$
$$\vdots$$

are functions of latitude ϕ_1 corresponding to length of meridional arc $B = x$. Provided that longitude λ is within $\lambda_0 \pm 3°30'$, three terms of the series are sufficient for computing C with an error $<0".001$.

Example 2-9. Compute grid convergence C at the point given in Examples 2-4 and 2-5.
   a. Given $\phi = 45° 53' 38".3864$, $\lambda = -77° 55' 03".8473$, $(\lambda_0 = -75°)$.

$$C = l \sin \phi \left[1 + \frac{l^2}{3} \cos^2 \phi (1 + 0.020\,44 \cos^2 \phi + 0.000\,09 \cos^4 \phi) \right.$$

$$\left. + \frac{l^4}{15} \cos^2 \phi (-1 + 3 \cos^2 \phi) \right]. \tag{2-22b}$$

Computing

$$l = \lambda - \lambda_0 = -0.050\,924\,0888$$

$$\sin \phi = 0.718\,053\,372$$

$$\cos \phi = 0.695\,99$$

one obtains

$$C = -0.036\,581\,68 = -2°\,05'\,45''.51.$$

b. Given $x = 5\,088\,227.618$, $y = -226\,452.508$. Compute first ϕ_1 and $P_1^{-1}y = b_1 y$ as in Example 2-5. Then solving

$$C = b_1 y \sin \phi_1 \left[1 - \frac{(b_1 y)^2}{3} (1 - 0.006\,81 \cos^4 \phi_1 - 0.000\,09 \cos^6 \phi_1) \right.$$

$$\left. + \frac{(b_1 y)^4}{15} (3 - \cos^2 \phi_1) \right] \tag{2-23b}$$

gives

$$b_1 y = -0.050\,957\,5207$$

$$\sin \phi_1 = 0.718\,505\,896$$

$$\cos \phi_1 = 0.695\,52$$

$$C = -0.036\,581\,68 = -2°\,05'\,45''.51.$$

Note the minus sign of C, which signifies an anticlockwise rotation angle from geographical north to grid north.

Scale Factor

Transverse Mercator scale factor. The scale factor of the Transverse Mercator projection, as defined previously by Eq. (2-1), is a function of latitude and distance from the central meridian. It is computed from geographical coordinates (ϕ, λ) by applying the general formula

$$m = 1 + a_8 l^2 + a_{10} l^4 + \cdots, \tag{2-24a}$$

where $l = \lambda - \lambda_0$ is the difference in longitude from central meridian λ_0 in radian measure and the coefficients

$$a_8 = \tfrac{1}{2} \cos^2 \phi (1 + e'^2 \cos^2 \phi)$$

$$a_{10} = \tfrac{1}{24} \cos^2 \phi [-4 + (9 - 28e'^2)\cos^2 \phi + 42e'^2 \cos^4 \phi + \cdots]$$

$$\vdots$$

are functions of latitude ϕ. The scale factor can also be computed from Transverse Mercator coordinates (x, y) from the formula

$$m = 1 + b_8 y^2 + b_{10} y^4 + \cdots, \tag{2-25a}$$

where the coefficients

$$b_8 = \tfrac{1}{2}R_1^{-2} = \tfrac{1}{2}c^{-2}(1 + e'^2 \cos^2 \phi_1)^2$$

$$b_{10} = \tfrac{1}{24}R_1^{-4}(1 + 4e'^2 \cos^2 \phi_1 + \cdots)$$

$$\vdots$$

are functions of latitude ϕ_1 corresponding to length of meridional arc $B = x$. Provided that longitude λ is within $\lambda_0 \pm 3°30'$, three terms of the series are sufficient for computing m with an error <0.01 parts per million.

Example 2-10. Compute scale factor m at the point given in Examples 2-4 and 2-5.

a. Given $\phi = 45° 53' 38''.3864$, $\lambda = -77° 55' 03''.8473$, ($\lambda_0 = -75°$). Then

$$m = 1 + \frac{l^2}{2} \cos^2 \phi \left[1 + 0.006\,815 \cos^2 \phi \right.$$

$$\left. + \frac{l^2}{12}(-4 + 8.81 \cos^2 \phi + 0.28 \cos^4 \phi) \right] \qquad (2\text{-}24b)$$

$$l = \lambda - \lambda_0 = -0.050\,924\,09$$

$$\cos \phi = 0.695\,9880$$

$$m = 1.000\,630\,21.$$

b. Given: $x = 5\,088\,227.618$, $y = -226\,452.508$. First compute ϕ_1 as in Example 2-5. Then, from

$$\frac{y}{R_1} = \frac{y(1 + 0.006\,815 \cos^2 \phi_1)}{6\,399\,903}$$

$$m = 1 + \frac{1}{2}\left(\frac{y}{R_1}\right)^2 \left[1 + \frac{1}{12}(1 + 0.03 \cos^2 \phi_1)\left(\frac{y}{R_1}\right)^2 \right], \qquad (2\text{-}25b)$$

$$y = -226\,452.5$$

$$\cos \phi_1 = 0.695\,5209$$

$$\frac{y}{R_1} = -0.035\,500\,39$$

$$m = 1.000\,630\,21.$$

Scale factor of modified systems. The scale factor of a modified Transverse Mercator grid (X, Y) is equal to the Transverse Mercator scale factor multiplied by central scale factor m_0

$$m = m_0(1 + b_8 y^2 + b_{10} y^4 + \cdots), \qquad (2\text{-}26a)$$

where $y = (Y - Y')/m_0$, and coefficients b_8, b_{10}, ... are functions of latitude ϕ_1 corresponding to length of meridional arc $B = (X - X')/m_0$.

Practical computation of scale factor from Transverse Mercator coordinates.
The effort of finding ϕ_1 renders Eq. (2-25a) impractical for routine computation of scale factors from Transverse Mercator coordinates x, y. The remedy is to tabulate coefficients b_8 and b_{10} as a function of x for the latitude range required or, preferably, to set up a numerical formula for the scale factor in direct terms of x and y with the aid of *midlatitude* ϕ_m.

Within latitude limits $\phi_m \pm 2°$, the first three terms of the series in Eq. (2-25a) can be replaced by the expression

$$m = 1 + y'^2(1 - D_1 x' + D_2 x'^2 + D_3 y'^2), \qquad (2\text{-}27a)$$

where

$$x' = \frac{x - x_m}{c} \qquad y' = \frac{y}{c V_m^{-2}\sqrt{2}}$$

and the coefficients have the values

$$D_1 = 4e'^2 \sin \phi_m \cos \phi_m V_m^{-5}$$
$$D_2 = 2e'^2(1 - 2 \cos^2 \phi_m)V_m^{-2}$$
$$D_3 = \tfrac{1}{6}(1 + 4e'^2 \cos^2 \phi_m).$$

As in the derivation of the midlatitude formulas (p. 21), ϕ_m denotes the latitude corresponding to the selected midvalue of meridional arc $B = x_m$.
A useful approximation of Eq. (2-27a) is

$$m = 1 + y'^2, \qquad (2\text{-}28)$$

which will give the scale factor with an error of ≤ 0.6 parts per million (within $\phi_m \pm 2°$, $\lambda_0 \pm 3°30'$).

Example 2-11. Clarke 1866 Spheroid, $x_m = 4\,900\,000$. Derive the midlatitude formulas for the (a) Transverse Mercator and (b) UTM scale factor.

$$V_m^2 = 1 + e'^2 \cos^2 \phi_m = 1.003\,498\,07 \qquad D_1 = 0.0135$$
$$\sin \phi_m = 0.697\,63 \qquad\qquad\qquad D_2 = -0.000$$
$$\cos \phi_m = 0.716\,45 \qquad\qquad\qquad D_3 = 0.169.$$

a. Transverse Mercator scale factor, x: $4\,700\,000 - 5\,100\,000$;

$$m = 1 + y'^2(1 - 0.0135x' + 0.169y'^2), \qquad (2\text{-}27b)$$

where

$$x' = \frac{x - 4\,900\,000}{6\,399\,903} \quad \text{and} \quad y' = \frac{y}{9\,019\,279}.$$

b. UTM scale factor, N: 4 700 000 – 5 100 000;

$$m = 0.9996 + 0.9996y'^2(1 - 0.0135x' + 0.169y'^2), \qquad (2\text{-}26\text{b})$$

where

$$x' = \frac{N - 4\ 900\ 000}{0.9996 \times 6\ 399\ 903} \quad \text{and} \quad y' = \frac{E - 500\ 000}{0.9996 \times 9\ 019\ 279}.$$

TRANSVERSE MERCATOR PROJECTION CORRECTIONS

Introduction

Trigonometric computations on the Transverse Mercator grid follow the simple rules of plane trigonometry. Angles and distances used in these computations should be the plane equivalents of the corresponding ellipsoidal angles and distances. This requires not only the reduction of measured quantities onto the surface of the ellipsoid but also the application of *projection corrections* from the ellipsoid to the plane.

The projection corrections are small. They may always be computed by using roughly approximate preliminary coordinate values for newly established stations; these can be derived from uncorrected measurements. [For graphical determination of projection corrections, see Ref. (1).]

Corrections to Horizontal Angles and Directions

Station-adjusted observations of horizontal angles or directions are, without further reduction, valid on the ellipsoid, except in the mountains (see p. 112). Because the total sum of angles in a geodetic triangle is greater than 180° by a small spherical excess and conformal mapping preserves the size of the angles, it is evident that the counterpart of the triangle on the grid must possess slightly curved sides to produce a sum of angles more than exactly 180°. But plane computations must be carried along exactly straight lines, which requires small corrections to the observed directions before they are used on the grid.

In Figure 2-5, the projected line of sight between stations P_1 and P_2 is represented by arc P_1P_2, whereas chord P_1P_2 is used in plane computations. For every line of sight, therefore, two kinds of grid azimuth must be considered: the *projected grid azimuth* of the arc, denoted by T, and the *plane azimuth* of the chord, denoted by t. Obviously, $t_{21} = t_{12} \pm 180°$, but in general, $T_{21} \neq T_{12} \pm 180°$.

As a rule, in conformal mapping, the projected line of sight bends toward an area of greater scale factor—on the Transverse Mercator plane, away from the central meridian. A point of reversing curvature will occur on lines crossing over the central meridian. The small angles between the chord and the arc,

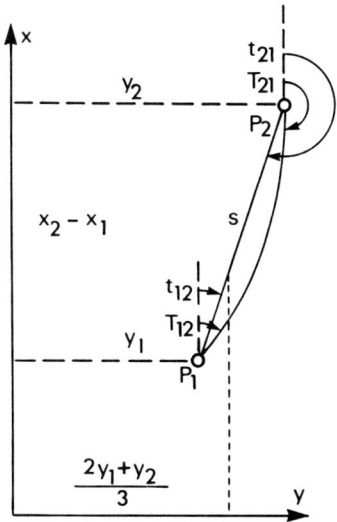

Figure 2-5. Plane azimuth t and projected grid azimuth T.

which are generally different at both ends of a line, are computed by the (approximate) formula

$$t_{AB} - T_{AB} = k(x_A - x_B)\frac{2y_A + y_B}{3} \tag{2-29}$$

or, in the case of a modified Transverse Mercator grid (X, Y),

$$t_{AB} - T_{AB} = \frac{k}{3m_0^2}(X_A - X_B)(2Y_A + Y_B - 3Y'). \tag{2-30}$$

The value k can be taken as a constant factor within latitudes $\phi_m \pm 2°$:

$$k'' = \frac{\rho''}{2R_m^2} = \left(\frac{321.14}{c}(1 + e'^2 \cos^2 \phi_m)\right)^2$$

or

$$k^{cc} = \frac{\rho^{cc}}{2R_m^2} = \left(\frac{564.19}{c}(1 + e'^2 \cos^2 \phi_m)\right)^2.$$

Within longitude zone $\lambda_0 \pm 3°30'$, the error in $t - T$ is likely to be less than $0''.1$ for lines not exceeding 50 km. [Reference (5) quotes more accurate formulas for longer lines.]

A set of observed horizontal directions $d_{12}, d_{13}, \ldots$ is converted into (observed) grid aximuths by equations

$$\begin{aligned} T_{12} &= d_{12} + z_T; & t_{12} &= d_{12} + (t_{12} - T_{12}) + z_T \\ T_{13} &= d_{13} + z_T; & t_{13} &= d_{13} + (t_{13} - T_{13}) + z_T \end{aligned} \tag{2-31}$$

$$\cdots\cdots\cdots\cdots\cdots \qquad \cdots\cdots\cdots\cdots\cdots\cdots\cdots$$

where orientation constant z_T is the same for every direction, but the differences $t - T$, variously known as *direction corrections, arc-to-chord corrections,* or $(t - T)$ *corrections*, will be different for each line of sight. The plane equivalent of a horizontal angle is obtained by considering the angle as the difference between the directions of its right and left arms.

Example 2-12. Convert the following set of four directions into plane azimuths.

Observed at P_1	Approximate distance (m)	Known coordinates (m)	
		x	y
P_2 0°00′00″.0	16 840	P_1 4 851 225.468	233 484.603
P_3 60 05 28 .4	28 780	P_4 4 857 475.489	204 140.594
P_4 156 02 51 .7	(30 000)	(Clarke 1866 Spheroid)	
P_5 185 52 38 .5	21 990		

Computation of orientation constant z_T and projected grid azimuths T.

$$\text{ctn } t_{14} = \frac{x_4 - x_1}{y_4 - y_1} = -0.212\,9914$$

$$t_{14} = 282° 01′ 25″.8.$$

For x: 4 700 000 − 5 100 000, from Example 2-11, $V_m^2 = 1 + e'^2 \cos^2 \phi_m = 1.003\,498\,07$. This gives $k'' = 25''.36 \times 10^{-10}$, with c in meters. Then

$$x_1 - x_4 = -0.062\,50 \times 10^5$$

$$\tfrac{1}{3}(2y_1 + y_4) = 2.237\,03 \times 10^5$$

$$t_{14} - T_{14} = 25''.36(-0.062\,50)(+2.237\,03) = -3''.5$$

$$z_T = t_{14} - (t_{14} - T_{14}) - d_{14} = 125° 58′ 37''.6.$$

	T	$\sin T$	$\cos T$
$P_1:P_2$	125° 58′ 37″.6	+0.809 25	−0.587 46
P_3	186 04 06 .0	−0.105 71	−0.994 40
P_4	282 01 29 .3	(−0.978 06)	(+0.208 34)
P_5	311 51 16 .1	−0.744 84	+0.667 24

Approximate coordinates and $t - T$ corrections. Because $t \cong T$,

$$x_2 = x_1 + s_{12} \cos t_{12} \cong 4\,841\,330; \quad x_1 - x_2 = 0.0989 \times 10^5$$

$$y_2 = y_1 + s_{12} \sin t_{12} \cong 247\,110; \quad \tfrac{1}{3}(2y_1 + y_2) = 2.3803 \times 10^5$$

$$t_{12} - T_{12} = 25''.36(+0.0989)(+2.3803) = +6''.0$$

$$x_3 = x_1 + s_{13} \cos t_{13} \cong 4\,822\,610; \quad x_1 - x_3 = 0.2862 \times 10^5$$

$$y_3 = y_1 + s_{13} \sin t_{13} \cong 230\,440; \quad \tfrac{1}{3}(2y_1 + y_3) = 2.3247 \times 10^5$$

$$t_{13} - T_{13} = 25''.36(+0.2862)(+2.3247) = +16''.9$$

$$x_5 = x_1 + s_{15} \cos t_{15} \cong 4\,865\,900; \quad x_1 - x_5 = -0.1467 \times 10^5$$

$$y_5 = y_1 + s_{15} \sin t_{15} \cong 217\,110; \quad \tfrac{1}{3}(2y_1 + y_5) = 2.2802 \times 10^5$$

$$t_{15} - T_{15} = 25''.36(-0.1467)(+2.2802) = -8''.5.$$

Result:

	T	$t - T$	t
$P_1 : P_2$	125° 58′ 37″.6	+6″.0	125° 58′ 43″.6
P_3	186 04 06 .0	+16.9	186 04 22 .9
P_4	282 01 29 .3	−3 .5	282 01 25 .8
P_5	311 51 16 .1	−8 .5	311 51 07 .6

Scale Correction

Because of the projection scale, trigonometric computations in plane coordinates have to deal with grid distances that generally differ from the corresponding geodetic distances measured along the surface of the ellipsoid. For a given line of finite length, *grid length s* is the product of *ellipsoidal length S* multiplied by average *scale factor* $\bar{m}$ of the line,

$$s = \bar{m}S = S + (\bar{m} - 1)S$$

$$S = \frac{s}{\bar{m}} = s - \left(\frac{\bar{m} - 1}{\bar{m}}\right)s \cong s - (\bar{m} - 1)s. \tag{2-32}$$

On the projection plane, both s and $\bar{m}$ should be referred to the projected arc rather than to the chord, but such refinement need not be considered if $s < 50\,\text{km}$. For longer lines, see Ref. (5).

The average scale factor of the chord occurs near its middle point, more precisely, at the point on the chord where

$$y^2 = \tfrac{1}{3}(y_1{}^2 + y_1 y_2 + y_2{}^2)$$

$$x = x_1 + \frac{(x_2 - x_1)(y - y_1)}{y_2 - y_1}, \tag{2-33}$$

x_1, y_1 and x_2, y_2 being the terminal coordinates. Alternatively, $\bar{m}$ can be computed by the formula

$$\bar{m} = \tfrac{1}{6}(m_1 + 4m_m + m_2), \qquad (2\text{-}34)$$

where m_1 and m_2 are the scale factors at the end points and m_m is the scale factor at the midpoint of the chord.

For the reduction of measured distances onto the ellipsoid, see pp.112 –119.

Example 2-13. Compute Transverse Mercator scale factor $\bar{m}$ for line P_1P_4 (approx. 30 km) in Example 2-12, given $x_1 = 4\,851\,225$, $y_1 = 233\,485$; $x_2 = 4\,857\,475$, $y_2 = 204\,141$.

a. At the point where $y^2 = \tfrac{1}{3}(y_1{}^2 + y_1y_2 + y_2{}^2)$, $y = 218\,977$, $x = 4\,854\,315$. Then, Eq. (2-27b), $\bar{m} = 1.000\,589\,57$.

b. $\bar{m} = \tfrac{1}{6}(m_1 + 4m_m + m_2)$; $y_m = 218\,813$; $x_m = 4\,854\,350$. From Eq. (2-27b), $m_1 = 1.000\,670\,30$, $m_2 = 1.000\,512\,38$, $m_m = 1.000\,588\,69$, and $\bar{m} = 1.000\,589\,57$.

Practical Application of Projection Corrections

Projection corrections can often be omitted in lower-order work, but this should not be done in a wholesale manner. Safe rejection criteria can be established for each type of survey by considering (a) the working distance from the central meridian, (b) the length of the lines of sight, and (c) the precision of the measurements. Table 2-1 shows typical values of the projection corrections. It should be noted that mere adjustment to superior control does not take adequate care of these corrections (scale in a traverse loop; bend of straight traverse).

Least-squares adjustment of first-order urban surveys can be done with advantage in plane coordinates, particularly if the network is not extensive or far from the central meridian. There is no question of ignoring any of the projection corrections; however, near the central meridian the simplest versions of the formulas are sufficient for first-order accuracy.

Table 2-1. Typical Values of Transverse Mercator Projection Corrections

Distance from central meridian	Direction corrections for north–south lines				Scale correction per 1000 m distance
	$s = 1$ km	$s = 10$ km	$s = 20$ km	$s = 30$ km	
10 km	0″.03	0″.3	0″.5	0″.8	1 mm
25 km	0″.06	0″.6	1″	2″	8 mm
50 km	0″.1	1″	3″	4″	3 cm
100 km	0″.3	3″	5″	8″	1 dm
200 km	0″.5	5″	10″	15″	0.5 m
350 km	0″.9	9″	18″	27″	1.5 m

As a final numerical example of this chapter, Example 2-14 has been chosen to demonstrate the use of projection corrections in transferring coordinate computations from the ellipsoid to the plane. The geodetic data have been taken from Ref. (6).

Example 2-14.   Given $\phi_A = 68°\,04'\,29''.1035$ N, $\lambda_A = 24°\,03'\,17''.8503$ E, α_{AB} $= 121°\,08'\,59''.10$, $S_{AB} = 74\,561.374$ (International Ellipsoid). Compute: ϕ_B, λ_B, and α_{BA}.

The central meridian is set through initial point A:

$$\lambda_0 = \lambda_A$$
$$l = 0; C_A = 0$$
$$T_{AB} = \alpha_{AB}.$$

From Eqs. (2-15a) and (2-10c), $x_A = B = 7\,554\,416.462$, $y_A = 0$.

The projection corrections are computed with approximate coordinates for terminal B

$$x_B \cong x_A + S_{AB} \cos T_{AB} = 7\,515\,848$$
$$y_B \cong y_A + S_{AB} \sin T_{AB} = \quad 63\,811.$$

Then, with $\phi_m = \phi_A$, Eqs. (2-28) and (2-33) for the scale factor, and Eq. (2-29) for direction corrections give

$$\bar{m} = 1.000\,016\,60$$
$$t_{AB} - T_{AB} = +2''.07$$
$$t_{BA} - T_{BA} = -4''.14.$$

The grid length and the plane azimuth of chord AB are then

$$S_{AB} = \bar{m}S_{AB} = 74\,562.612$$
$$t_{AB} = T_{AB} + (t_{AB} - T_{AB}) = 121°\,09'\,01''.17.$$

The final plane coordinates of terminal B,

$$x_B = x_A + S_{AB} \cos t_{AB} = 7\,515\,846.325$$
$$y_B = y_A + S_{AB} \sin t_{AB} = \quad 63\,811.657,$$

are transformed into geographical coordinates by Eqs. (2-16a),

$$\phi_B = 67°\,43'\,19''.1536 \text{ N}$$
$$\lambda_B = 25°\,33'\,45''.8684 \text{ E}.$$

The back azimuth is

$$\alpha_{BA} = (t_{AB} + 180°) - (t_{BA} - T_{BA}) + C_B$$
$$= 302°\,32'\,48''.32,$$

where $C_B = 1°\,23'\,43''.01$ is the grid convergence at terminal B computed from Eq. (2-22a) or (2-23a).

ESTABLISHMENT OF LOCAL PLANE COORDINATE GRID

Experience shows that a local coordinate system, once brought into operation, will resist all future change. There are countless examples of badly designed or erroneous systems that have been perpetuated for the simple reason that revision of the existing coordinate values would be prohibitively costly and inconvenient. In the design of a new system, careful thought should be given to its long-term use.

The Transverse Mercator projection leaves free choice in the selection of two parameters: the longitude of the central meridian and the central scale factor. Apart from the constants of the reference ellipsoid, three more independent parameters are involved in the implementation of a geodetic datum that will fix the coordinates of one initial point, and the azimuth of one initial line, on the grid.

Choice of Central Meridian

Ordinarily, it is best to select the central meridian for a local plane coordinate system at or near the center of the area it is intended to serve, although other considerations may predominate in regions where a suitable system, such as a modified TM system in 3° zones, has already been firmly established. The important point to consider is the scale factor:

Transverse Mercator scale factor	Distance from central meridian (km)
1.0000	0
1.0001	90
1.0002	128
1.0004	180
1.0006	221
1.0008	255
1.0010	285

Although any one of the scale factors may be brought equal to unity by choosing a suitable central scale, the east–west rate of change in the scale cannot be altered, and this rate increases rapidly with distance from the central meridian. Should, for instance, the use of the grid in detail surveys demand that a true distance of 100 m is not distorted in the projection by more than ±1 cm (as might be accepted as a reasonable stipulation), the effective grid area will be reduced from an original width of 180 km to barely 41 km when the average working distance from the central meridian grows to 200 km.

Central Scale Factor; Cities of High Elevation

The distortion of true lengths, popularly called the "scale error," increases on the Transverse Mercator projection with the square of the distance from the central meridian. On grid systems with zones several degrees wide in longitude, the maximum scale error would become intolerable unless reduced by applying a suitable central scale factor less than unity.

A central scale factor greater than unity is usually applied in localities significantly above the sea level. In such a case, a constant enlargement of scale by

$$m_0 = 1 + 0.000\ 000\ 157\ h, \tag{2-35}$$

where h denotes the mean elevation in meters, helps to offset the reductions of horizontal lengths to the ellipsoid level. For example, if $h = 2000$ m, $m_0 = 1.000\ 314$, which results in an increase of grid distances by 3 cm for each 100 m of horizontal length, thereby bringing the horizontal distances computed from grid coordinates into reasonable agreement with the corresponding true distances measured on the ground.

Geodetic Datum

After selection of the central meridian and the central scale factor, the next step in setting up an urban plane coordinate system consists of the computation of grid coordinates for the primary geodetic stations to which the local first-order control net is being tied. These coordinates will be kept *fixed* in the grid adjustment of the local networks, into which the observed angles and distances are entered in their projected values (see also pp. 133–134). The result is a system of grid coordinates in which the geodetic datum of the primary stations has been preserved, and which effectively represents geodetic positions on the ellipsoid in terms of plane coordinates. Such a system is "local" only because of its local central meridian; it is in fact linked to any other similar system by conversion through geographical coordinates.

Some common sources of error that are sometimes overlooked in the process of transferring the geodetic datum include misinterpretation or confusion among the ground markings on a primary geodetic station and reliance on any primary station that, since its establishment, may have suffered from movement on unstable ground. Large errors, of course, will show up in the net adjustment, but as an added precaution, it is always advisable to measure independently a few selected distances between the fixed stations, reduce them onto the plane, and compare the results with grid distances computed from the coordinates.

In regions lacking primary networks, the above procedure is not possible, and the plane coordinate system must be set up on a *local datum* by means of the following procedure.

1. One station of the first-order urban network, which must possess at least one measured side, is given arbitrary grid coordinates X, Y. The point of origin ($X = 0$, $Y = 0$) should fall southwest, well outside the grid limits.
2. The network is oriented on the projection plane by observing the astronomical azimuth (e.g., from the Sun or Polaris) or gyroscopic azimuth of one of its sides, allowing for grid convergence if the point of observation is outside the central meridian.
3. Projection corrections based upon map latitude are applied to the measured directions (angles) and distances in the network.
4. Grid coordinates are computed in rigorous adjustment of the oriented network on the projection plane.

A local system thus established is a conformal plane coordinate system, although its coordinates cannot be immediately converted into other systems such as the UTM. But as soon as the primary geodetic net becomes available in the region, the whole system can be transformed to the geodetic datum by translation of the origin, slight rotation, and change of scale, thereby avoiding a laborious repetition of all the network computations.

Summary of Recommendations

In most countries, federal planners have gradually come to recognize the merits of the Transverse Mercator projection in providing a countrywide, uniform grid system made up of identical north–south projection zones. In the United States, for instance, the old State Plane Coordinate System (SPC), introduced in 1933, which consists of a combination of different conformal projections, is soon to be abandoned in favor of a plane coordinate grid in (modified) Transverse Mercator zones. Much can be said for a well-organized federal plane coordinate system that has been properly designed to handle all the surveying and mapping and is tied to the national geodetic control network. However, a certain drawback inherent in the system cannot be overlooked when dealing with the specific requirements of urban surveying.

It is obvious that a fixed grid system with equally spaced central meridians must necessarily lead to a situation in which some cities, through no fault of their own, will be bisected by the boundary line between two adjacent zones and other cities will be uncomfortably close to the boundary line. Although, from the technical standpoint, the zonal boundaries do not present any real difficulty in dealing with plane coordinates, it would appear unreasonable to expect any city to maintain coordinate files for perhaps thousands of points in two entirely different coordinate systems and to accept the inconvenience and cost of such a duplication. In any case, the bulk of the city control is needed only for local purposes. In the opinion of the authors, the plane coordinate system for each urban community, or group of communities, should be designed so as to fulfill the particular needs of that community, even if this means a departure from an otherwise accepted regional coordinate system.

To obtain the best benefits from the Transverse Mercator projection as an urban plane coordinate grid, the central meridian should be chosen so that the grid limits are within 100 km of the meridian. The application of a central scale factor should be avoided unless it is necessary for the convenience of detail surveys, as in locations high above sea level.

References

1. Saastamoinen, J. A Canadian grid system in 3° Transverse Mercator zones, *The Canadian Surveyor*, June 1964.
2. Jordan, W., Eggert, O., and Kneissl, M. *Handbuch der Vermessungskunde*, 10th ed., J. B. Metzlersche Verlagsbuchhandlung, Stuttgart, Vol. 4-1, 1958.
3. Schmid, E. The general term in the expansion for meridian length, *The Canadian Surveyor*, June 1971.
4. Lauf, G. B., and Young, F. Conformal transformations from one map projection to another, using divided difference interpolation, *Bulletin Géodésique*, 1961.
5. Bomford, A. G. Transverse Mercator arc-to-chord and finite distance scale factor formulae, *Empire Survey Review*, 1962.
6. Korhonen, J. Coordinates of the stations in the first order triangulation of Finland, *Publications of the Finnish Geodetic Institute*, 1967.

Additional Readings

Bomford, G. *Geodesy*, 3rd ed., Oxford University Press, 1971.

Clark, D. *Plane and Geodetic Surveying*, Vol. 2, revised by J. E. Jackson, Constable and Co. Ltd., London, W.C.2.

Hotine, M. The orthomorphic projection of the spheroid, *Empire Survey Review*, Vols. 8 and 9, Nos. 62, 63, 64, 65, 66, 1946–47.

Jordan, W., Eggert, O., and Kneissl, M. *Handbuch der Vermessungskunde*, 10th ed., Vol. 4-2, 1959.

Maling, D. H. *Coordinate Systems and Map Projections*, George Philip & Son, Ltd., London, W.C.2, 1973.

Richardus, P., and Adler, R. K. *Map Projections for Geodesists, Cartographers and Geographers*, North-Holland/American Elsevier, New York, 1972.

Thomas, P. D. *Conformal Projections in Geodesy and Cartography*, U.S. Coast and Geodetic Survey Special Publication No. 251, 1952.

Chapter 3

Horizontal Control Surveys

INTRODUCTION

Purpose and Scope of Horizontal Control

The purpose of control surveys is to provide a uniform framework of reference for the coordination of all surveying activities within a given area. Control surveys consist of horizontal and vertical controls, which require fundamentally different methods of establishment, although some control points may be common to both control nets.

A horizontal control system for an urban area should be considered a *public utility*. Although it is possible to build a city with little horizontal control—and many cities provide notable examples—the lack of this utility will eventually lead to rising costs and difficulty with all surveying within the city, whether it be legal, technical, or topographical surveys or surveys for underground installations. The average city dweller may not readily appreciate the need for horizontal control or, more specifically, the need for spending tax money for such a purpose, although an incidence of sewage backup may be convincing of the necessity for vertical control. Nevertheless, the economic benefits from horizontal control are no less real, and the effort and money invested are soon recovered in the benefits. In recent years, this has become increasingly evident, since much of the earlier difficulty of establishing horizontal control has been removed by modern methods of electromagnetic distance measurement.

The scarcity, until recently, of control surveys in American cities demonstrates that an orderly system of horizontal control, however desirable, does not come about on its own merits. Only proper recognition of the technical and economic value of an integrated survey will lead civic administrations of urban communities to set up surveying departments properly equipped with qualified staff to establish and maintain an adequate horizontal control net. Although required capital outlays are relatively low, it is important that these surveys are conducted under the supervision of engineers with experience in geodetic work.

No two cities are alike in their topography, layout, or administration, and rigid rules cannot be formulated for any "best" method of horizontal control. But to fulfill its basic purpose, the horizontal control network should be built up and maintained with two basic requirements in mind: The accuracy of the control should be superior to that of the surveys it coordinates, and equally important, additional control should always be ready for the user when and where it is needed. These are no small requirements; only competent planning and foresight can meet them adequately.

Basic Geodetic Concepts

The following definitions are adopted in this chapter for fundamental geodetic surfaces:

> A *horizontal surface* is any continuous surface that encloses the earth at every point perpendicular to the plumb line, i.e., perpendicular to the direction of gravity.
>
> *Mean sea level* is a particular horizontal surface referenced by fundamental bench marks and specified in the official vertical datum of a country.
>
> The *geoid* is a particular horizontal surface which, for practical reasons, is defined as coincident with mean sea level.
>
> The *reference ellipsoid* is a mathematical surface that closely fits the geoid and is specified in the official horizontal datum of a country.
>
> The *projection plane* is a representation of the reference ellipsoid on a plane surface, as described in Chapter 2.

The global run of every horizontal surface, such as the geoid, is affected by the uneven distribution of mass in the earth's crust (Fig. 3-1). In a world geodetic system, a single reference ellipsoid can be fitted so that the undulations of the geoid with respect to the ellipsoid are less than 100 m; an example of a better fit

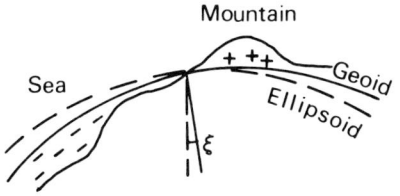

Figure 3-1. Attraction due to the excess mass of mountains raises the geoid above the ellipsoid in the continents; ξ, deflection of the vertical.

achieved in continental systems is shown in Figure 3-2, which shows the North American geoid.

The angle between the plumb line and the normal to the ellipsoid is called the deviation or *deflection of the vertical*, the north–south component of which is customarily denoted by ξ, the east–west component by η. These angles typically amount to a few seconds of arc, although deflections as large as 30″ are not unusual in mountainous regions. On a geoid chart, the deflections of the vertical are easily visualized as slope angles of the "terrain" shown by the contour lines of the geoid. Conventionally, ξ and η are given *positive* algebraic signs if the geoid slopes down relative to the ellipsoid toward *north* and *east*, respectively.

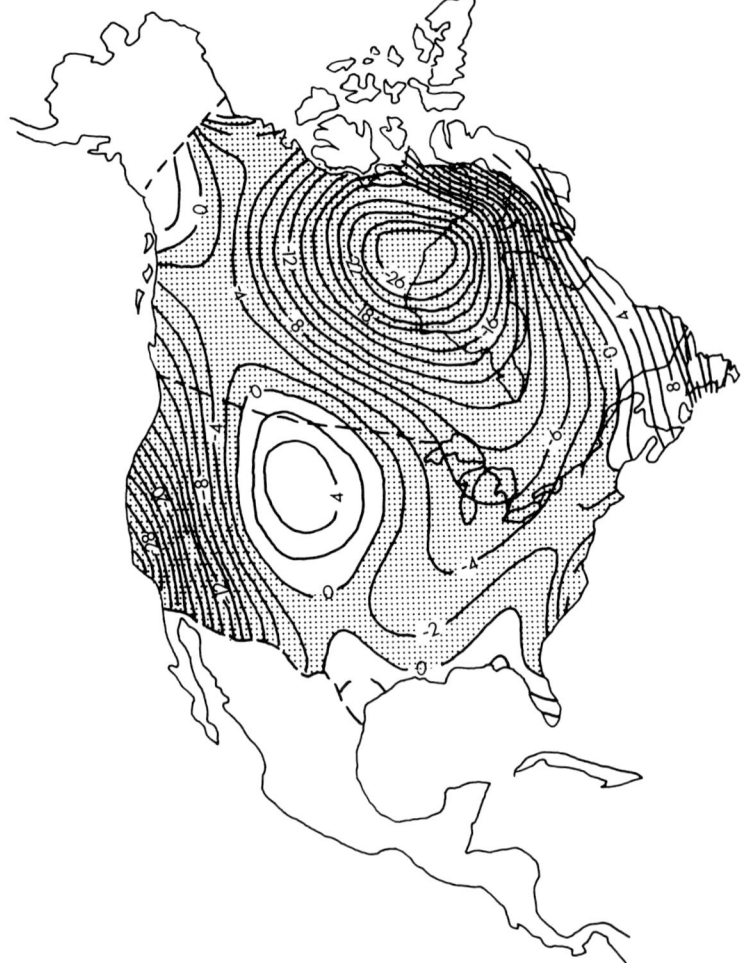

Figure 3-2. Geoid chart of North America. Contours in meters above the Clarke 1866 Spheroid, NAD 1927 (1).

Units of Measurement

Standard units of length. It is not surprising that the early length standards were correlated to lengths or widths of human limbs. Despite lack of uniformity, the reference to dimensions of human limbs conveyed at least the *order* of the magnitude in a convenient way. However, this system led to a great number of standard units that, with time, were independently defined in each autonomous area by its ruler. For instance, in Germany prior to the adoption of metric units in the second half of the nineteenth century, more than 100 different foot standards were in use!

Since length units were needed not only for surveying tracts of land but also in trade and industry, a great variety was used in most countries. For example, the basic English length units were:

Mile	Furlong	Chain	Land rod	Yard	Foot	Link	Inch
1	8	80	320	1760	5280	8000	63 360
	1	10	40	220	660	1000	7920
		1	4	22	66	100	792
			1	$5\frac{1}{2}$	$16\frac{1}{2}$	25	198
				1	3	$4\frac{6}{11}$	36
					1	$1\frac{17}{33}$	12
						1	7.92

The same units may have different absolute values in different English-speaking countries. Consequently, care must be taken when converting dimensions from this system to the metric system.

Early attempts were also made to derive the basic length units from the dimensions of the earth, but only in modern times has this concept been more widely introduced. Following various determinations of meridian lengths in the eighteenth century, the French Academy of Science recommended in its first report, October 1790, that a decimal system of length units (in addition to money and weights) be introduced. The new length unit, the meter, was defined as

$$\frac{1}{10\,000\,000} \text{ length of a meridian quadrant of the earth.}$$

After a new measurement of the meridian arc Dunkirk–Barcelona in 1792–1798, the initial length of the meter was slightly modified, and three platinum reference standards were built and stored in Paris as the primary reference standards. In France, the meter was introduced as a legal unit in 1795, and in 1840 it became the only official unit.

In 1875, 18 countries signed a treaty, the Metric Convention, establishing in Sèvres, a suburb of Paris, an International Bureau of Weights and Measures, which assumed the custody of metric standards and produced a number of new primary and secondary reference standards for distribution among the member countries. Shortly afterward, the standard meter was defined in terms of the wavelength of red cadmium light in order to make it independent of possible changes in the prototypes owing to structural changes of the metals used. The meter, as expressed in wavelengths of light from a specified source, under precisely specified conditions, can be reproduced at any competent physical laboratory with a superior accuracy, an important advantage over a standard in the form of a metal bar.

In 1960, the meter was redefined in terms of the wavelength of the orange light of crypton:

$$1 \text{ m} = 1\,650\,763.73\,\lambda.$$

Other length units based on the meter are:

$$
\begin{array}{ll}
1000 \text{ m} & = 1 \text{ km (kilometer)} \\
0.1 \text{ m} & = 1 \text{ dm (decimeter)} \\
0.01 \text{ m} & = 1 \text{ cm (centimeter)} \\
0.001 \text{ m} & = 1 \text{ mm (millimeter)} \\
0.000\,001 \text{ m} & = 1 \text{ } \mu\text{m (micrometer).}
\end{array}
$$

In 1959, those countries using the English units adopted a common conversion factor, 1 in. = 2.54 cm, for industrial purposes. This gives

$$1 \text{ international foot} = 0.3048 \text{ meters.}$$

By comparison, the British imperial foot is about 1 part per million shorter, and the U.S. customary foot about 2 parts per million longer.

Units of area. These are derived units, of which the following are commonly used in surveying:

$$1 \text{ acre} = \tfrac{1}{640} \text{ square miles} = 4840 \text{ square yards} = 43\,560 \text{ square feet}$$

$$1 \text{ hectare} = 100 \text{ ares} = 10\,000 \text{ square meters.}$$

Therefore, 1 acre = 4 046.9 square meters, approximately.

Angular units. Angles can be measured in degrees, grads, radians, mils, or, occasionally, in time units.

If a full circle is divided into 360 equal parts, the resulting angular unit is called the degree. In this system

$$1° \text{ (degree)} = 60' \text{ (minutes)} \quad \text{and} \quad 1' = 60'' \text{ (seconds).}$$

This system is known as the sexagesimal system of angular units, and the units are referred to as sexagesimal minutes and sexagesimal seconds.

If a full circle is divided into 400 parts (or the quadrant is subdivided into 100 parts), one refers to the centesimal system consisting of grads, centesimal minutes, and centesimal seconds.

$$1^g \text{ (grad)} = 100^c \text{ (centesimal minutes)} \quad \text{and} \quad 1^c = 100^{cc} \text{ (centesimal seconds).}$$

The circumference of a circle with radius 1 is equal to 2π, and π is 3.141 592 653 6. Hence,

$$1^{rad} = 57.295\ 78° = 63.661\ 98^g.$$

The mil is one-thousandth part of a radian.

Finally, owing to the association of angular magnitude with the rotation time of the earth, the subdivision of the circle in time units should be mentioned:

$$15° = 1^h,\ 15' = 1^{min} \text{ (time minute)},\ 15'' = 1^s \text{ (time second)}.$$

In surveying, the sexagesimal system has been used throughout the world. However, sexagesimal units are somewhat awkward in arithmetic operations because of the need for continued conversion of smaller units into larger, and vice versa. Conversions are also a frequent source of error and lower the efficiency of field work, where rapid checks must be performed, often mentally. As the result, in many countries, particularly in Central and Eastern Europe, the surveying instruments equipped with centesimal subdivision are used almost exclusively. This also facilitates the subsequent computational and office work for which the trigonometric and other auxiliary tables in the centesimal system are available. In this book, both systems of units will be used in order to familiarize the reader with them.

DENSITY AND ACCURACY REQUIREMENTS OF HORIZONTAL CONTROL NETWORKS

The integration of positional and geometrical information can be made in a number of ways, depending upon available technology, manpower, and local needs. At one end of the spectrum is an integrated system based entirely on large-scale graphical maps (from photogrammetric or field surveys), correlated by the grid of the coordinate system defined on the ground by the control net. The maps are supplemented by files of field sketches containing the results of original surveys. The eventual setting out or relocation of some details in the field is done either from measurements taken directly from graphical maps or on the basis of original survey documents searched from files.

At the other end of the spectrum is a computerized, integrated system based entirely on coordinates of all points. In this system, horizontal control is always used for positioning and relocation work, including property and engineering surveys. Graphical displays serve only for indexing data or providing an overview.

Combinations of these two extreme cases are possible and are in existence in some cities. Owing to the steadily increasing cost of manual labor and the decreasing cost of computerization, use of systems based entirely on numerical values seems to be inevitable in the future.

The density and accuracy requirements of horizontal control become much higher in numerical systems because, in extreme cases, the coordinates may become the only evidence for the positioning and relocation of details, including property boundaries.

Spacing of the horizontal control points should allow the surveyor to tie detailed surveys with one or two instrument setups at the most. Otherwise, regular use of the control could prove uneconomical. The use of EDM instruments permits long-range direct measurements, but city traffic and other obstacles in commercial and densely populated areas limit the sights to from 100 to 200 m. Therefore, the maximum distance between adjacent control points in these areas should be of the order of 400 m. On the other hand, considering the unfavorable influence of centering errors on error propagation in traverses with short sights, the distances between control points should be not shorter than approximately 100 m.

Ideally, the whole city should be covered with a uniform and simultaneously adjusted network of densely located points. Uniform coverage may be difficult to achieve at once for economic reasons. Different parts of the city may have to be densified with horizontal control at different periods of time. To provide uniformity in successive densification of control, a higher-order control net must be established first as a framework for the densification surveys. The higher-order control points should be of high enough density to encourage surveyors to tie their surveys into it in suburban and less densely populated areas of the city. Distances of from a few hundred meters to approximately 2 km seem to be adequate, with maximum distances about 3 km. If such a higher-order network could be established and adjusted simultaneously over an area that could be considered as the projected city limit, the two-order network could serve as a basis for the integrated city survey system. However, the 2-km density of points may be uneconomical in the noncultivated areas surrounding some cities, for instance, in densely wooded regions, even if future growth of the city is expected in that direction. In addition, the city network should be an integral part of the regional or national control net. In such cases, an even higher-order net with distances of 5 to 15 km between the points should be considered as the framework for the previously described two-order control scheme.

Therefore, the integrated survey system in urban areas should be based on a horizontal control framework established in three stages:

> Stage I: first-order control with spacing of up to 15 km,
>
> Stage II: second-order control with spacing up to 3 km,
>
> Stage III: third-order control with spacing up to 400 m.

Certain limiting factors independent of the accuracy of surveying instruments may serve as a constraint on accuracy in the foreseeable future. Limiting factors include the instability of survey markers due to climatic effects and ground movements, the centering accuracy of survey instruments, and limitation in the physical identification of surveyed points.

A positional error in the order of 1 cm in terms of the semimajor axis of the standard-error ellipse may be considered a reasonable limiting accuracy for relocating a point by using independent coordinate surveys. This corresponds to a tolerance limit of 25 mm at the 95 % confidence level. This value agrees, for instance, with working standards of land surveyors in New York City who are requested to locate corners of properties in the Manhattan area to within 1 in. (25 mm).

Even higher accuracies may be needed in special construction surveys, when fitting together prefabricated construction elements, for instance. This "internal" accuracy, encountered in engineering projects and the subject of separate surveys, is not generally a concern of city survey systems for which primary interest is restricted to the relative position of individual construction projects and their relation to other details such as property boundaries.

The accepted value of 25 mm as the maximum positional error in a relocation survey consists of three partial errors: Errors of the control network if the original and relocation surveys are tied to different points of the network. Errors of the original connecting survey. Errors of the connecting surveys in the relocation procedure. If each of these factors has approximately the same influence, the accuracy of the control surveys would be in the order of $25/\sqrt{3} = 14$ mm in terms of the semimajor axis of the relative-error ellipses at the 95 % confidence level.

If 200 m is accepted as an average spacing between the control points, the required relative accuracy becomes $14/200\,000 \cong 1/14\,000$ for the lowest-order control. The higher-order control should be more accurate, so that no significant distortions are introduced into the lower-order control while holding the higher-order points fixed during the adjustment process. The Canadian Federal Surveys and Mapping Branch (2) recommends a factor of 2.5 for the increase in the accuracy from the lower- to the higher-order network. The authors have not checked the validity of this factor. Should it prove acceptable, the accuracy of the control surveys in urban areas would then be:

> First order: 1/88 000
> Second order: 1/35 000
> Third order: 1/14 000,

in terms of the semimajor axes of relative-error ellipses at the 95 % confidence level.

The above-listed accuracies for control networks are recommended for cities which intend to adopt the purely numerical system of the integrated survey system based on a mathematical description (coordinates) of boundaries as the primary evidence in legal (property) surveys. The cities which, within the foreseeable future, intend to continue to use physical monumentation and the original survey documents as the primary evidence (see Chapter 7) may lower the above accuracy requirements by 2 to 3 times for the second- and third-order control networks. In this case, a photogrammetric aerotriangulation (see Chapter 8) may prove to be more economical in the control densification than the ground survey. The first-order control, however, should be always of the highest possible accuracy, as specified above.

NETWORK DESIGN

Configuration of Geometric Nets

First-order horizontal control. The basic methods of establishing horizontal control—*triangulation, trilateration,* and *traverse*—are well known and do not call for explanation; the only question is how they are best put to use in the urban survey. Pure triangulation schemes with base lines and base extension nets belong to the past, and trilateration without measured angles has been found inadequate, although it may have to be considered in cities with heavy smog. The present practice favors triangulation, strengthened with measurement of distances, various schemes of traverse, and hybrid systems.

As a rule, first-order control should be established as a homogeneous, simultaneously adjusted network of points. It should cover the entire urban area and its expected expansion for the next 20 to 25 years.

The ideal configuration for the first-order urban net consists of an area network of equilateral triangles, but this can seldom be realized in practice. Figure 3-3 illustrates a typical design, in which the ideal configuration is approached by using a braced quadrilateral as net element. Additional diagonals connecting intervisible stations can be included in the basic figure; long diagonals that add little to the strength of the network should be omitted.

In terms of national geodetic networks, the city control net may be classified as second order, obtained by densification from the first-order net in which the average distance between adjacent stations is from 30 to 50 km. Existing geodetic stations within the urban area also become stations of the city net, and connecting ties should be established with outside geodetic control (Fig. 3-4).

Although national geodetic networks have always been measured to the highest standards of accuracy, much of the work was done before the era of electromagnetic distance measurements and without the support of the superior control that has only recently become available from satellite geodesy. First-order urban networks now being established may therefore surpass former geodetic work in accuracy. The city net should nevertheless be tied to national geodetic control, even if it is adjusted independently of it.

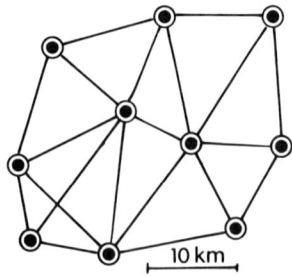

Figure 3-3. Part of the first-order network of Stuttgart, German Federal Republic (3).

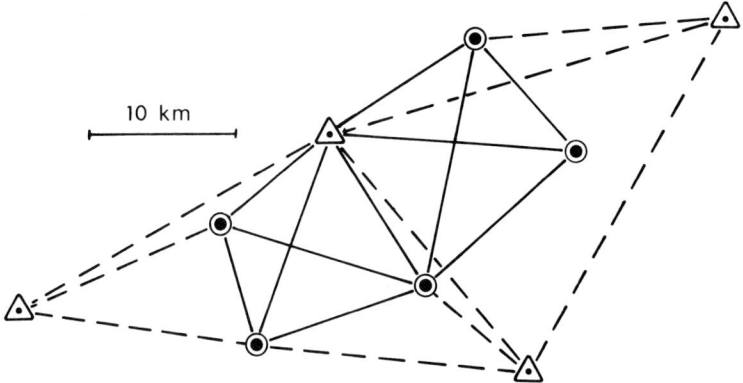

Figure 3-4. First-order network tied to the national geodetic control.

Under exceptionally difficult conditions, the first-order urban net can be measured with precise traverses following the most favorable transportation routes, usually railroads and highways. The number of legs in each traverse should be kept as low as the terrain allows, and the individual traverses should be interconnected to form a rigid network as in Figure 3-5.

Traverse nets are more economical and faster to measure than triangulation. They require, however, a very careful error preanalysis and design to meet the accuracy requirements.

Second-order nets. Before the advent of portable electronic distance-measuring (EDM) equipment and simultaneous adjustment of large networks, second-order control was established in two stages. To start with, the first-order net was densified locally by filling in a few "major stations" of second order, as in Figure 3-6. The increase in the number of fixed points made the triangulation of the remaining second-order points relatively more easy. This procedure led to inhomogeneity in the accuracy of the network. Today there is little excuse for carrying out the densification by this "patching" method. Traversing with EDM, preferably with junction points at each station and supplemented from

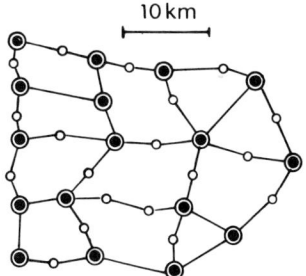

Figure 3-5. First-order city control by precise traverse.

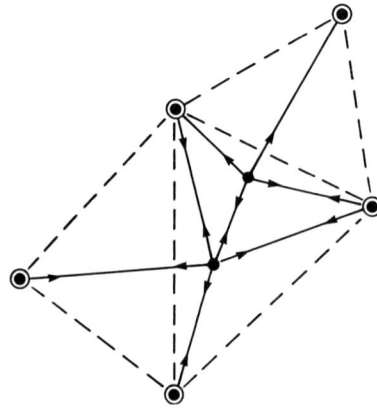

Figure 3-6. Old-type densification by triangulation (the "patching method").

time to time by diagonals, should be used in the densification. The adjustment of the second-order network should be carried out simultaneously on as large a number of points as possible, with ties to all surrounding higher-order points.

Densification by single points or single traverses as need arises should be avoided because it may lead to the situation shown in Figure 3-7, in which two traverses, A–B and C–D, have been measured and adjusted separately to different first-order points. The points I and II of the two traverses are only 1.5 km apart. If the first-order points, A, B, C, and D, which are approximately 10 km apart, satisfy the relative accuracy requirement of 1/88 000, the relative positional error between them is 10 km/88 000 = 113 mm. This error will propagate to points I and II even if the traverses are errorless, and the relative error between them may be as large as 113 mm/1.5 km = 1/13 000 instead of the required 1/35 000.

An idealized model of the densification to the second order is shown in Figure 3-8. The second-order network is adjusted simultaneously in large groups of points that fill up "unit blocks" of the first-order frame-net.

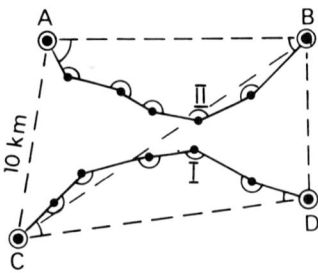

Figure 3-7. An ill-designed densification.

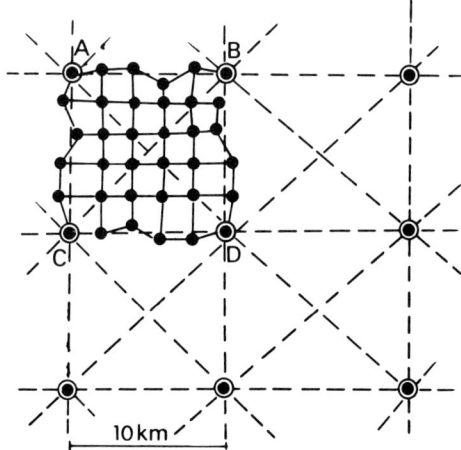

Figure 3-8. Idealized model of a densification network. The second order is simultaneously adjusted within the unit "block" of the first-order points, A, B, C and D.

Third-order nets. Third-order horizontal control is the end product of the densification process. It comprises the main body of control points intended for local use in detail surveys. The third-order stations should be established on carefully selected sites, such as street intersections, where they are most convenient for the users.

Third-order control, filled into the network of first-order and second-order stations, is best established by the traverse method, preferably in larger blocks, each containing several junction points. No traverse, however short, must be left open, i.e., without a closure to a first-order station, second-order station, or a junction point. Figure 3-9 shows a portion of the second-order and third-order networks in the city of Fredericton in Canada.

Fourth-order points. Control points that do not qualify for any of the previous classes may be described as fourth order. These may include coordinated boundary monuments, manholes, hydrants, and stations from which occasional detail surveys are carried out. Conspicuous structures unsuitable as ordinary control points or outside the working area, such as sight-seeing and telecommunication towers, which are, however, valuable in providing reference azimuths, are usually determined by the intersection method as fourth-order points.

Reconnaissance and Preliminary Design

In a given topographical situation, any number of alternative solutions are usually found for the realization of a horizontal control network. The task of the designer of a network is to determine, from among the available alternatives,

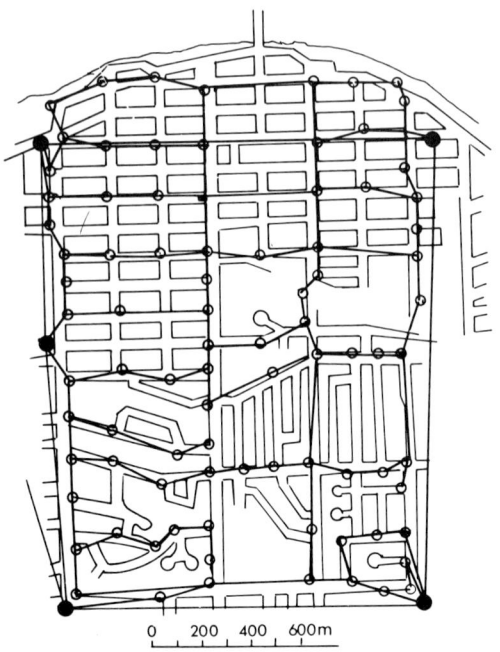

Figure 3-9. Part of the second- and third-order control in Fredericton, Canada (4).

the most economical and practical solution that will produce the required network of control points to the desired standards of accuracy and density.

The design should be preceded by a thorough office and field reconnaissance. Reconnaissance requires a good knowledge of all the phases of survey work, and it should be left to experienced hands. Oversights and errors due to superficial reconnaissance are difficult or impossible to correct at a later stage; the time spent on good planning never goes unrewarded.

Time spent in field reconnaissance can be greatly reduced by making preparatory investigations in the office. These preparations include:

acquisition of complete information on previously established geodetic control, such as coordinates and stability of the existing points, and accuracy parameters preferably in the form of the full variance–covariance matrix of the network;

study of the region's geology and soil stability for ground monumentation of survey markers;

inspection of available topographical maps and listing of building heights in the area for determination of intervisibility; and

acquisition of information on the ownership of buildings suitable for rooftop or wall monumentation, followed by requests for permission for monumentation.

Office reconnaissance based on the use of topographical maps can frequently be carried out to the point where a definite plan for the network design emerges. In field reconnaissance, the design is confirmed, and the exact location and type of monument is decided for each station. The intervisibility between the pertinent stations can be checked fairly accurately by computing the elevation of the line of sight at suspected obstruction points by the formula

$$h(\text{m}) = \frac{s_1 h_2 + s_2 h_1}{s_1 + s_2} - \frac{s_1 s_2}{15}, \tag{3-1}$$

where h_1 and h_2 are the terminal elevations in meters, and s_1 and s_2 are the distances from the terminals in kilometers. The formula takes into account the effects of the curvature of the earth and vertical refraction. For example, if $h_1 = 305$ m, $h_2 = 275$ m, $s_1 = 8.0$ km, and $s_2 = 3.3$ km, the line of sight would be cut off by an obstruction with elevation $h \geq 282$ m. Reference (5) gives a graphical solution for convenient checking of large numbers of lines.

Another factor to be watched in selecting lines of sight is *lateral refraction*. Horizontal differences in air temperature are the principal cause for lateral refraction; a light ray tends toward warmer air, where the propagation speed is greater. Temperature differences along the path are obviously harmless since they do not change the direction of the ray; however, horizontal temperature gradients at right angles to the path cause lateral refraction.

The heat-island structure of most metropolitan centers, consisting of built-up areas interspersed by parklands, is particularly affected by lateral refraction, which is believed to be one of the most important single sources of error in the measurement of horizontal directions. Assuming that a uniform horizontal temperature gradient dT/dx persists over the whole length S of the line of sight, angle γ of lateral refraction can be calculated by the approximate formula

$$\gamma'' = 8''.0 \left(\frac{pS}{T^2}\right)\left(\frac{dT}{dx}\right), \tag{3-2}$$

where p is the barometric pressure in millibars, T is the temperature in kelvins ($T = 273.15 + t°\text{C}$), and dT/dx is measured horizontally at right angles to the line of sight.

A gradient of only $0.001°\text{C/m}$, if it persists over the whole length of a 10-km line, will cause a lateral refraction of $0''.9$. But temperature gradients may reach values of $0.3°\text{C}$ and even higher (6) in urban areas if the lines of sight pass close to the walls of buildings exposed to the sun. The values of γ may then become equal to $5''$ for distances of 200 m. It should be emphasized, however, that the distribution of the horizontal gradients of temperature is unpredictable and difficult to measure. Therefore, Eq. (3-2) may serve only as an estimation of expected errors, not for the correction of measurements. To diminish the errors caused by lateral refraction, the following points should be considered in selecting the lines of sight:

First-order lines should be raised as high as possible, with no portion of the line running less than 10 m above the ground.

Grazing lines over hillsides must be avoided, and no line should pass near a smokestack or a wall of a tall building.

Heat islands, or cold islands, should be bisected centrally, rather than crossed off the center (the rule is reversed if distances are being measured instead of directions).

Rooftop pillars should possess an unobstructed view in all directions; no lines should pass near walls or posts, and lines should be placed clear of outlets of hot air.

A traverse following a straight roadway should preferably have stations alternating between the left and right sides of the road; traversing on city streets should, if possible, be directed along the center line between buildings.

Once the locations of points have been selected and intervisible lines checked in the field, the preliminary design of the control network can be prepared. This will include:

1. a map at a scale of, say, 1:25 000, marked with all existing and proposed points of the geodetic and first-order city network and showing all considered intervisibility lines.
2. a map or sketch at a scale of 1:10 000, or a similar scale, marked with all existing and proposed points of the densification network and showing all the possible connecting lines.
3. the exact location and type of monumentation of survey stations.
4. the proposed survey instruments to be used and preliminary specifications for field observations.
5. the proposed method of tying the city network to the national geodetic network, if applicable.

In points 1 and 2 of the preliminary design, the general rules governing the required density and configuration of the networks should be followed as much as possible (see pp. 47–53).

The preliminary design leaves alternatives open for the final selection of the lines to be observed between the points and final decisions on the requirements of measurement accuracies and field procedures. These decisions are extremely important because the final design of the geometry of the network and of survey procedures should provide the most economical solution and meet the required accuracy of the network, with a minimum of lines observed, using the simplest possible survey techniques. Therefore, the final design of the network must be based on rigorous preanalysis and optimization.

Accuracy Preanalysis and Optimization

Review of definitions and notations used in the accuracy analysis. Only a brief review of the principles of the accuracy preanalysis is given here since it is assumed that readers are familiar with the basic theory of error propagation and accuracy analysis. Readers who lack the background are referred to Additional Readings for a list of texts that deal with the subject matter in more detail.

The accuracy of a surveying network is fully defined if errors of relative positions between any two points in the network are known at a certain *confidence level* (probability level). Usually a 95 % probability that the error of the relative position does not exceed the given value is accepted as the confidence level. The *relative position* of one point with respect to another point is the difference in the adjusted positions of the two points as defined by the differences of the correspondingly adjusted coordinate values. If the positional accuracy of a point is related to the reference points that are treated as fixed (errorless) in the process of the network adjustment, the term *absolute* positional accuracy is applied.

Variances and *covariances* of coordinates of the analyzed network are a basis for calculations of the absolute and relative positional errors. They allow the calculation of following quantities that are frequently used in describing the positional accuracies:

Standard deviations of coordinates or differences in coordinates.

Semimajor and semiminor axes of *standard-error ellipses* (absolute or relative).

Standard deviations of distances, directions (azimuths), or angles calculated from adjusted coordinates.

The most general description of the positional accuracy of a point is given in a graphical form by an *error curve*.

Brief definitions of the above terms and quantities are given below.

The *variance* (denoted by σ^2 or m^2) of a coordinate value or of an observation is a statistical measure of reliability of the value. It is also called mean square error.

The *covariance* of a pair of coordinate values belonging to one or two points (denoted by $\sigma_{X_i Y_i}$, $\sigma_{X_i X_j}$ or by $m_{X_i Y_i}$, $m_{X_i X_j}$) is a measure of statistical dependence of the two values. The same applies to the covariance of two observations. If two coordinates or two observation values are uncorrelated, their covariance equals zero. In most cases, coordinates of points that are calculated using the same set of observations are correlated.

The *standard deviation* (denoted by σ or m) of a coordinate value, a difference in coordinates, or an observation is the positive square root of the corresponding variance. It is also called standard error or root mean square (rms) error. Standard deviation represents a probability (confidence level) of 68 % that the difference between the given (calculated) coordinate and its true value is within the interval $+\sigma$ to $-\sigma$. To increase the confidence level to 95 or to 99%, the standard deviation must be multiplied by 1.96 or 2.58, respectively. Standard deviations of X- and Y-coordinates of a point describe the positional accuracy of the point only in directions of the X- and Y-axes of the coordinate system. Usually, we like to know the maximum and minimum standard deviations (σ_{max} and σ_{min}) for the point and their directions. These can be calculated from

$$\sigma_{max}^2 = \tfrac{1}{2}(\sigma_X^2 + \sigma_Y^2 + \sqrt{(\sigma_X^2 - \sigma_Y^2)^2 + 4\sigma_{XY}^2}), \qquad (3\text{-}3)$$

$$\sigma_{min}^2 = \tfrac{1}{2}(\sigma_X^2 + \sigma_Y^2 - \sqrt{(\sigma_X^2 - \sigma_Y^2)^2 + 4\sigma_{XY}^2}), \qquad (3\text{-}4)$$

and

$$\tan 2\phi = \frac{\sin 2\phi}{\cos 2\phi} = \frac{2\sigma_{XY}}{\sigma_Y^2 - \sigma_X^2}, \qquad (3\text{-}5)$$

where ϕ ($\leq 180°$) is the azimuth of the direction of σ_{max} (Fig. 3-10). The direction of σ_{min} is $\phi + 90°$. The quadrant of 2ϕ is determined by the algebraic signs of the numerator and the denominator in Eq. (3-5), i.e., by the signs of $\sin 2\phi$ and $\cos 2\phi$.

The *standard-error ellipse* of a calculated absolute position is an ellipse described by σ_{max} and σ_{min} as the semimajor axis a ($a = \sigma_{max}$) and the semi-minor axis $b = \sigma_{min}$. If drawn around a point, it is usually interpreted as depicting the region in which we have 39% confidence that it contains the position of the corresponding point determined from errorless observations. To increase the confidence to 95% or 99%, the semiaxes a and b of the error ellipse must be multiplied by 2.45 and 3.03, respectively. If an error ellipse is calculated on the basis of variances and covariances of differences ΔX and ΔY of coordinates of two points i and j, it is called a *relative standard-error ellipse*, which describes the relative positional accuracy between the points i and j. The calculations are then in the form

$$a^2 = \tfrac{1}{2}(\sigma_{\Delta X}^2 + \sigma_{\Delta Y}^2 + \sqrt{(\sigma_{\Delta X}^2 - \sigma_{\Delta Y}^2)^2 + 4\sigma_{\Delta X \Delta Y}^2}) \qquad (3\text{-}6)$$

$$b^2 = \tfrac{1}{2}(\sigma_{\Delta X}^2 + \sigma_{\Delta Y}^2 - \sqrt{(\sigma_{\Delta X}^2 - \sigma_{\Delta Y}^2)^2 + 4\sigma_{\Delta X \Delta Y}^2}) \qquad (3\text{-}7)$$

and

$$\tan 2\phi = \frac{2\sigma_{\Delta X \Delta Y}}{\sigma_{\Delta Y}^2 - \sigma_{\Delta X}^2}, \qquad (3\text{-}8)$$

where

$$\sigma_{\Delta X}^2 = \sigma_{X_j}^2 + \sigma_{X_i}^2 - 2\sigma_{X_i X_j} \qquad (3\text{-}9)$$

$$\sigma_{\Delta Y}^2 = \sigma_{Y_j}^2 + \sigma_{Y_i}^2 - 2\sigma_{Y_j Y_i} \qquad (3\text{-}10)$$

$$\sigma_{\Delta X \Delta Y} = \sigma_{X_j Y_j} - \sigma_{X_j Y_i} - \sigma_{X_i Y_j} + \sigma_{X_i Y_i}. \qquad (3\text{-}11)$$

The *error curve*, called also *pedal curve*, if drawn around a point, describes positional standard deviations σ_ε in any desired direction (Fig. 3-10), at an angle measured clockwise from the semiaxis a of the error ellipse.

$$\sigma_{\varepsilon_i}^2 = a^2 \cos^2 \varepsilon_i + b^2 \sin^2 \varepsilon_i \qquad (3\text{-}12)$$

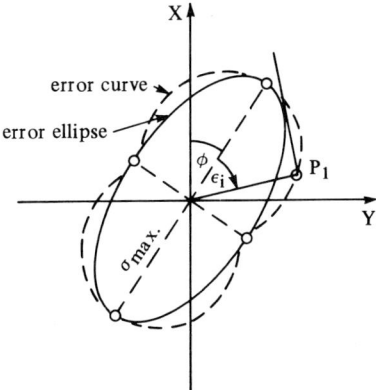

Figure 3-10. Error curve and error ellipse.

Graphically, the error curve is found by determining its points as an apex P of a right angle between the direction line of ε and the tangent line to the error ellipse.

The *variance–covariance matrix* of coordinate values of the points in the network is an ensemble of variances and covariances of the individual values obtained by applying general rules of error propagation (variance–covariance law) to the calculation of coordinates. The variance–covariance matrix, denoted by Σ_x, is square and symmetrical. For instance, for two points i and j in a network, Σ_x will be a 4×4 matrix of the form

$$\Sigma_x = \begin{pmatrix} \sigma_{X_i}^{2} & \sigma_{X_i Y_i} & \sigma_{X_i X_j} & \sigma_{X_i Y_j} \\ \sigma_{X_i Y_i} & \sigma_{Y_i}^{2} & \sigma_{Y_i X_j} & \sigma_{Y_i Y_j} \\ \sigma_{X_i X_j} & \sigma_{Y_i X_j} & \sigma_{X_j}^{2} & \sigma_{X_j Y_j} \\ \sigma_{X_i Y_j} & \sigma_{Y_i Y_j} & \sigma_{X_j Y_j} & \sigma_{Y_j}^{2} \end{pmatrix}$$

with variances on the main diagonal and covariances at the intersections of the ith and jth rows and columns.

Error propagation and positional accuracy. If a set of n unknowns u_i (vector matrix U) is computed from a set of k known variables z_i (vector matrix Z) that have known variances and covariances (matrix Σ_z), according to the general formula, known as the propagation of variance–covariance matrices, we can calculate the variance–covariance matrix Σ_U of the unknowns from

$$\Sigma_U = B \Sigma_Z B^{T}, \tag{3-13}$$

where B is the so-called design matrix. Elements of the B matrix are calculated as partial derivatives of the functions

$$u_i = u_i (z_1, z_2, \ldots, z_k),$$

where $i = 1, 2, \ldots, n$ at the approximate points of variables. Thus

$$
B = \begin{vmatrix}
\dfrac{\partial u_1}{\partial z_1} & \dfrac{\partial u_1}{\partial z_2} & \cdots & \dfrac{\partial u_1}{\partial z_k} \\[2ex]
\dfrac{\partial u_2}{\partial z_1} & \dfrac{\partial u_2}{\partial z_2} & \cdots & \dfrac{\partial u_2}{\partial z_k} \\[2ex]
\vdots & \vdots & & \vdots \\[2ex]
\dfrac{\partial u_n}{\partial z_1} & \dfrac{\partial u_n}{\partial z_2} & \cdots & \dfrac{\partial u_n}{\partial u_k}
\end{vmatrix}.
$$

We have, therefore, in more detail

$$
\Sigma_U = \begin{pmatrix}
\sigma_{u_1}^2 & \sigma_{u_1 u_2} & \cdots & \sigma_{u_1 u_n} \\
\sigma_{u_2 u_1} & \sigma_{u_2}^2 & \cdots & \sigma_{u_2 u_n} \\
\vdots & \vdots & & \vdots \\
\sigma_{u_n u_1} & \sigma_{u_n u_2} & \cdots & \sigma_{u_n}^2
\end{pmatrix} = B \begin{pmatrix}
\sigma_{z_1}^2 & \sigma_{z_1 z_2} & \cdots & \sigma_{z_1 z_k} \\
\sigma_{z_2 z_1} & \sigma_{z_2}^2 & \cdots & \sigma_{z_2 z_k} \\
\vdots & \vdots & & \vdots \\
\sigma_{z_k z_1} & \sigma_{z_k z_2} & \cdots & \sigma_{z_k}^2
\end{pmatrix} B^T .
$$

If we have only one unknown, say u_1, as a function of several known variables, say z_1, z_2, and z_3, the above general law of variance–covariance propagation leads to the well-known formula of error propagation,

$$
\sigma_{u_1}^2 = \left(\frac{\partial u_1}{\partial z_1}\right)^2 \sigma_{z_1}^2 + \left(\frac{\partial u_1}{\partial z_2}\right)^2 \sigma_{z_2}^2 + \left(\frac{\partial u_1}{\partial z_3}\right)^2 \sigma_{z_3}^2
$$
$$
+ 2\frac{\partial u_1}{\partial z_1}\frac{\partial u_1}{\partial z_2}\sigma_{z_1 z_2} + 2\frac{\partial u_1}{\partial z_1}\frac{\partial u_1}{\partial z_3}\sigma_{z_1 z_3} + 2\frac{\partial u_1}{\partial z_2}\frac{\partial u_1}{\partial z_3}\sigma_{z_2 z_3}, \qquad (3\text{-}14)
$$

where the partial derivatives are calculated for approximate values of $z_1, z_2,$ and z_3.

In the design of a horizontal control network, we can usually estimate a priori the accuracy of planned observations (angles, distances, azimuths, etc.), and we know approximate coordinates of the points, for instance, from a large-scale layout of the network. Each observation l_i can be expressed as a function of coordinates of k points to which the observation is related,

$$
l_i = l_i (X_1, Y_1, X_2, Y_2, \ldots, X_k, Y_k) \qquad (3\text{-}15)
$$

with $i = 1, 2, \ldots, n$, where n is the number of observations.

If we denote by A the matrix of partial derivatives of the functions in Eq. (3-15), calculated for approximate values of the coordinates, we can calculate, by using the general rule of error propagation, the expected variance–covariance matrix of the coordinates from

$$
\Sigma_x = \sigma_o^2 (A^T P A)^{-1} = \sigma_o^2 Q, \qquad (3\text{-}16)
$$

where

$$A = \begin{vmatrix} \dfrac{\partial l_1}{\partial X_1} & \dfrac{\partial l_1}{\partial Y_1} & \dfrac{\partial l_1}{\partial X_2} & \dfrac{\partial l_1}{\partial Y_k} \\[2mm] \dfrac{\partial l_2}{\partial X_1} & \dfrac{\partial l_2}{\partial Y_1} & \dfrac{\partial l_2}{\partial X_2} & \dfrac{\partial l_2}{\partial Y_k} \\[2mm] \vdots & \vdots & \vdots & \vdots \\[2mm] \dfrac{\partial l_n}{\partial X_1} & \dfrac{\partial l_n}{\partial Y_1} & \dfrac{\partial l_n}{\partial X_2} & \dfrac{\partial l_n}{\partial Y_k} \end{vmatrix} \tag{3-17}$$

The value $\mathbf{P}$ denotes the matrix of *weights* of observations if the latter are of different accuracies. The weights are arbitrary quantities that are inversely proportional to the variances of observations and are calculated as ratios $p_i = \sigma_o^2/\sigma_i^2$ where σ_o^2 is a variance of the observation for which weight $p_0 = 1$. The value of σ_o^2, called also a variance factor, may be arbitrarily taken as any number but is usually taken as $\sigma_o^2 = 1$. In general terms, the $\mathbf{P}$ matrix is calculated as

$$\mathbf{P} = \sigma_o^2 \mathbf{\Sigma}_L^{-1} \tag{3-18}$$

where $\mathbf{\Sigma}_L$ is a variance–covariance matrix of observations. If the observations are not correlated (the usual case), the $\mathbf{P}$ matrix is a diagonal matrix of the form

$$\mathbf{P} = \sigma_o^2 \begin{vmatrix} \dfrac{1}{\sigma_{l_1}^2} & & & \\[2mm] & \dfrac{1}{\sigma_{l_2}^2} & & \\[2mm] & & \ddots & \\[2mm] & & & \dfrac{1}{\sigma_{l_n}^2} \end{vmatrix}$$

If all the observations are of equal accuracy σ_l, it is convenient to accept $\sigma_o^2 = \sigma_l^2$, the $\mathbf{P}$ matrix then becomes a unit matrix, and Eq. (3-16) becomes

$$\mathbf{\Sigma}_x = \sigma_l^2 (\mathbf{A}^T \mathbf{A})^{-1}. \tag{3-19}$$

For individual types of observations, we can write the following equations and their differentials with partial derivatives calculated for approximate values of the coordinates:

1. *Distance observation* between points i and j,

$$s_{ij} = \sqrt{(X_j - X_i)^2 + (Y_j - Y_i)^2}, \tag{3-20}$$

and

$$ds = \frac{X_j - X_i}{s} dX_j + \frac{(Y_j - Y_i)}{s} dY_j - \frac{X_j - X_i}{s} dX_i - \frac{Y_j - Y_i}{s} dY_i. \tag{3-21}$$

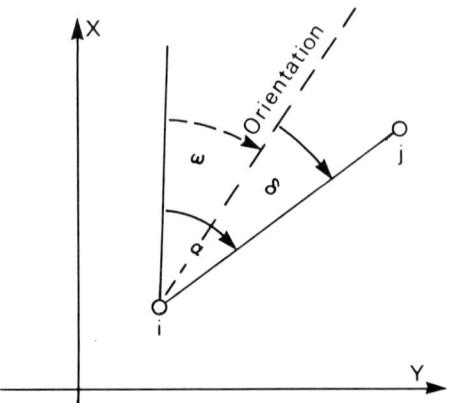

Figure 3-11 Direction observation.

2. *Azimuth observation* between points i and j,

$$\alpha = \arc\tan \frac{Y_j - Y_i}{X_j - X_i} \tag{3-22}$$

and

$$d\alpha = \frac{-(Y_j - Y_i)}{s^2} dX_j + \frac{(X_j - X_i)}{s^2} dY_j + \frac{(Y_j - Y_i)}{s^2} dX_i \tag{3-23}$$

$$- \frac{(X_j - X_i)}{s^2} dY_i.$$

3. *Direction observation* (Fig. 3-11),

$$\delta = \alpha - \omega = \arc\tan \frac{Y_j - Y_i}{X_j - X_i} - \omega, \tag{3-24}$$

where ω is orientation angle of the zero readout of the horizontal circle. From Eq. (3-23), we have

$$d\delta = \frac{-(Y_j - Y_i)}{s^2} dX_j + \frac{(X_j - X_i)}{s^2} dY_j + \frac{(Y_j - Y_i)}{s^2} dX_i$$

$$- \frac{(X_j - X_i)}{s^2} dY_i - d\omega. \tag{3-25}$$

The orientation angle ω adds one more unknown (one more column in matrix **A**) with the coefficient equal to -1 for each group of directions measured.

4. *Angle observation* between L (left) and R (right) arms from station C (central), as shown in Figure 3-12,

$$\beta = \alpha_R - \alpha_L = \arc\tan \frac{Y_R - Y_C}{X_R - X_C} - \arc\tan \frac{Y_L - Y_C}{X_L - X_C}, \tag{3-26}$$

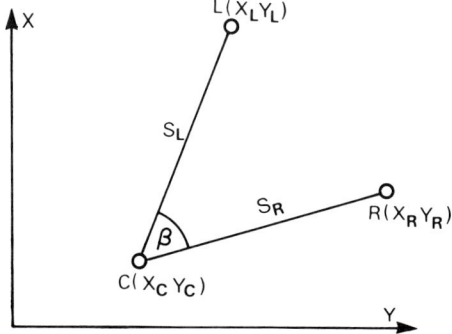

Figure 3-12. Angle observation.

and

$$d\beta = \frac{X_R - X_C}{s_R^2} dY_R - \frac{Y_R - Y_C}{s_R^2} dX_R - \frac{X_L - X_C}{s_L^2} dY_L + \frac{Y_L - Y_C}{s_L^2} dX_L$$

$$+ \left(\frac{X_L - X_C}{s_L^2} - \frac{X_R - X_C}{s_R^2}\right) dY_C - \left(\frac{Y_L - Y_C}{s_L^2} - \frac{Y_R - Y_C}{s_R^2}\right) dX_C.$$

(3-27)

The coefficients of Eqs. (3-23), (3-25), and (3-27) should be multiplied by one radian in seconds ($\rho'' = 206\ 265''$) if the variances and covariances of the corresponding observations in the weight matrix are in sexagesimal seconds. The linear units of the coordinates, distances, and their variances should be properly chosen to make the coefficients of angular and linear observations numerically convenient for computations.

The coefficients of Eqs. (3-21), (3-23), (3-25), and (3-27) are elements of the matrix **A** in Eq. (3-16). Note that if an observation is made to a fixed (treated as errorless) point in the analyzed network, the partial differentials in respect to its coordinates equal zero and the corresponding columns in the matrix **A** are omitted.

Example 3-1 illustrates a practical application of Eq. (3-16) in the positional accuracy preanalysis of a point P (Fig. 3-13) whose coordinates are supposed to be determined on a basis of three angles and three distances measured from three known control points assumed as fixed and errorless.

Example 3-1. Positional Error Analysis.
Planned measurements: distances s_1, s_2, and s_3 with standard deviation $\sigma_s = 10$ mm and angles β_1, β_2, β_3 with $\sigma_\beta = 2''$. Observations will be uncorrelated.

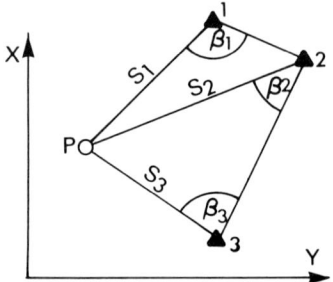

Figure 3-13. Positioning of point P.

Approximate coordinates of the fixed points and of the new point P (taken from a large-scale map) are:

Pt	Y(m)	X(m)
1	600	800
2	900	700
3	600	100
P	200	400

Units of seconds and centimeters are used in the analysis. The matrix $\mathbf{A}$ will have two columns (two unknown coordinates) and six rows (six observations). Using Eq. (3-27) for angles and Eq. (3-21) for distances, we obtain

$$\mathbf{A} = \begin{matrix} Y_P & X_P & \\ \begin{pmatrix} -2.58 & 2.58 \\ -1.07 & 2.49 \\ -2.48 & -3.30 \\ -0.71 & -0.71 \\ -0.92 & -0.39 \\ -0.80 & 0.60 \end{pmatrix} & \begin{matrix} \beta_1 \\ \beta_2 \\ \beta_3 \\ s_1 \\ s_2 \\ s_3 \end{matrix} \end{matrix}$$

Taking $\sigma_o^2 = 1$, $\sigma_\beta^2 = 4$, and $\sigma_s^2 = 1$ (cm^2), we obtain a 6×6 diagonal weight matrix

$$\mathbf{P} = \begin{pmatrix} 0.25 & \cdot & \cdot & \cdot & \cdot & \cdot \\ \cdot & 0.25 & \cdot & \cdot & \cdot & \cdot \\ \cdot & \cdot & 0.25 & \cdot & \cdot & \cdot \\ \cdot & \cdot & \cdot & 1 & \cdot & \cdot \\ \cdot & \cdot & \cdot & \cdot & 1 & \cdot \\ \cdot & \cdot & \cdot & \cdot & \cdot & 1 \end{pmatrix}$$

and

$$\mathbf{A}^T \mathbf{PA} = \begin{pmatrix} 5.48 & 0.10 \\ 0.10 & 6.95 \end{pmatrix}$$

The variance–covariance matrix is obtained from Eq. (3-16)

$$\mathbf{\Sigma}_X = \begin{pmatrix} 0.18 & -0.00 \\ -0.00 & 0.14 \end{pmatrix} = \sigma_o^2 \mathbf{Q}$$

and the following standard deviations for coordinates of point P will be obtained

$$\sigma_Y = \sqrt{0.18} = 4 \text{ mm} \quad \text{and} \quad \sigma_X = \sqrt{0.14} = 4 \text{ mm}$$

Semiaxes of the standard-error ellipse, calculated from Eqs. (3-3) and (3-4), are

$$a = \sigma_{max} = 4 \text{ mm} \quad \text{and} \quad b = \sigma_{min} = 4 \text{ mm}$$

The positional accuracy of point P at 95% probability in terms of the semi-major axes may be expected to be $2.45 \times 4 = 10$ mm with respect (relative) to the points 1, 2, and 3 that have been treated as fixed in the error analysis.

If points 1, 2, and 3 in the above example belonged, for instance, to a second-order control network that was previously adjusted simultaneously within a first-order network, we might be interested to learn the positional accuracy of our new point P with respect to the first-order points. In this case we have to know the $\mathbf{\Sigma}_{X_{1,2,3}}$ for points 1, 2, and 3 from the previous analysis, and the new variance–covariance matrix $\mathbf{\Sigma}_{X_P}$ for point P with respect to the first-order points will be obtained (7) from

$$\mathbf{\Sigma}_{X_P} = \sigma_o^2 (\mathbf{P}_{X_{1,2,3}} + \bar{\mathbf{A}}^T \mathbf{P} \bar{\mathbf{A}})^{-1}, \tag{3-28}$$

where

$$\mathbf{P}_{X_{1,2,3}} = \mathbf{\Sigma}_{X_{1,2,3}}^{-1},$$

enlarged by as many zero columns and rows as new unknowns are added to the network; $\bar{\mathbf{A}}$ is the design matrix with partial derivatives in respect not only to point P but also in respect to points 1, 2, and 3, which are treated in this case as unknowns; $\mathbf{P}$ is the same matrix (weights of the new observations), and σ_o is the same.

We have, therefore, in Example 3-1, the $\mathbf{P}_X$ matrix with a dimension of 8×8 (six columns and rows for old coordinates plus two zero columns and two zero rows to accommodate the two new unknowns). Matrix $\bar{\mathbf{A}}$ would be 6×8 matrix for six observations and eight unknowns (which are also treated in this approach as quasi-observations).

The solution given by Eq. (3-28) would give the 8×8 matrix $\mathbf{\Sigma}_{X_P}$ which would give all the necessary information for calculating the positional error of P with respect to the first-order points with the same result as the simultaneous adjustment of the coordinates of P with the second-order network to which points 1, 2, and 3 belong.

Further discussion of error analysis and more examples are given on pp. 121–131.

Optimization and final design of the network. The procedures shown in the simple Example 3-1 may be easily extended to a preanalysis of control networks of any size with large computers. Programs are available, and the cost of the computer analysis is minimal. For instance, when using the IBM-370/158 computer (50 000 computations per second), calculation of the full variance–covariance matrix for a network consisting of 100 new points (200 unknowns) takes less than 3 min of the effective run time, which costs $32 at the rates ($650 per hour) charged in 1976 by the University of New Brunswick. A UNB program (8) gives absolute and relative error ellipses for 36 stations, on the above-mentioned computer, in 33 s at a cost of $9.

City survey offices, therefore, have no excuse for not designing a control network based on rigorous error analysis. The same can be said about the adjustment of the networks (see discussion beginning on p. 121).

The final design of a network must be preceded by an optimization process giving answers to the following questions arising from the preliminary design: Which of the possible intervisibility lines between proposed stations should be utilized in the final design of the configuration of the network? What type and what accuracy of observations are necessary to satisfy the positional accuracy requirements?

Optimization is not an easy task; it requires good experience from the designer. The problem of optimization has been the subject of many publications and discussions on the development of mathematical tools that will facilitate the determination of the influence of individual observations on changes in the positional accuracy of the whole network. Publications such as those in Refs. (9), (10), and (11) are just a few examples of new approaches in the field of optimization. To date, however, the only practical way of performing the optimization of control networks is by the trial-and-error method of running the accuracy preanalysis of the simulated network several times, by changing each time the number, type, and accuracy of observations and analyzing which alternative gives the best results in the most economical way or the best compromise between accuracy and cost.

A detailed discussion on the subject of optimization is beyond the scope of this book, and only a brief numerical example is given to demonstrate the power of the optimization.

Example 3-2. Let us consider a design of the network shown in Figure 3-14. All possible intervisibility lines are shown in the figure. Points 1 and 2 belong

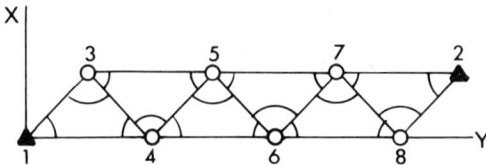

Figure 3-14. Triangulation vs. triangulateration.

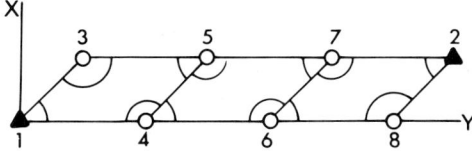

Figure 3-15. Quadratic loop traversing.

to a higher-order control; they will be kept fixed (errorless) when determining the coordinates of remaining new points. Distances between neighboring points range from 1.4 to 2 km. Required accuracy in terms of the semimajor axes of relative standard-error ellipses is 1:90 000. The preliminary design proposes to measure angles with a 1″ theodolite in four sets, estimating $\sigma_\beta = 2″$ and distances with a short-range EDM instrument that gives standard deviation $\sigma = 10$ mm for distances of up to 2 km. The following five alternatives have been considered for the final design.

Case 1. All the angles and distances measured (triangulateration) as marked in Figure 3-14.

Case 2. Only angles measured (triangulation) as marked in Figure 3-14.

Case 3. Only distances measured (trilateration).

Case 4. Quadratic loop traversing (Fig. 3-15).

Case 5. Straight loop traverse (Fig. 3-16).

Approximate coordinates of the points are:

Station	X(m)	Y(m)
1	0	0
2	1000	7000
3	1000	1000
4	0	2000
5	1000	3000
6	0	4000
7	1000	5000
8	0	6000

Full variance–covariance matrices and absolute and relative standard-error ellipses have been calculated for all five alternatives by means of, as in Example 3-1, Eqs. (3-3) to (3-11) and (3-16) to (3-27).

Table 3-1 lists the results of calculations of semimajor axes of absolute- and relative-error ellipses.

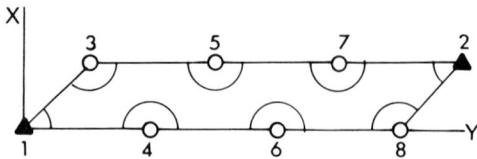

Figure 3-16. Straight loop traverse.

The whole analysis took only 11 s of the total computer time, including compiling, using the UNB (8) program and IBM 370/158 computer.

Comparison of the results from the five cases leads to the following conclusions.

1. As to absolute positional accuracy, the straight traverse loop with only 16 observations gave the best results, followed closely by the quadratic loops (22 observations) and triangulateration (31 observations). The pure triangulation (18 measured angles) and trilateration (13 distances) gave more than doubly poor results.
2. As to relative positional accuracy, the quadratic loop traverses and triangulateration gave practically the same results. The straight traverse loop did not satisfy the requirement of 1:90 000 relative accuracy, because the error ellipses of the lines 4–5 and 5–6 exceeded the limit (18 mm/1.4 km = 1:78 000). Similarly, pure triangulation and trilateration have to be discarded, unless the accuracy of measurements in those networks is increased.

Summarizing, the quadratic loop traverses would probably be accepted for the final design unless it proves more economical to perform more precise angle measurements in the triangulation in order to avoid the use of two instruments, the theodolite and EDM equipment as it is necessary in traversing.

Table 3-1. Standard-Error Ellipses (Semimajor Axis a in Millimeters)

Absolute-error ellipses a(mm)						Relative-error ellipses a(mm)					
		Case:							Case:		
Pt	1	2	3	4	5	Line	Distance (km)	1	2	3	4	5
3	12	26	21	10	10	4–5	1.4	8	17	16	8	17
4	16	31	28	14	13	4–6	2.0	11	19	21	11	10
5	19	39	33	17	16	4–7	3.2	13	21	23	12	15
6	19	39	33	17	16	4–8	4.0	14	25	24	13	12
7	16	31	28	14	13	5–6	1.4	8	15	15	10	18
8	12	26	21	10	10	5–8	3.2	15	27	25	14	16

If the urban network is designed according to the previously specified required densities and configurations, a maximum of about 100 stations will be subjected to a simultaneous adjustment at one time within the blocks of higher order. Therefore, as already indicated, the computer cost for the proper optimization based on several runs of the preanalysis is negligible. Thus we do not need to obey blindly old specifications such as "Do not design angles smaller than 30°," or "Do not run a traverse with a ratio of neighboring sides smaller than 1:4." Each surveyor may design the shape of the network and accuracy of measurements according to local conditions and the best economical solution as long as it can be proven by preanalysis that the general accuracy specifications are satisfied and the surveyor understands what accuracy of measurement can be achieved with the instrumentation available.

MONUMENTATION OF CONTROL POINTS

Location of Control Points

Three types of monumentation are used in urban control networks: ground monumentation, wall monumentation, and rooftop markers.

The first-order network is usually monumented on the roofs of buildings; ground stations are used only in suburban districts. Roof monumentation is also used occasionally in the second-order network, but with the use of traversing as the basic approach in second-order work, preference is given to suitably monumented ground stations. Third-order control points are as a rule ground stations, but wall monumentation is becoming more and more popular.

When choosing the type of monumentation, the following factors should be considered: stability and permanency of the markers, accessibility, and intervisibility to other points.

Generally, the ground markers are subjected to frequent destruction during reconstruction of streets. Between 5 and 20% of ground monuments are destroyed every year in urban areas. Intervisibility between the ground-monumented control points is very often lost permanently by new construction along the line of sight or blocked temporarily by parked cars causing delay in the surveys. City traffic and construction excavations cause ground movements that lead to the displacement of monuments. Snow and ice during the winter season in high geographical latitudes make it difficult to locate ground markers.

Some of these problems can be overcome by monumenting the markers on the walls and roofs of buildings (see pp. 72–77).

Whatever type of monumentation is used, each monument has to be referenced to at least three witness markers established within a few meters of the monument. Distances to the witness markers are taped and are shown on topographical descriptions in the form of sketches, which have to be prepared for each control point. Figure 3-17 shows a typical topographical description used in the city of Ottawa, Canada.

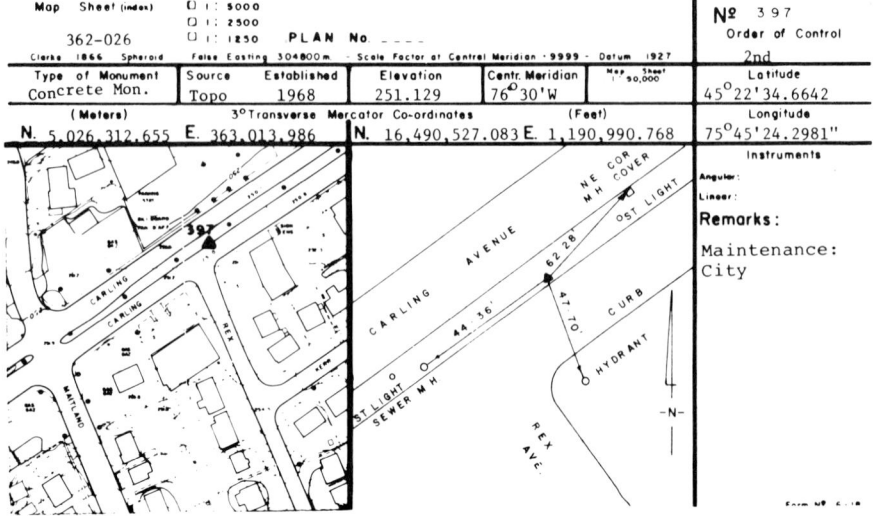

Map Sheet (index)	☐ 1 : 5000						№ 397
	☐ 1 : 2500						Order of Control
362–026	☐ 1 : 1250	PLAN No. _ _ _ _					2nd
Clarke 1866 Spheroid	False Easting 304800 m	- Scale Factor at Central Meridian · 9999 -		Datum 1927			
Type of Monument	Source	Established	Elevation	Centr. Meridian	Map Sheet 1' 50,000		Latitude
Concrete Mon.	Topo	1968	251.129	76°30'W			45°22'34.6642
(Meters)		3°Transverse Mercator Co-ordinates			(Feet)		Longitude
N. 5,026,312.655	E. 363,013.986		N. 16,490,527.083	E. 1,190,990.768			75°45'24.2981"

Instruments
Angular :
Linear :

Remarks :

Maintenance:
City

Figure 3-17. Topographical description, City of Ottawa (courtesy of the National Capital Commission).

Ground Monumentation

Data on the geology of the area and soil stability, collected during reconnaissance, is indispensable for proper selection of types of ground monuments and avoiding sites where the soil is subject to horizontal movement.

In those few cities where bedrock occurs near the surface, permanent monu-'mentation of control points does not present any difficulty. Iron or bronze bolts, or tablets cemented into bedrock, can be expected to remain stable over long periods of time, although small shifts of the order of a few millimeters will occur because of thermal expansion and contraction in the individual rock slate. The worst soil for monumentation is clay, in which the points move with water conditions and frost action. The best foundations known are large boulders or concrete blocks embedded in thick layers of coarse gravel, such as moraine. In shallow overburden, concrete monuments should be connected with bedrock by reinforcing steel bars. Examples of typical survey markers, as recommended in Ref. (2), are shown in Figures 3-18, 3-19, and 3-20.

The first-order and second-order points, if not monumented in the bedrock, are usually doubly marked by monumenting a smaller marker concentrically under the main one.

Third-order stations are frequently monumented directly in the curbs of sidewalks by drilling a small hole and cementing into it a small bronze tablet with a centering mark. Permanency of this type of monumentation is questionable, although experience shows that the curb markers may at times show better stability than deep-ground markers if the latter are not properly designed according to soil and climatic conditions.

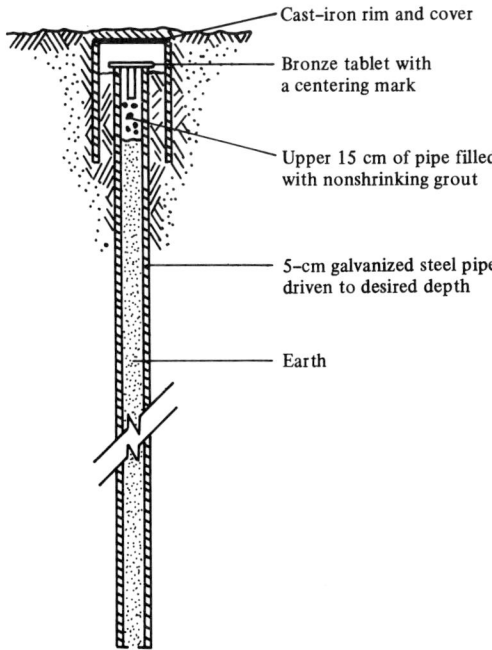

Cast-iron rim and cover

Bronze tablet with
a centering mark

Upper 15 cm of pipe filled
with nonshrinking grout

5-cm galvanized steel pipe
driven to desired depth

Earth

Figure 3-18. Survey marker recommended for muskeg, bog, swamp, very fine silts, construction fill, and clays of medium to high plasticity.

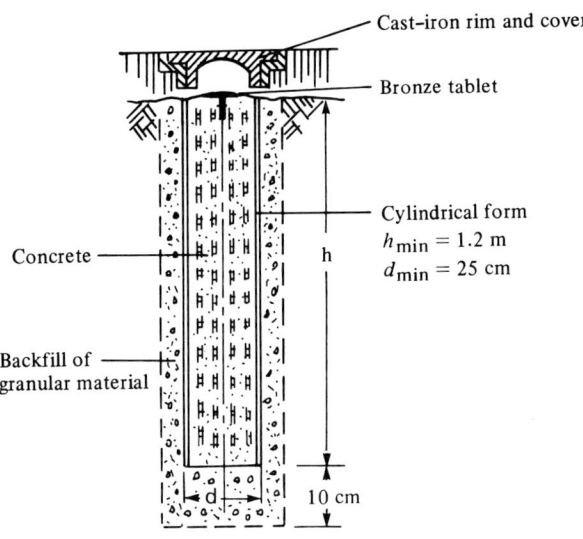

Cast-iron rim and cover

Bronze tablet

Cylindrical form
$h_{min} = 1.2$ m
$d_{min} = 25$ cm

h

Concrete

Backfill of
granular material

d

10 cm

Figure 3-19. Survey marker recommended for granular soils, tills, silts, clays of low plasticity, and construction fill.

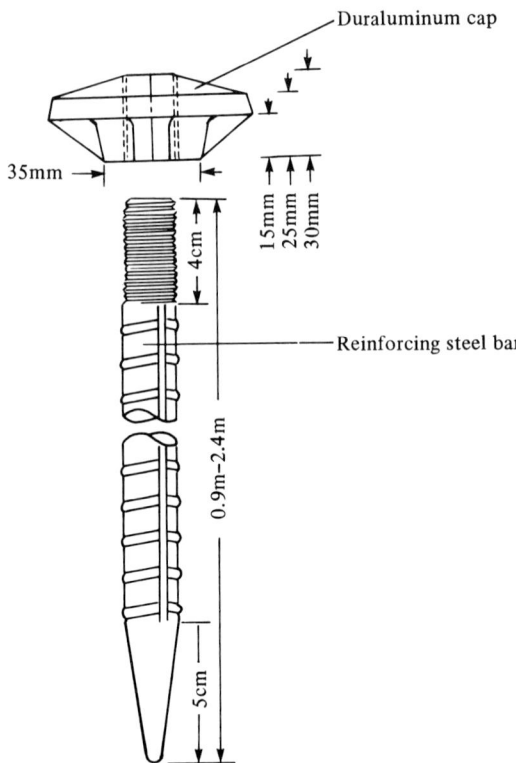

Figure 3-20. Survey marker recommended for permafrost, weathered and soft rocks (set in drilled holes), and highly organic wet soils.

Wall Markers

The idea of wall monumentation was developed a long time ago and has been tested in various European countries (12) and further improved in Canada (4). Different approaches were developed that could be classified into two groups as to the use of the markers. In the first group, the wall markers serve only as reference points for reconstructing the original location of traverse stations. In the second group, the wall markers serve directly as coordinated control points.

Several solutions have been proposed (12) for reconstructing the original positions of traverse stations. One of the simplest is shown in Figure 3-21. A portable centering arm, which fits into the wall marker, determines uniquely the location of the survey station. The same type and length of the centering arm must be used in original traversing and in the subsequent use of the control network. A disadvantage of this method is the necessity of locating traverse stations very close to the walls of buildings.

The use of wall markers directly as coordinated control points has been tested in Canada (4). Traverses in this approach are run with forced centering,

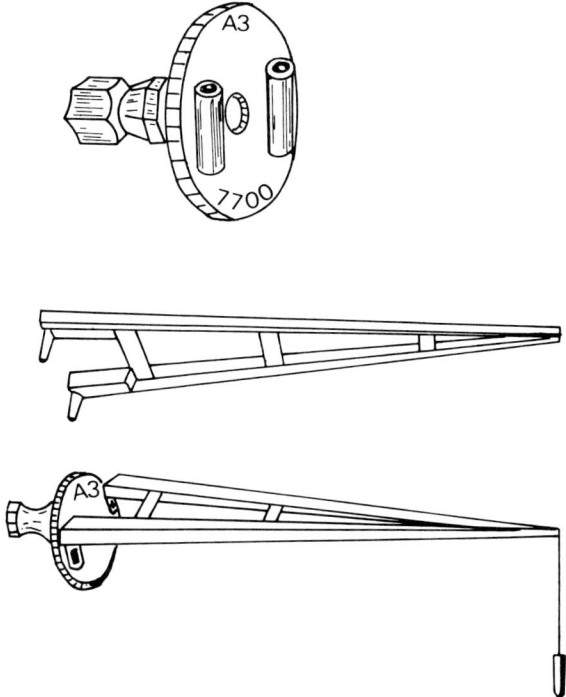

Figure 3-21. Wall marker and the portable centering arm (12).

without permanently marking the traverse stations on the ground. Instead, three or more small markers are monumented and targeted on the buildings in the vicinity of each traverse station. The wall points are established as the traverse progresses by measuring distances and directions to each wall marker as, for instance, in the traverses in Figure 3-22. If a new traverse, 3–7 in Figure 3-22, has to be connected to traverse 1–5 in the vicinity of station 3, e.g., in position 3a, the connecting measurements are made to the nearest wall points by angular

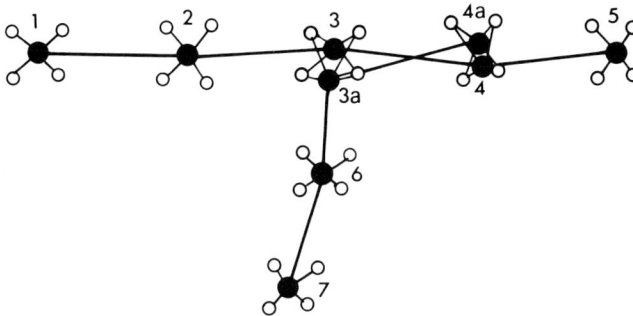

Figure 3-22. Traversing with wall monumentation.

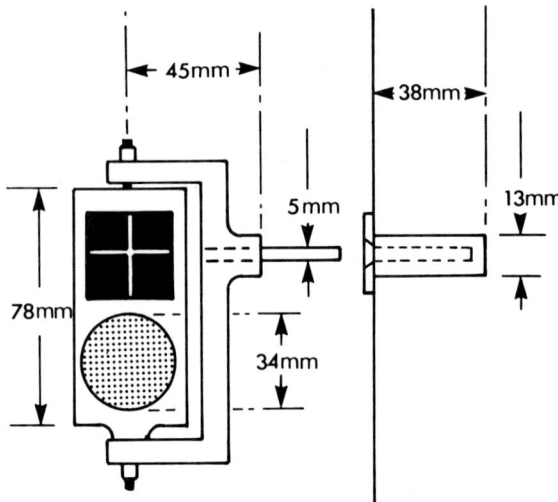

Figure 3-23. Design of the wall marker and target.

resection, linear intersection, or a combination of angle and distance measurements. The orientation of the new traverse may be done either by sighting directly to any of the previously positioned wall markers at stations 2 or 4 or by positioning a new target, e.g., at 4a.

Once the traverse network covers the area of a higher-order control, the network is simultaneously adjusted. The adjustment includes the coordinates of all the traverse stations and the wall points that have been used in connecting surveys. Coordinates of other wall points are calculated by a conventional polar method after the adjustment of the traverse stations.

Figures 3-23 and 3-24 show a design of the wall markers and targets that were used in a test network in Fredericton, Canada (4) (see Figure 3-9). The target may be rotated about its vertical axis and is equipped with a small plastic reflector allowing for distance measurements from traverse stations using laser EDM instruments. The experience of the test network shows that four markers (one traverse station) can be monumented in less than $\frac{1}{2}$ hour with an electric rotary hammer and epoxy glue. To cement the plugs in a horizontal position, a spirit level on a short rod, inserted to the plug, is used during setting time. A small brass cover protects the opening of the marker when the latter is not in use.

The only problem with the shown targets is that the coordinated point is not the hole of the marker but the center of the cross on the removable target. In effect, all users of the control points must have the same type of targets or must remember that the control point is 45 mm in front of the marker.

The use of wall monumentation has the following advantages compared to conventional ground monumentation:

1. Buildings are generally more permanent than curbs or sidewalks since they are not subjected to such frequent reconstruction or destruction.

2. Buildings with solid deep foundations are subjected less to horizontal displacement than ground monuments.
3. When using the densification control, the location of survey instruments is not restricted to centering above specific ground marks. The instrument may be set up at the most convenient place as long as at least two wall points are visible from the survey station.
4. Markers on the walls are subjected to little if any coverage by snow and ice and they may be sighted above the traffic if placed high enough above the ground.

Traversing with wall monumentation imposes additional field work and additional computations during the establishment of control, as well as later in successive surveys. This additional work is partly compensated for by the fact that the surveyor saves time by not having to center the instruments above an existing ground mark. Another compensation comes from the possibility of setting up the survey instruments in the most convenient location at the time of the survey, thus avoiding difficulties due to temporary or permanent obstacles on the lines of sight.

The additional time required for taking two sets of measurements to four wall points, for instance, is practically no longer than the time necessary for referencing the ground monuments in conventional surveys.

Figure 3-24. Targeting of the wall control point.

Connecting surveys to the wall points may be made by an angular resection to a minimum of three wall points, by distance measurements to a minimum of two points, or by a combination of the two methods.

In some cases, coordinates of the new position of the instruments must be calculated directly in the field. This does not impose any difficulties with programmable pocket calculators (13).

Rooftop Monumentation

Flat roofs of large buildings are ideal for the monumentation of first- and second-order control points. Although accessibility and connecting survey to the lower-order points may be a problem, roof monumentation is indispensable in obtaining intervisibility lines between points in the densely builtup areas of the cities. Use of EDM has at least partially solved the problem of connecting surveys, which formerly required indirect distance determination from the roof to the ground stations by using, for instance, angular intersections from ground-established base lines. Tops of very high buildings may demonstrate movements of several centimeters due to wind and thermal expansion. The surveys between the roof points should, therefore, be performed on calm and cloudy days, and generally buildings over 20 storeys high should not be used for roof monumentation.

Roof markers are usually placed on the top of walls expanded above the roof. Figure 3-25 shows a typical roof monumentation in the form of a steel rod cemented about 50 cm into the wall. A steel plate for survey instruments or a target fit concentrically onto the rod. Figure 3-26 shows typical roof monumentation in the city of Zurich. A concrete plate is placed loosely (unattached) on the roof. Its stability is assured only by the heavy weight of the plate and is checked within one millimeter before each survey by measuring short distances

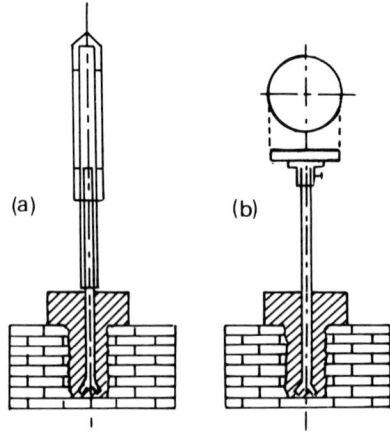

(a)   (b)

Figure 3-25. Rooftop marker with (a) target and (b) instrument table (12).

Figure 3-26. Rooftop concrete block with a marker and openings for tripod legs.

to reference nails cemented to the surrounding wall. Three holes on the plate for tripod legs facilitate the setting up and centering of the survey instruments or targets above the marker cemented into the concrete.

ELECTROMAGNETIC DISTANCE MEASUREMENTS (EDM)

Basic Principles and Classification

The EDM instruments that are used in control surveys and in detailed mapping may be classified into two groups according to the type of electromagnetic radiation that carries measuring signals:

Microwave EDM instruments that generally employ radio waves of the wavelength $\lambda_o = 3$ cm. Only one instrument in this group (Tellurometer Model MRA-4) uses a different wavelength of $\lambda_o = 8$ mm;

Electrooptical EDM instruments that utilize visible and near-infrared radiation to carry the measuring signals. In this group, we have either instruments with helium–neon (He–Ne) lasers with $\lambda_o = 0.63$ μm or with gallium–arsenide (Ga–As) diodes that produce invisible radiation with $\lambda_o \cong 0.9$ μm. Only one instrument (Mekometer) utilizes another source of radiation, a xenon flash tube that produces a mixture of wavelengths with a mean λ_o of 0.43 μm. Older types of instruments used tungsten lamps or mercury arc lamps, but they have been replaced by the He–Ne laser, which, thanks to its coherent radiation, may be utilized over long distances at night and in the daylight owing to a high power density and the possibility of filtering off the ambient light.

Generally, the shorter the carrier wavelength, the better the accuracy of the EDM. Therefore, microwave instruments are generally less accurate than the electrooptical instruments. On the other hand, the longer the wavelength, the better the penetration through haze and fog. Therefore, microwave instruments may be utilized over long distances in bad atmospheric conditions. Electrooptical instruments are useful only when visibility is good.

Infrared instruments are characterized by very weak radiation, and their maximum range is generally shorter than that of other instruments—of the order of 1 to 3 km, depending on the model—unless an infrared laser is used as, for instance, in the HP-3820 electronic tacheometer.

All EDM instruments that are used in surveying utilize a modulated radiation for distance measurements. The wavelength of the modulated signal is called a pattern wavelength; it is used as a unit of the measurement. Different instruments use different pattern wavelengths, ranging from a few meters to 40 m depending on the model, with the exception of the Mekometer, which uses $\lambda = 60$ cm.

Table 3-2 gives the main characteristics of some EDM instruments that were on the market in 1977. The stated accuracies are listed in terms of standard

Table 3-2. Sample of EDM Instruments (1977)

Model	Manufacturer	Range (km)	Accuracy[1] $\pm a$ (mm)	$\pm b$ (ppm)	Remarks
		Microwave EDM instruments			
MRA-5	Tellurometer (Pty. Ltd), S. Africa	70	15	3	Antenna can be separated up to 25 m
CA-1000	Tellurometer (Pty. Ltd), S. Africa	30	15	5	Lightweight, 3.5 kg with battery
MRA-4	Tellurometer (Pty. Ltd), S. Africa	40	3	3	Out of production
SIAL MD	Siemens-Albis, Switzerland	150	10	3	Reduced ground swing, separated antenna
		He–Ne laser EDM instruments			
600	AGA, Sweden	30	5	2	
710	AGA, Sweden	5 (with 6 prisms)	5	2	Automatic tacheometer, autom. recording (optional)
78	AGA, Sweden	8 (with 6 prisms)	10	2	

Table 3-2. (*continued*)

Ranger IV	Laser Systems & Electronics (for K & E), USA	12 (with 7 prisms)	5	2	16.2 kg, 5- to 20-min warm-up
Ranger V	Laser Systems & Electronics, USA	25 (with 7 prisms)	10	2	16.2 kg, 5- to 20-min warm-up
Rangemaster II	Laser Systems & Electronics, USA	60 (with 30 prisms)	5	2	18.0 kg, 5- to 20-min warm-up
Infrared EDM instruments					
DI-3S	Wild–Heerbrugg, Switzerland	1.5 (with 3 prisms)	5	5	Aiming unit mounted on theodolite, plungeable with the telescope, semi-auto. slope correction
TAC-1 Tachymat	Wild–Heerbrugg, Switzerland	0.7 (with single prism)	5	5	Automatic tacheometer, cassette tape recording, compact
MA 100	Tellurometer (Pty. Ltd.)	1.5 (with triple prism)	1.5	2	
CD-6	Tellurometer (Pty. Ltd.)	0.7 (with single prism)	5	5	Mounted on theodolite, 3.7 kg
DM501	Kern, Switzerland	1.5 (with 3 prisms)	5	5	Mounted on theodolite, plungeable with the telescope, 1.6 kg, compact.
HDM-70	Cubic Corp., USA	1.5 (with triple prism)	5	10	Aiming unit (1 kg) mounted on theodolite, plungeable with the telescope, slope-horiz. corr.
HP3820A	Hewlett–Packard, USA	5 (with 6 prisms)	5	5	Fully auto. tacheometer, compact, electronic data recording
12 A	AGA, Sweden	2 (with 3 prisms)	5	5	Mounted on theodolite, 2.5 kg
Eldi-1	Carl-Zeiss, W. Germany	2 (with 1 prism)	5	2	8 kg
Eldi-3	Carl-Zeiss, W. Germany	0.4 (with 1 prism)	5	2	Mounted on theodolite, 3.8 kg
Reg-Elta 14	Carl-Zeiss, W. Germany	0.7 (with 3 prisms)	10	2	Automatic tacheometer, auto. data recording
Beetle 500S	Precision Int., USA	0.5 (with 3 prisms)	10		Mounted on theod., semiautom. slope reduction, 2.5 kg
Beetle 1000S	Precision Int., USA	1 (with 3 prisms)	10		Mounted on theod., semiauto.slope reduction, 2.5 kg
Xenon-flash EDM instruments					
ME 30000 Mekometer	Kern, Switzerland	2	0.2	2	

[1] As claimed by manufacturers. See discussion on pp. 84–85.
Standard deviation $\sigma_s = \sqrt{a^2 + b^2 \times S^2}$.

deviations as claimed by the producers. A discussion of EDM accuracy can be found on pp. 84–90.

All EDM instruments use the same principle of distance measurement. The modulated signal is continuously transmitted from one end of the measured distance and is reflected or retransmitted back at the other end. A difference in phase between the reference (transmitted) and the returning modulated signal is measured in the transmitting instrument.

If an exact integer number m of half-wavelengths is contained in the distance, the phase difference is zero. In all other cases, the difference in phase is translated into a fraction U of the half-wavelength and is displayed in linear units.

The distance S between the transmitter and the reflector is therefore equal to

$$S = U + \frac{m\lambda}{2}. \tag{3-29}$$

To find the number m, the measurement must be repeated with two or more different wavelengths.

The pattern wavelength λ is a function of the modulation frequency f and of the velocity v of propagation of electromagnetic waves,

$$\lambda = \frac{v}{f}. \tag{3-30}$$

In vacuo, the velocity v of propagation is constant for all electromagnetic waves and is equal to

$$c = 299\ 792.5\ \text{km/s}.$$

In the atmosphere, the velocity v of propagation is always smaller than c and may be calculated from

$$v = \frac{c}{n}, \tag{3-31}$$

where n is the refractive index of air, which is a function of the density of air and carrier wavelength.

The value of n ranges from $n = 1$ *in vacuo* to about $n = 1.0003$ for average atmospheric conditions. Its exact value may be determined on the basis of meteorological measurements of the temperature, barometric pressure, and humidity of air along the measured line. Therefore, the value of λ of the modulated signal is unknown during the measurement unless n is known, and then

$$\lambda = \frac{c}{nf} \tag{3-32}$$

can be calculated. The frequency f of modulation can be stabilized and is usually known with a high degree of accuracy.

The manufacturer usually gives the value of $\lambda = \lambda_1$ for specified atmospheric conditions, i.e., for a certain value of $n = n_1$. Thus

$$\lambda_1 = \frac{c}{n_1 f}. \tag{3-33}$$

Therefore, the distance that is recorded by the EDM instrument is equal to

$$S_1 = U_1 + m\frac{\lambda_1}{2}.$$

where U_1 is a fraction of $\frac{1}{2}\lambda_1$.

If, during the measurements, $n = n_2 \neq n_1$, the correct value of λ equals

$$\lambda_2 = \frac{c}{n_2 f}, \tag{3-34}$$

and the actual distance is

$$S = U_2 + m\frac{\lambda_2}{2}. \tag{3-35}$$

From Eqs. (3-34) and (3-33),

$$\lambda_2 = \lambda_1 \frac{n_1}{n_2}, \tag{3-36}$$

and finally, the corrected distance may be calculated as equal to

$$S = U_1 \frac{n_1}{n_2} + \frac{m\lambda_1 n_1}{2n_2} = S_1 \frac{n_1}{n_2}. \tag{3-37}$$

Equation (3-37) gives the basic formula for correcting the measured distance according to actual atmospheric conditions.

Because the electronic centers of EDM instruments usually do not coincide exactly with the plumbing center, a zero correction Z_o has to be determined and added to the calculated distance. Further corrections that must be made to reduce the measured distance are discussed on pp. 112–119.

The final reduced distance S_o is calculated as

$$S_o = S_1 \frac{n_1}{n_2} + Z_o + \Delta S, \tag{3-38}$$

where S_1 is the measured distance,

n_1 is the refractive index accepted for laboratory calibration,

n_2 is the refractive index during the field measurements (to be determined by the observer),

Z_o is the zero correction, and

ΔS is a compound correction due to reductions.

Determination of EDM Corrections

Refractive index. The group refractive index for the visible and near-infrared modulation radiation in dry air at 0°C, 760 mm of mercury pressure ($= 1013.250$ mbar) and with 0.03% of carbon dioxide, may be calculated from the Barrel and Sears formula,

$$n_o = 1 + \left(287.604 + \frac{4.8864}{\lambda_o^2} + \frac{0.068}{\lambda_o^4} \right) 10^{-6}, \qquad (3\text{-}39)$$

where λ_o is the wavelength of the carrier radiation in micrometers. For instance, for a He–Ne laser, $\lambda_o = 0.6328 \ \mu m$ and $n_o = 1.000\ 300$. If the temperature t, barometric pressure p, and humidity of the air differ from the standard conditions, then the group refractive index is calculated from

$$n - 1 = \frac{0.269\ 578(n_o - 1)}{273.15 + t} p - \frac{11.27}{273.15 + t} 10^{-6}e, \qquad (3\text{-}40)$$

where e is the partial pressure of water vapor in millibars, t is in degrees Celsius, and p is in millibars.

The value of e is determined from measurements of a difference in temperature between dry t_D and wet t_w bulb thermometers and by using an approximate formula

$$e = E - 0.7(t_D - t_w) = E - 0.7 \ \Delta t, \qquad (3\text{-}41)$$

where E is the saturated water vapor pressure in millibars.

The values of E are tabulated in meteorological tables. Sample values are given in Table 3-3, including the change ΔE per 1°C in the vicinity of the indicated temperatures.

This table shows that humidity has very little influence on electrooptical measurements. In extreme conditions, as for instance at temperature $t = 30°C$ and humidity 100%, i.e., for $t_w = t_D$, there is $e = E = 42.4$ mbar. If, in these conditions, the influence of e is neglected by placing $e = 0$ in Eq. (3-40), the error of the calculated n would be only 1.6 ppm. Therefore, the second term in Eq. (3-40) is usually neglected in practice.

Table 3-3. Saturated Water Vapor Pressure

	t_w				
	−10°C	0°C	10°C	20°C	30°C
E (mbar)	2.6	6.1	12.3	23.4	42.4
$\Delta E/1°C$	±0.23	±0.44	±0.8	±1.4	±2.4

The refractive index for microwaves can be calculated from the Essen and Froome formula, which may be written in the form

$$n - 1 = \frac{77.624}{T} 10^{-6} p + \left(\frac{0.372}{T^2} - \frac{12.92}{T} 10^{-6} \right) e, \qquad (3\text{-}42)$$

where $T = 273.15°C + t$, t is in °C, and p and e are in millibars.

Calculation of n from Eqs. (3-40) or (3-42) is needed only in high-precision distance measurements. In routine surveys, nomograms for the refractive-index corrections, usually supplied by the manufacturers with EDM instruments, can be used. The nomograms are prepared on the basis of Eqs. (3-40) or (3-42) and the accepted calibration value of the refractive index. The nomograms give corrections of the measured distance that are slightly less accurate (about 2 ppm) than the results of rigorous calculations. Some EDM instruments have a built-in automatic correction system that requires only that the observer feed into the instrument the results of the meteorological measurements. The Mekometer 3000 has built-in compensators that automatically correct the modulated wavelength for local atmospheric conditions; the operator does not have to measure the temperature and barometric pressure at the station of the instrument. It must be remembered, however, that the fully automatic correcting systems give accurate results only when atmospheric conditions along the measured line are the same as those at the instrument station. In other cases, manual calculations of additional corrections are always necessary.

Zero correction. The electric centers of EDM instruments usually do not coincide with the centering marks used for plumbing the instruments above the survey station. The internal distance traveled by electromagnetic waves in the instruments is usually longer than the direct distance between the point of arrival of the signal and the centering mark. This difference may be quite large and in some models of Tellurometer or Geodimeter is about 30 cm.

Manufacturers of EDM instruments always supply information about the value of the zero corrections that should be added to the measured distance to compensate for the difference. Most new instruments are calibrated in such a way that the zero correction is equal to zero. It has been found, however, that the value of the zero correction may change after prolonged use of instruments. The change is usually small in electrooptical instruments (a maximum of a few millimeters), but in microwave instruments it may be of the order of several centimeters (14). Therefore, the zero correction should be frequently checked by measuring several distances on a calibration base line. Distances from 50 to 500 m are recommended for the calibration of electrooptical instruments and from 200 to 1000 m for microwave instruments. If known distances are not available, the value of the zero correction may be found by using a method of subdivided distances. The method requires a straight line in a flat area with a few marked points, say points A, B, C, and D as shown in Figure 3-27. The total distance AD and the subdistances AB, BC, and CD are

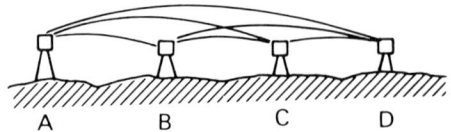

Figure 3-27. Determination of the zero correction using the method of subdivided distances.

measured with the instruments to be calibrated. Assuming that the zero correction Z_0 is the same (constant) for each measured distance, its value can be calculated by comparing the total distance and the sum of the subdivided distances, which means

$$AD - (AB + BC + CD) = 2Z_0.$$

If the distances in Figure 3-27 were measured in all combinations, the value of Z_0 could also be calculated from the comparisons

$$AC - (AB + BC) = Z_0$$

$$BD - (BC + CD) = Z_0.$$

In practice, it is recommended that a line be used with a minimum of four subdivisions measured in all combinations and that the value of Z_0 be calculated by using the least-squares adjustment method. Of course, the measured distances should be corrected for the refractive index, slope, etc., prior to the calculation of Z_0.

Some EDM instruments demonstrate a relationship between the value of Z_0 and the value of the phase difference U, showing a cyclic change when changing the distance over the range of the modulated half-wavelength. The cyclic change may be discovered by measuring a known distance and changing it in steps, equal to, for instance, $\frac{1}{20}$ of the half-wavelength.

If repeated measurements of different distances show a pronounced relationship between the values of Z_0 and U, different zero corrections should be applied in field measurements for different values of U by means of a tabulated or a graphical plot of the changeable Z_0. In most cases, however, the changes of Z_0 are within the accuracy of its determination, and an average value of Z_0 should be applied as a constant correction regardless of the value of U contained in the measured distance.

Accuracy Analysis of EDM

Error equation. If Eqs. (3-33) and (3-37) are placed into Eq. (3-38), the reduced distance S_0 can be expressed in the form

$$S_0 = U_1 \frac{n_1}{n_2} + m \frac{c}{2n_2 f} + Z_0 + \Delta S. \qquad (3\text{-}43)$$

Variance $\sigma_{S_o}^2$ of the distance S_o may be obtained by differentiating Eq. (3-43) and applying the ground rules of error propagation. This gives

$$\sigma_{S_o}^2 = \sigma_U^2 + \left(\frac{m}{2nf}\right)^2 \sigma_c^2 + \left(\frac{m}{2nf^2}\right)^2 \sigma_f^2 + \left(\frac{m}{2n^2f}\right)^2 \sigma_n^2 + \sigma_{Z_o}^2 + \sigma_{AS}^2. \quad (3\text{-}44)$$

Equation (3-44) may be simplified by an approximation, $2S = m\lambda = mc/(nf)$, and the result is

$$\sigma_{S_o}^2 = \sigma_U^2 + S^2\left[\left(\frac{\sigma_c}{c}\right)^2 + \left(\frac{\sigma_f}{f}\right)^2 + \left(\frac{\sigma_n}{n}\right)^2\right] + \sigma_{Z_o}^2 + \sigma_{AS}^2. \quad (3\text{-}45)$$

In the above equations, σ_U represents the standard deviation of the whole value of $U_1(n_1/n_2)$. Since U_1 usually corresponds to a very short distance, the influence of the errors in n_1 and n_2 on the accuracy of U_2 may be neglected.

The accuracy of EDM is usually given in the technical literature in the general form

$$\sigma_S^2 = a^2 + b^2S^2 \quad (3\text{-}46)$$

or, simplified,

$$\sigma_S = \pm a \pm bS. \quad (3\text{-}47)$$

Equation (3-45) may be reduced to Eq. (3-46) by substituting

$$\sigma_U^2 + \sigma_{Z_o}^2 = a^2 \quad (3\text{-}48)$$

$$\left(\frac{\sigma_c}{c}\right)^2 + \left(\frac{\sigma_f}{f}\right)^2 + \left(\frac{\sigma_n}{n}\right)^2 = b^2, \quad (3\text{-}49)$$

where σ_c is the error of the velocity of propagation of light in a vacuum,
  σ_f is the error of modulation frequency,
  σ_n is the error of the refractive index,
  σ_u is the error of the phase-difference determination, and
  σ_{Z_o} is the error of the calibration correction (zero error).

The error σ_{AS} of the geometrical reductions is not included in the Eqs. (3-46) and (3-47).

It can now be seen why the accuracies of EDM instruments, as shown in the Table 3-2, consist of two numbers; the first number represents the value of a in millimeters, and the second the value of b in parts per million.

Equations (3-48) and (3-49) show all the sources of error in EDM. Each source will be discussed separately.

Velocity error. The value of c was accepted in 1957 as being equal to 299 792.5 km/s, with a standard deviation of $\sigma_c = 0.4$ km/s. Recent results confirm the value of c and give the standard deviation as $\sigma_c \leq 0.1$ km/s. This corresponds to a relative error of 0.3 ppm. The error is negligibly small for surveying application of the EDM. Its influence is of a constant nature and introduces a constant scale change in the absolute distance determination.

Modulation-frequency errors. Frequency of oscillations is measured in units of 1 hertz (Hz), and

$$1 \text{ Hz} = 1 \text{ cycle/s}$$
$$1000 \text{ Hz} = 1 \text{ kHz}$$
$$10^6 \text{ Hz} = 1 \text{ MHz}$$
$$10^9 \text{ Hz} = 1 \text{ GHz}.$$

To achieve, for instance, a modulation wavelength of 10 m, oscillations of the signal with a frequency of about 30 MHz (Eq. 3-30) must be produced.

The modulation frequency can be calibrated with an accuracy of about 0.1 ppm and can be stabilized during the use of the EDM instrument within a few hertz if the oscillation circuit, which includes crystals of quartz, is kept at a constant temperature. If the temperature is not controlled, a drift of the frequency will occur, producing errors of up to 10 ppm or more. Most EDM instruments have built-in ovens with a thermostat and require a warm-up time. Even with the oven, the frequency may drift because of aging of the control crystals. Some microwave instruments show a drift of about 50 Hz per year, which, for a modulation frequency of 10 MHz, for instance, would produce an error of 5 ppm. It is therefore recommended that the frequencies be checked at least once a year, or even every month if the instrument is used in extreme weather conditions. Methods of frequency stabilization and of calibration may differ from one model of EDM instrument to another. The users of EDM instruments should consult the manufacturer about requirements for frequency calibration of their instruments.

Refractive-index errors. The influence of errors σ_p, σ_t, and σ_e in measurements of the barometric pressure p, temperature t, and water vapor pressure e may be calculated by applying the law of error propagation to Eqs. (3-40) and (3-42), which gives, for electrooptical distance measurements,

$$\sigma_n{}^2 = \left[-\frac{0.269\,578(n_0 - 1)}{(273.15 + t)^2} p \right]^2 \sigma_t{}^2 + \left[\frac{0.269\,578(n_0 - 1)}{273.15 + t} \right]^2 \sigma_p{}^2. \quad (3\text{-}50)$$

For instance, for $\sigma_t = 1°C$ and $\sigma_p = 3.5$ mbar, σ_n/n varies between 1.2 and 1.8 ppm for temperature ranges from $-30°C$ to $+30°C$ and for p varying from 980 to 1070 mbar. In average conditions, a $1°C$ error produces an error of 1 ppm in n, and 1 mbar error produces an error of 0.3 ppm. As already stated, the influence of water vapor pressure may be neglected in electrooptical measurements if the instruments are calibrated in average humidity conditions.

Error propagation in microwave measurements gives

$$\sigma_n{}^2 = \left(\frac{77.624}{T} 10^{-6} \right)^2 \sigma_p{}^2 + \left(-\frac{77.624p}{T^2} 10^{-6} + \frac{12.92e}{T^2} 10^{-6} - \frac{0.74e}{T^3} \right)^2 \sigma_T{}^2$$

$$+ \left(\frac{0.372}{T^2} - \frac{12.92}{T} 10^{-6} \right)^2 \sigma_e{}^2. \quad (3\text{-}51)$$

The above formula leads to the following conclusions: An error of 1°C in T will produce an error σ_n/n ranging from 0.8 to 2 ppm for extreme values of p, T, and e, with a value of $\sigma_n/n = 1.2$ ppm in the average atmospheric conditions; and an error of 1 mbar in p will produce an error $\sigma_n/n = 0.3$ ppm within the temperature range $-30°C$ to $+30°C$.

The influence of errors in the determination of e is very critical in microwave measurements. For instance, for $t = 0°C$, $\sigma_n/n = 5\sigma_e$ in parts per million and for $t = 30°C$ $\sigma_n/n = 4\sigma_e$ in parts per million.

The value of σ_e may be calculated by applying the error propagation rule to Eq. (3-41), which gives

$$\sigma_e^2 = \sigma_E^2 + (0.7\sigma_{\Delta t})^2. \tag{3-52}$$

For instance, for $t_w = 30°C$ and $\sigma_{t_w} = 1°C$, the error in E is equal to 2.4 mbar, as shown in Table 3-3. If, in these conditions, the error in reading Δt from the wet and dry thermometers is equal to $\sigma_{\Delta t} = 0.5°C$, from Eq. (3-52) we have

$$\sigma_e = \sqrt{(2.4^2 + 0.35^2)} = 2.4 \text{ mbar,}$$

and the error of the distance determination due to the error σ_e becomes

$$\frac{\sigma_S}{S} = \frac{\sigma_n}{n} = 4 \times 2.4 = 9.6 \text{ ppm.}$$

In summary, the errors in meteorological measurements produce the following errors in the distance determination in average atmospheric conditions:

1. An error of 1°C in temperature produces an error of 1 ppm;
2. An error of 1 mbar in barometric pressure produces an error of 0.3 ppm;
3. An error of 0.5°C in $(t_w - t_D)$ and an error of 1°C in t produces an error of up to 10 ppm in microwave measurements.

In practice, meteorological conditions are only measured at each end of the measured distance, or even at one end only. In many cases, particularly in measurements in diversified topographical conditions, the mean temperature along the measured line may differ by several degrees Celsius from the end measurements. Surveys in urban areas, where heat radiation from paved areas and buildings may drastically change from one point to another, require special attention and care in the determination of the meteorological data. A repetition of measurements in different atmospheric conditions (different values of n) with, for instance, one set of measurements in daylight and another at night, may improve the accuracy. Measurements of meteorological conditions at intermediate points along the measured distance will also improve the results if high accuracy is sought, as for instance in the primary control net for a city. The thermometers, barometers, and psychrometers should be of good quality and should be frequently checked and calibrated.

Error of the phase-difference determination. As mentioned before, the measurement of a phase difference between the transmitted and returned modulated signal is the basis of all EDM instruments listed in Table 3-2. There are many possible ways to determine the phase difference, by either null-point or digital methods. All give a resolution of 1/1000 of the cycle or even better. This yields, for example, an error of 10 mm in a single distance determination with the Kern DM501, which utilizes the modulation half-wavelength of about 10 m, or about 0.3 mm in measurements with the Mekometer, which uses the modulation wavelength of 0.6 m. Phase measurements are usually repeated several times during distance determination, and the phase error is decreased by taking the mean of the results.

Phase measurements in microwave instruments may be seriously affected by ground reflections of a portion of the signal traveling between the two end stations. The cone angle α of radiation of microwaves may be calculated approximately from

$$\sin\left(\frac{\alpha}{2}\right) = \frac{\lambda}{D},$$

where λ is the carrier wavelength and D is the diameter of the microwave reflector.

For instance, for $\lambda = 3$ cm and $D = 36$ cm, as is the case in some models of Tellurometers, the cone angle of radiation is about $9°$. Therefore, unless the distance is measured across a deep valley, a portion of the signal may be reflected from the ground and may interfere with the direct measuring signal at the remote station. The interference may produce a systematic phase shift or even a cancellation of the direct signal, making it impossible to measure the distance. The latter case—cancellation of the signal—may occur if the measurements are performed above extended reflecting bodies such as calm water, paved areas, or flat metal roofs, which may produce mirror-type reflections in which the reflected signal is almost the same strength as the direct signal and may differ $180°$ in phase. Many factors affect the strength and the phase difference between the direct and reflected signals. Two factors, namely, the angle of incidence of the reflected signal and the carrier frequency of the microwave instruments, can be controlled to a certain extent by observers during the distance measurement. The angle of incidence may be changed according to the height of the instrument above the survey marker. In many cases, a difference of only a few decimeters may produce a large change in the phase and strength of the reflected signal. The same effect may be achieved by a change of the carrier wavelength by altering the so-called cavity tuning. To reduce the influence of ground reflections, it is recommended that phase measurements be repeated 10 to 20 times, with the cavity tuning (the carrier wavelength) being changed in equal steps each time. If the repeated measurements give an unsatisfactory spread of the results, the whole procedure should be repeated using different heights of the instruments. To evaluate whether the spread of the results is satisfactory, the mean result from all cavity tunings is calculated, and deviations of individual results are plotted. Thus a ground swing curve is obtained, as shown in Figure 3-28. If the ground swing is close to a full cycle of a sinusoidal curve and the maximum

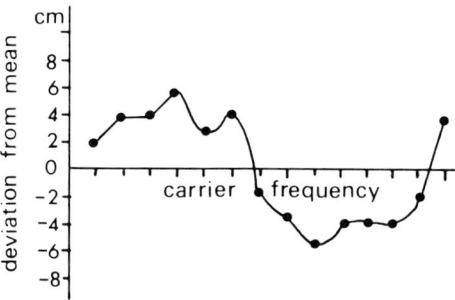

Figure 3-28. Typical ground swing curve of a "perfect" measurement.

spread of the results does not exceed five times the carrier wavelength (15 cm for instruments with $\lambda = 3$ cm), it may be said that ground reflections have little or no influence and the mean distance is almost perfect.

   If the ground swing curve shows large irregularities and particularly sudden "jumps" at certain cavity tunings (Fig. 3-29), the measurements may be affected by strong reflections, and the mean distance probably has a large systematic error. In such cases, the measurements should be repeated with different heights of the instruments or an attempt should be made to tilt the instruments upward slightly to change the reflection parameters. If, by changing the ground-reflection conditions, the ground swing curve still shows systematic irregularities, the distance measurements may be affected by internal reflections. This may be the case when the parabolic reflector is bent or otherwise distorted, which causes a back-reflection of the transmitted signal.

   The influence of ground reflections on the standard deviation of the measured distance does not exceed a value of 1.5 to 3 cm for instruments with $\lambda = 3$ cm if the measurements are repeated on 20 different cavity tunings and the ground swing curve has a satisfactory shape. In extremely bad conditions, strong reflections may affect the mean distance by errors of the order of 10 cm or more.

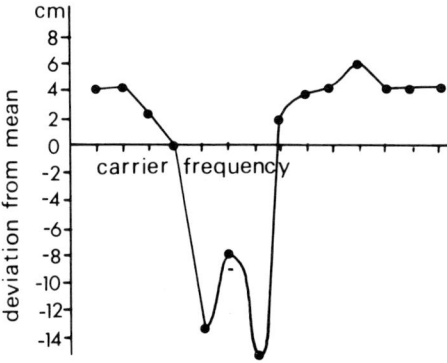

Figure 3-29. Ground swing curve indicating a strong reflection.

Measurements with the Tellurometer MRA-4, which utilizes an 8-mm carrier wavelength, are affected by ground reflections 4 to 5 times less than measurements with the 3-cm wavelength.

Conclusions and recommendations. The total error of EDM may be expressed in the form

$$\sigma_S{}^2 = a^2 + b^2 S^2 + \sigma_{\Delta S}{}^2. \qquad (3\text{-}53)$$

Values of the parameters a and b for different EDM instruments are shown in Table 3-2. It must be realized that it is very difficult, although not impossible, to achieve the listed accuracies.

As to electrooptical instruments, the listed values for the parameter a are quite reasonable if the instruments are frequently checked for the Z_0 correction. Determination of the parameter b is dependent mainly on knowledge of the refractive index of air and the stability of the modulation frequencies. The thermometers and barometers should be carefully checked and calibrated. Experience shows that the claimed value of 2 ppm for the parameter b for some instruments is usually exaggerated in routine applications of the electrooptical instruments even when all field procedures are performed according to specifications. A value of $b = 4$ ppm is more realistic unless the measurements are repeated several times in different atmospheric conditions with a carefully calibrated instrument and the mean value is taken as the final result of the measured distance.

Microwave instruments, which are vulnerable to ground reflections and are affected by uncertainties in the determination of the relative humidity of the air, may yield much larger deviations from the listed values of the parameters a and b. In the evaluation of surveying results, their use requires much more skill and experience than do electrooptical instruments. The instruments with 3-cm microwaves may give a value for a between 2 and 3 cm if the Z_0 correction is frequently checked separately for each pair of instruments and if the ground reflections are minimized by proper field procedure and skillful interpretation of the ground swing curves. The value of b, although claimed by manufacturers to be of the order of 3 ppm, is in reality much larger. Experience shows that a value of $b = 6$ ppm for routine surveys seems to agree with reality unless the surveys are repeated several times in *uncorrelated* conditions.

A good review of sources of error and recommended field procedures, particularly in the use of microwave instruments, is given by Bomford (15) with an extensive bibliography on EDM up to 1970.

Choice of EDM Instruments

EDM in first-order control. As stated before, spacing of first-order control points in urban areas should be of the order of 10 km, with a relative accuracy of about 4 ppm in terms of the semimajor axis of the standard-error ellipses. Theoretically, most of the microwave and medium-range laser instruments should be able to satisfy the range and accuracy requirements. Practically, however, difficulties may be encountered with microwave instruments owing to the strong effect of ground reflections in urban areas and uncertainties in the determination of the refractive index. Therefore, laser instruments are recommended. If smog and poor visibility over large cities make the use of electrooptical instruments impossible, trilateration based on microwave instruments is the only other choice for primary control work. In such cases, instruments that utilize the shortest possible carrier wavelength should be used, as for instance, Tellurometer MRA-4 with 8-mm radiation, in order to minimize parameter b of the error equation by repeating measurements in different atmospheric conditions and frequent checking of the modulation frequencies. It should be remembered that, from the geometric point of view, trilateration nets are generally weaker (16) than other types of networks; this places higher demands on the accuracy of distance measurements and the design of the shape of the primary net.

EDM in densification-control surveys. Spacing of stations in the second-order control usually ranges from a few hundred meters to about 2 km, rarely to 3 km, and is about 200 m in the third-order networks.

Densification surveys are mostly performed on tripods close to the ground, along streets and expressways. City traffic interrupts the electromagnetic radiation. Some of the older types of EDM instruments may show wrong results when the electromagnetic beam is interrupted. Most of the new EDM instruments may be used in traffic without harm to the survey results as long as the reflecting station is exposed long enough to allow "counting" of the modulated wavelengths.

The counting time may differ from about 1 to 20 s depending on the instrument. Some instruments have to start counting from the beginning each time the beam is interrupted; other instruments stop counting when the reflecting station is temporarily screened and continue counting within the "windows" of the traffic. Since this mode of operation has an important bearing on the economy of city surveys, the effect of traffic should be carefully considered before the instrument is purchased.

In some cities, high tripods (2.5 m) are used in city surveys to allow for measurements above the traffic. This requires some type of a portable platform for the operator.

Monumentation of second-order control on the roofs of buildings and monumentation of third-order control in the walls of buildings solve, at least partially, difficulties due to traffic.

Use of microwave instruments in establishing a detailed control net is out of the question because of unfavorable conditions for minimizing parameter a of the error equation (tunnel-type reflections (14) along the streets) and the short distances involved.

Electrooptical instruments with accuracy parameters $a \leq 10$ mm and $b \leq 5$ ppm should be used in second-order control surveys, and instruments with $a \leq 5$ mm and $b \leq 10$ ppm should be used in third-order control. All distances should always be measured in both directions, and great care should be taken with the differences in elevation for the reductions of slope distances.

Distances and angles in control surveys should be measured separately; therefore, there is little advantage in using EDM instruments mounted on theodolites. Accuracy of angle measurements deteriorates when an additional strain is applied on the axial system of the theodolite. Generally, all combinations of EDM and theodolites are more useful for detailed surveying and mapping than for urban control surveys.

Laser instruments may use small plastic reflectors over distances of up to about 150 m instead of the conventional large cube reflectors. This is advantageous in third-order control surveys, particularly in traversing with wall monumentation.

It should be remembered that in some countries the use of laser instruments may be prohibited for safety reasons. Most cities do not yet have regulations governing the allowable power densities of the electromagnetic radiation of the surveying instruments, but the authorities should be consulted on the use of laser instruments. Generally, a radiation with a power density up to 1 μW/cm^2 at the eye entrance may be accepted as harmless for a prolonged viewing in the

Figure 3-30. Electronic tacheometer HP-3820A (courtesy of Hewlett–Packard).

laser beam. Most EDM laser instruments do not produce higher power density at distances exceeding a few hundred meters.

Use of electronic tacheometers. The development of electronic tacheometers such as HP-3820 (Fig. 3-30), Tachymat TAC-1, AGA-710, or the older Reg Elta14 offers new possibilities in the automation of urban surveying and mapping. Simultaneous measurement of angles and distances with an automatic recording of the results allows determination of position of up to 60 points per hour from one instrument station. Higher accuracy of measurements of the newer models also permits the use of the instruments in third-order control surveys.

This opens a possibility for a simultaneous densification of the control together with detailed surveying and mapping (see Chapter 5). The electronically recorded in-the-field measurements may be fed directly to a computer for numerical processing of survey data and automatic map drawing.

MEASUREMENT OF HORIZONTAL ANGLES

Instruments, Signals, and Centering Devices

There has been no significant improvement for over more than a century in the accuracy of angle-measuring devices used in field surveying, and at the present time nothing suggests any change in the foreseeable future. Electronic readout systems available in some new instruments allow for the automation of angle recording and make the measurements less strenuous for the observer than optical systems, but the latter are still superior in accuracy. Therefore, the optical theodolite is likely to remain essentially unchanged for some time.

For the establishment of urban horizontal control, two types of instruments can be considered: the precision theodolite and the 1″ theodolite. Table 3-4 gives general characteristics of the two types.

Precision theodolites such as Kern DKM-3, Wild T-3, or RankMicroptic-3 are used only in first-order control surveys. The 1″ theodolites, as, e.g., Wild

Table 3-4. General Characteristics of Theodolites

Specification	Precision theodolite	1″ theodolite
Weight of instrument	10 to 15 kg	3 to 5 kg
Free objective diameter	60 to 70 mm	35 to 45 mm
Magnification	24 to 45×	24 to 35×
Diameter of horizontal circle	90 to 140 mm	70 to 80 mm
Circle graduation interval	4′ to 10′	10′ to 20′
Micrometer graduation interval	0″.2 to 0″.5	1″
Sensitivity of plate bubble	5″ to 10″/2 mm	20″ to 30″/2 mm

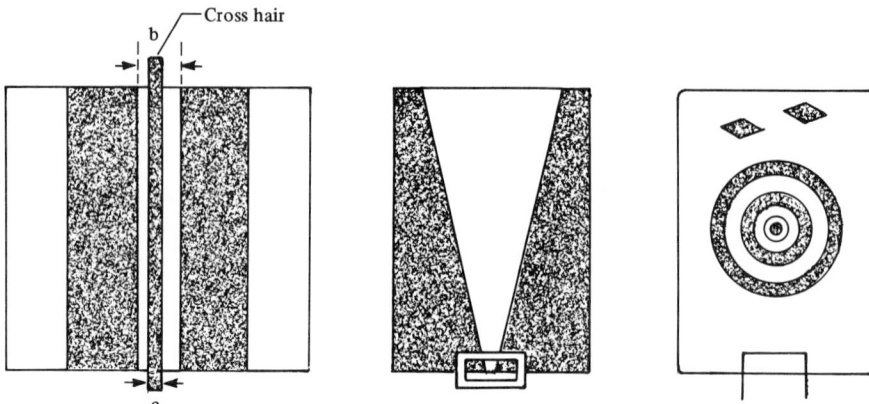

Figure 3-31. Different types of plate signals.

T-2, Kern DKM-2, or Zeiss-Jena Theo 010, are used in second- and third-order densification. Handled with care, a theodolite will require little maintenance and will provide trouble-free service almost indefinitely. Protection against shocks and jolts in transportation, cleanliness in handling, storage in dry condition, and regular oiling service about once every 2 years are the main requirements. Dust should only be removed from the optical surfaces with a soft brush, and the objective lens must never be touched by fingers or left without its protective cover. Excess moisture from a wet instrument enclosed in its hermetically sealed carrying case will slowly penetrate and condense on the inner glass surfaces and leave stains that ruin the reading optics. Regular oil changes are necessary because of the extremely narrow tolerances (1 μm) of the moving parts.

Targets for daylight observations are either of a cylindrical shape or in the form of plate signals. The first type, used frequently over long distances (first-order lines) has a drawback in phase error arising from nonsymmetric illumination, which affects the accuracy of pointing. Plate signals such as shown in Figure 3-31 avoid this error if they are not shaded by outside objects. According to (7), the rectangular pattern of the signal is the best because it offers the largest comparison area for symmetrical pointing of the cross hair. The dimensions of the target should be $b = 1.2(D/M)10^{-3} + c$, where D is the sighted distance, M is the magnification of the telescope, and c is the width of the cross hair at the signal distance. For a single cross hair, c is usually between $10^{-5} D$ to $2 \times 10^{-5} D$. The triangular and circular patterns have the advantage in that they may be sighted from varying distances. A combination of black and yellow colors for the targets is recommended as best for observing.

Figure 3-32 shows a common type of signal for daylight and night traversing.

Battery-operated point sources of light serve as signals for night measurements over long distances. The current devices work on the same principle as

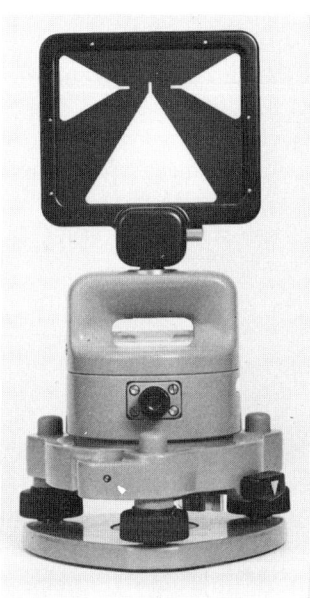

Figure 3-32. Traverse signal of commonly used type. Additional plate allows sightings of up to 5 km (courtesy of Wild Heerbrugg Ltd).

automobile headlights but differ in detail of construction. One model has a 4-V/1.5-W light bulb with a short filament accurately centered at the focal point of a 140-mm diameter paraboloidal reflector of 60-mm focal length. The bulb socket and the aiming device of the system have been made adjustable. The optimum brightness and size of signals are selected according to line lengths and visibility conditions with a rheostat and a series of aperture disks.

A centering inaccuracy of 1 mm at the signal or the instrument will cause a maximum error of 0″.2 in the measured direction at a distance of 1 km, and proportionately more on shorter lines. Accurate centering is therefore more critical in third-order traverses than in a first-order network, not so much in centering over the markers but in setting the theodolite and the signals exactly over the same points. This is mechanically forced in most makes of traverse theodolites, and the equipment consists of a fixed tribrach on top of which the instrument and the signals are interchangeable within an accuracy of 0.2 mm.

Plummets consisting of a small telescope and a spirit level are available for fast and accurate centering in observation towers and for various other applications, but special equipment for this purpose is not necessary in ordinary work. A plate for centering on rooftop pillars should be purchased as an accessory to the triangulation theodolite.

Most theodolites offer an optical plumb as standard accessory as a centering device for third-order traversing on tripods. Experience in the use of these devices has not always been encouraging. The optical plumb is only as good as its adjustment, which must be checked frequently; in some makes, this cannot be done in a practical way during the course of measurement. A centering rod equipped with a circular level is convenient if the points have been marked by round drill holes about 3 mm in diameter. With its tip resting in the hole, the rod can be rotated, and any maladjustment of the bubble is discovered at once.

Additional detail on the features and construction of modern theodolites and centering devices are given, with other publications, in Refs. (17) and (18).

Field Observations

Observation methods. Network observations are usually made by the *direction method* or the *angle method*. The *direction* of a line is given by the angle it forms with a fixed reference line. In horizontal control surveys, this angle is always counted in a clockwise direction from the reference line, a convention that is incomparably better than the mixed assortment of bearings so popular in land surveying in some countries. Thinking in terms of directions is convenient and simple: Addition or subtraction of 360° does not affect the direction of a line, whereas 180° will reverse it; the size of any angle (up to 360°) is obtained by subtracting the direction of its left arm from the direction of its right arm; and so on.

In using the direction method in the configuration shown in Figure 3-33, the work proceeds as follows. When the theodolite has been set up and roughly leveled at station P_0, the observer first locates all the signals so that they can be found without delay when making measurements. If conditions are right for measurement, the instrument is leveled accurately, the work is started by pointing to station P_1 along the right arm of the largest angle in the configuration, and the horizontal circle is read. There is no need to preset the circle to any fixed initial reading. The telescope is then turned clockwise, making the pointings and readings of stations P_2, P_3, and P_4 in consecutive order, which completes the first half-set. The telescope is then turned 180° around its horizontal and vertical

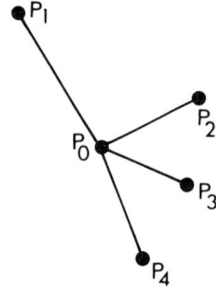

Figure 3-33. A group of four directions.

axes, and a second half-set is obtained from pointings in the reversed order P_4, P_3, P_2, and P_1. Only the minutes and seconds are read and recorded in the second half-set; the degrees can be omitted. The two half-sets complete the first direction set. The procedure is repeated as many times as required; the horizontal circle is rotated between each individual set by $180°/n$, where n is the total number of sets measured. The readings on the optical micrometer are similarly changed so that they spread evenly along the entire length of the micrometer scale.

While the observations proceed, the means are computed in the angle book as shown in Example 3-4.

Example 3-4. Recording of a Direction Set (Centesimal Units)

Sta	I	II	Mean	Reduced directions
1	$300^g.1151$	.1161	.1156	$0^g.0000$
2	5.5549	.5559	.5554	105.4398
3	58.2020	.2027	.2024	158.0868
4	111.8061	.8068	.8064	211.6908

In the final column, the set has been referenced to the first direction by adding orientation constant $400^g - 300^g.1156 = 99^g.8844$ to each mean direction. Any one of the four directions may be chosen as reference direction.

Some observers extend the direction set completely around the horizon by including the first station twice in the set, but this is not believed to offer any particular advantage.

In the *angle method* of observation, individual angles between neighboring lines are usually measured independently (Fig. 3-34) in a required number of

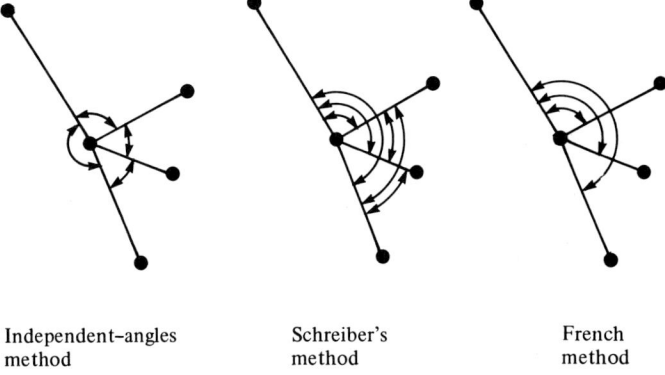

Independent–angles method   Schreiber's method   French method

Figure 3-34. Variants of angle method of observations.

sets. The closure of the sum of all the angles measured around the horizon to 360° provides an additional check of the field work; otherwise, the procedure is the same as for direction groups containing two lines each. The method of independent angles should be used on stations where the instrument support lacks perfect stability or when, because of adverse weather conditions, visibility to targets of individual lines is changing. Several other variants of the angle method of observation have been developed in different countries. The best known are Schreiber's method of measuring angles in all combinations of lines (Fig. 3-34) and the French method of overlapping angles with a common reference direction (Fig. 3-34).

The two latter variants will not be discussed in more detail because they do not seem to offer any advantage over the direction or independent-angles methods.

Errors and design of angle measurements. Measurement of horizontal angles with a theodolite provides a good example of error elimination by observational procedure; the effects of most of the axial errors and eccentricities of the instrument are eliminated from the mean of two telescope positions, and rotation of the horizontal circle between the sets reduces the influence of errors in the circle graduation.

Perfectly graduated circles cannot be manufactured. In modern theodolites, the graduations are marked on glass circles automatically and then etched. Accidental and systematic errors of the graduation lines in a well-made circle should not exceed 1 or 2″ of arc in the 1″ theodolites and are usually smaller than 0″.5 in precision theodolites. Circle graduations may be calibrated by a method introduced in 1913 by Heuvelink (19), but without special equipment, the procedure is tedious. The graduation errors, however, are of little consequence because of measurement of angles or directions in different segments of the circle effectively eliminates them even when the number of observed sets is small.

Other errors may be divided into centering, leveling, pointing, and reading errors and errors due to instability (twist or sinking) of the tripod or of the observation tower and atmospheric refraction. The last source of error, atmospheric refraction, has been already discussed on pp. 55–56.

Centering errors. Four methods of centering are considered: optical plumbs, plumbing rods, string plumb bobs, and, finally, automatic centering (self-centering).

Plumbing with an optical plumb gives a standard deviation of centering $\sigma_e = 0.5$ mm per meter of the height of the instrument above the survey marker if the plumb is well adjusted and the marker has a well-defined center. Rod plummets such as in Kern systems offer approximately the same accuracy. The string plumb bobs give $\sigma_c = 1$ mm per meter in windless weather or even the same accuracy as optical plumbs, particularly when centering under the survey markers, as in mining surveys. In air currents, however, they give unpredictably large errors.

Optical plumbs, particularly those that are built in tribrachs, require frequent checking and adjustment. Errors of 1 cm or more may occur owing to transportation shocks between survey stations.

Self-centering systems such as those used in traversing equipment allow for an exchange of theodolite with the targets with a standard deviation less than 0.1 mm.

Influence of the centering errors on an angle measurement may be calculated as

$$\sigma_{\beta_c}^{\ 2} = (\rho'')^2 \left[\frac{\sigma_{c_1}^{\ 2}}{D_1^{\ 2}} + \frac{\sigma_{c_2}^{\ 2}}{D_2^{\ 2}} + \frac{\sigma_{c_3}^{\ 2}}{D_1^{\ 2} D_2^{\ 2}} (D_1^{\ 2} + D_2^{\ 2} - 2D_1 D_2 \cos \alpha) \right], \quad (3\text{-}54)$$

where σ_{c_1} and σ_{c_2} = centering errors of the targets
$\quad\quad D_1$ and D_2 = distances to the targets
$\quad\quad \sigma_{c_3}$ = centering error of the theodolite
$\quad\quad \rho''$ = 206 265" or, from an approximate formula,

$$\sigma_{\beta_c} = \frac{2\sigma_c}{D} \rho'' \quad\quad\quad (3\text{-}55)$$

if the σ_c of the theodolite and of the targets are approximately the same and the distances to the targets are approximately equal.

The error can be decreased by $\sqrt{n}$ if the angle is remeasured in n sets and the instruments are recentered between the sets. Recentering should be preceded by a rotation of the whole theodolite and of the targets in steps of $180°/n$.

Leveling error. Sensitivity of tubular spirit levels of the medium- and high-accuracy theodolite is usually between 10 and 40" per division.

A careful observer and a well-adjusted spirit level may give the standard deviation σ_L of leveling equal to about 0.2 of the sensitivity if the vial of the level is protected against sunlight or other heat sources. A temperature difference of 1°C between the ends of a 100-mm-long vial will give about 5" error in leveling.

An error of v'' in leveling of the theodolite will introduce an error σ'' in horizontal direction α,

$$\sigma'' = v'' \sin(\alpha - \alpha_o)\tan h, \quad\quad\quad (3\text{-}56)$$

where h is the elevation angle of the target and α_o is the direction of the vertical plane of mislevel. This error cannot be eliminated by observational procedure. The measurement of direction sets containing inclined lines of sight, which often occur on traverse closure stations, therefore necessitates a painstakingly careful leveling of the instrument. An additional level (striding level) of high sensitivity must be used on the rotation axis of the telescope.

Pointing error. The diffraction limit (resolving power) of optical systems gives the maximum accuracy of pointing equal to about $10''/M$ where M is magnification of the telescope. This accuracy is further decreased by improper design of the target, poor visibility conditions, thermal turbulence of air, or focusing error.

The effect of focusing on pointing usually does not exceed 1″ in precision theodolites with an internal focusing system. Its influence is cancelled in angle measurements if distances to two targets are approximately equal or are no shorter than the order of 1 km. If distances are very short, angles with a large ratio of the larger to the shorter distances (the ratio should not exceed 3) should be avoided if a very high precision of angle measurements is required.

Generally, with a properly designed target, under average visibility and thermal turbulence conditions, the standard deviation of a single pointing over short distances is equal to

$$\sigma_p = \frac{30''}{M} \quad \text{up to } \sigma_p = \frac{60''}{M} \tag{3-57}$$

for distances larger than a few hundred meters.

If an angle is measured in n sets, there are $2n$ pointings at each target. The angle between two targets is calculated from

$$\beta = \frac{\sum_{i=1}^{2n} (\delta_2 - \delta_1)_i}{2n}, \tag{3-58}$$

where δ_2 and δ_1 are readings of the horizontal circle in the direction of targets 2 and 1, respectively, in one half-set. Applying the error-propagation law and assuming that the pointing errors $\sigma_{p_1} = \sigma_{p_2} = \sigma_p$, we have

$$\sigma_{\beta_p} = \frac{\sigma_p}{\sqrt{n}}. \tag{3-59}$$

Reading error. The following standard deviations of a single readout of the horizontal circle may be expected:

for theodolites with optical micrometers and with the smallest division of $d'' = 1''$ or $0.5''$,

$$\sigma_r = 2.5 \, d''; \tag{3-60}$$

for theodolites with a microscope and with an estimation of the fraction of the smallest division,

$$\sigma_r = 0.3 \, d'', \tag{3-61}$$

where d'' is the nominal angular value of the smallest division (typically from $10''$ to $1'$);

for vernier theodolites with two verniers,

$$\sigma_r = 0.3 \, d'', \tag{3-62}$$

where d'' is the angular value of the vernier division. Those values may be slightly enlarged by random eccentricity (wandering) of the horizontal circle and by the graduation errors of the horizontal circle.

If an angle β is measured in n sets, we have $2n$ readings in each direction, and the total influence of the reading error may be calculated as

$$\sigma_{\beta_r} = \frac{\sigma_r}{\sqrt{n}}. \tag{3-63}$$

Instability of tripods and observation towers. Thermal expansion due to sunlight or other heat sources may produce a systematic twist of tripods of up to several seconds and of observation towers up to several minutes of arc. Another source of instability may be the sinking of tripod legs in the soil, ice, or pavement, or movements of towers due to wind forces.

Shading of instruments and tripods against heat radiation is strongly recommended in precise angle measurements and a frequent checking of leveling and centering is necessary.

If instability is suspected, the following measuring procedures are recommended to ensure good accuracy of measurement:

> measurements of each set should be made as quickly as possible,
>
> sequence of pointing at two targets should follow the pattern 1–2–2–1 in each set,
>
> instruments should be releveled and recentered between the sets,
>
> if more than two directions are to be observed from the same station, each angle between pairs of targets should be measured separately (angle method instead of the direction method), and the closure should be checked to 360°.

Example of the design of angle measurements. Consider, as an example, a closed loop traverse consisting of four stations to be measured for an engineering project in a moderately flat area.

Distances between the traverse stations are approximately equal to 300 m. Specifications for the measurements allow a maximum misclosure of the four angles of 12″. How should the measurements be made and with what type of theodolite?

Considering the 12″ maximum errors as equal to 3σ, the permissible standard deviation for the closure of the traverse is 4″. If the same precision σ_β of measurements at each station is assumed, then according to the rule of the error propagation, $\sigma_\beta\sqrt{4} = 4''$, and the permissible standard deviation of the angle measurements at each station becomes

$$\sigma_\beta \leq \frac{4}{\sqrt{4}} = 2''.$$

Assuming that a properly adjusted theodolite and properly designed targets are used in the measurements, the total error will be influenced only by errors of reading, pointing, and centering. The random errors of instrument leveling may be neglected in flat country. Therefore,

$$\sigma_\beta = \sqrt{\sigma_{\beta_r}^{\,2} + \sigma_{\beta_p}^{\,2} + \sigma_{\beta_c}^{\,2}} \leq 2''.$$

For the purpose of the preanalysis, it is assumed that

$$\sigma_{\beta_r} = \sigma_{\beta_p} = \sigma_{\beta_c} = \frac{2''}{\sqrt{3}} = 1''.16.$$

Eqs. (3-60) and (3-63) give

$$\sigma_{\beta_r} = \frac{2.5d}{\sqrt{n}} = 1''.16,$$

and $d = (1.16/2.5) = 0''.5$ for $n = 1$ or $d = 1''$ for $n = 4$.

Then, from Eqs. (3-57) and (3-59) for average conditions,

$$\sigma_{\beta_p} = \frac{45}{M\sqrt{n}} = 1''.16,$$

with $M = 39$ for $n = 1$ or $M = 20$ for $n = 4$. From Eq. (3-55),

$$\sigma_{\beta_c} = \frac{2\sigma_c}{D} \rho'' = 1''.16,$$

with $\sigma_c = 0.9$ mm for $D = 300$ m.

Conclusions: The angles should be measured either in one set with a theodolite that has the magnification $M = 39$ and micrometer readout $d = 0''.5$ or in four sets with a $1''$ theodolite and a minimum magnification of $M = 20$. In both cases, an optical plummet or a centering rod should be used. For instance, the Kern DKM-3 theodolite ($M = 45$, $d = 0''.5$) in one set or a Wild T-2 ($M = 28$, $d = 1''$) in four sets could be used.

Generally, as a check, angles should always be measured in at least two sets. Therefore, the choice would be to measure the angles either with a precision theodolite in two sets or with a $1''$ theodolite in four sets.

Handling the theodolite. Electromagnetic distance measurements of high accuracy can be made from relatively lightly constructed observation towers, and no special skill is required in the observation work. The handling of a theodolite is different. A rigid stand for the instrument, which in towers means a double construction separating the instrument support from the observer, is required. The observer also must possess considerable skill that can be developed only through practice. Anyone can learn to operate a theodolite on a few days' instruction, but to entrust network observations to a novice after such brief learning would be foolish.

In viewing through the telescope, contrary to common practice, both eyes should be kept open and focused to infinity. Turning the instrument by hand requires firm, but subtle, movements; the clamps and screws must be turned without the slightest sidewise pressure, and the final move must be made consistently in the same clockwise or anticlockwise direction. Precision theodolites allow the selection of different magnifications by changing the eyepiece; smaller magnifications are used in conditions of weak light. The signal image must fall onto the plane of the cross hairs without parallax, and the pointings with the vertical wire made to a point near its center. Most observers have a personal bias in pointing. This is harmless if all the signals are similar—if, for instance, lights observed do not appear greatly different in size and brightness.

For an individual to record his or her own observations is both impractical and inconducive to accurate work; moreover, the eyes are strained with constant refocusing between the telescope and the field book. An experienced observer loses no time in carrying out pointings and readings, and the recorder is fully occupied in repeating and writing down the readings, calculating the means in the angle book, and taking care that the instrument and its stand are shaded from direct sunlight. In addition, the observer needs a recorder's help in centering the instrument on towers (p. 104).

Time and network schedules. Climate and weather impose severe restrictions on theodolite work. The observing teams must get accustomed to flexible working hours—which can be less irregular in traverse measurement with short sights than in triangulation—in order to take advantage of good observing conditions whenever they occur.

Thermal convection and turbulence in the atmosphere make all observation work, except centerings, generally impossible during hot periods of the day. Cool climates occasionally allow midday observation of long lines on cloudy days. Thus the main observing periods are restricted to: early morning work shortly after sunrise, late afternoon work until evening, and night work with lights after sunset. Good conditions seldom occur, so that skillful time management is necessary, especially in high-order daylight work.

A working program, popularly known as a *menu*, is prepared beforehand for each station occupied. The menu is a list of the different direction groups to be observed, with approximate azimuths and distances for each line determined from the layout. The groups are compiled in strict accordance with the classification of lines into different orders; the order of any one line is determined by its *lower-order* end station, regardless of the direction in which the line is observed. For example, a second-order line may connect two second-order stations as well as a first-order station with a second-order point. Lines of the same order are grouped together to form individual direction sets, including in each set one *reference line* of higher order, if such line exists.

Network specifications usually set a limit to the number of lines allowed in direction groups of different orders. Such regulations are useful as general guidelines and should be followed, but the observer must also consider local conditions of observation, such as stability of the instrument stand, when deciding how many lines to include in a group. Subdivision of a direction group will increase the amount of observation work; for instance, a group of 11 directions may be replaced by 2 subgroups of 6 directions each. If there are no reference lines of higher order, one of the directions must be observed in both subgroups.

Network specifications may contain not only rules on the total number of sets required in the observations of different order but also, depending on refraction, instructions for the distribution of the work. Mean directions derived from day and night observations or from observations spread over several days are likely to be more reliable than means from an equal number of sets measured once only. The observer should prorate observations as much as possible. If the total amount of work makes two visits to the same station necessary, first-order lines should not be completed on one day and second-order lines on the other, but one-half of both observations should be made on each day.

Eccentric observations. Observations on a horizontal control station must frequently be made from an eccentric instrument location. Similarly, signals cannot always be erected so that the observed point will lie exactly on the same vertical line with the station marker. Resorting to eccentric observation need not involve any loss of accuracy in the measurements; if the conditions of observation can be improved by moving the theodolite, say, farther away from an obstructing structure, the observer should not hesitate to do so. Computations and accuracy analysis of corrections for eccentric observations are discussed on pp. 110–112.

In observation towers, the positions of the instrument, the station marker, and the signal are usually determined by projecting the latter two vertically onto the instrument table and measuring the horizontal distances and directions of the projected points from the instrument. In direction measurement of short eccentricity lines, the projected points can be transferred to focusing distance of the theodolite with a piece of taut string. The directions of eccentricity lines can be observed and recorded in the course of main angle observations. Two sets should be measured to provide a check.

The projecting, or plumbing, of the marker and signal centers on the instrument table can be done with a theodolite from two points chosen at right angles and some distance away from the tower. The theodolite is set up on a tripod and leveled carefully. The station marker, or string of a plumb bob centered on it, is pointed with the vertical wire of the cross-hair, the telescope is turned about its horizontal axis, and the pointing is transferred to the instrument table. A straight line can be drawn on the table in line with the vertical wire. The recorder's assistance is, of course, necessary in marking points for this line, which can be done with a long pencil held in vertical position. A second parallel line is obtained from a pointing made in the other telescope position. The projected position of the station marker falls on a straight line drawn midway between these two; the theodolite is moved to another location to determine it as the intersection point of two lines.

The above procedure fails if the projected points fall outside the instrument table, in which case the eccentricities can be determined indirectly as shown in Figure 3-35. The indirect method is also, more often than not, the only practical solution for centering the directions measured from church towers. If large eccentricities are involved, base AB should be made roughly parallel to CD.

Preliminary Computations

Station adjustment of complete sets of observed directions. Field-book recordings mark the beginning of a computation process that gradually translates the observations into coordinates of control points. The observer is responsible for the first stage in this process, the *station adjustments*.

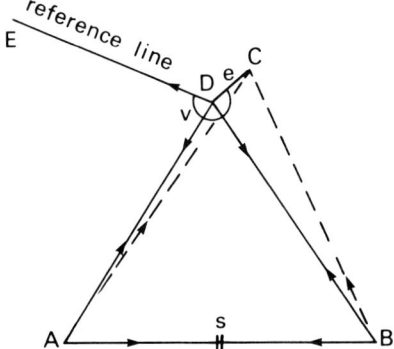

Figure 3-35. Measurement of eccentricity (v, e) indirectly with the aid of a taped horizontal base; $\overline{AB} = s$. Observed directions are shown by arrows.

For a group of n directions $\delta_1, \delta_2, \ldots, \delta_n$ observed in s complete sets, the total number of observations is equal to ns. If the direction to station 1 (Fig. 3-36) is selected as the reference ($0^g00^\circ00^{cc}$) direction, all the ns observations may be adjusted in terms of $n - 1$ unknown angles $X_2, X_3, \ldots, X_n$, and in terms of s orientation unknowns ω (one unknown for each set). By using the parametric least-squares adjustment, we shall have ns observation equations of the form (for ith set):

$$v_1^{\,i} = \omega^i - \delta_1^{\,i}$$

$$v_2^{\,i} = \omega^i + X_2 - \delta_2^{\,i}$$

$$\vdots \qquad \vdots \qquad \vdots \qquad \vdots$$

$$v_n^{\,i} = \omega^i + X_n - \delta_n^{\,i},$$

where $\delta_1^{\,i}, \delta_2^{\,i}, \ldots, \delta_n^{\,i}$ are observed directions in ith set.

Applying the least-square condition that $[vv] = $ minimum, one obtains the adjusted angles $\overline{X}_2, \overline{X}_3, \ldots, \overline{X}_n$ and adjusted orientation unknowns $\overline{\omega}^1, \overline{\omega}^2, \ldots, \overline{\omega}^s$ from

$$\mathbf{X} = (\mathbf{A}^T\mathbf{A})^{-1}\mathbf{A}^T\mathbf{L},$$

where $\mathbf{X}$ is the solution vector matrix, $\mathbf{A}$ is the matrix of coefficients of the observation equations, and $\mathbf{L}$ is the vector of observed directions (reduced to the reference direction).

The adjusted angles $\overline{X}_i$ are directly equal to the correspondingly adjusted directions $\overline{\delta}_i$ because to be able to orient the group of adjusted directions, the adjusted value of the reference direction $\overline{\delta}_1$ is again accepted as $0^g00^\circ00^{cc}$.

By substituting the adjusted values of unknowns into the observation equations, the residuals v are calculated; this allows for an estimation of the variance value of a single observation (in one set)

$$\hat{\sigma}_o^{\,2} = \frac{[vv]}{df}$$

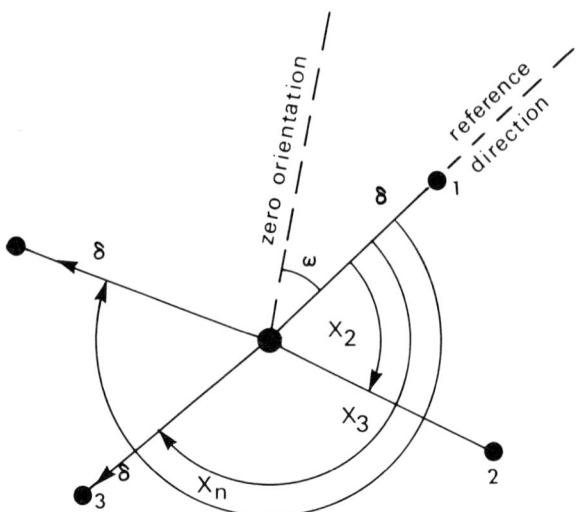

Figure 3-36. Station adjustment.

where df is the number of degrees of freedom. Since the total number of observations is ns and the number of unknowns is $n - 1 + s$, one can write

$$df = ns - (n - 1 + s) = (n - 1)(s - 1), \qquad (3\text{-}64)$$

and finally

$$\hat{\sigma}_o^2 = \frac{[vv]}{(n - 1)(s - 1)}. \qquad (3\text{-}65)$$

In practice, the whole procedure of the rigorous adjustment and calculation of residuals may be done in a much simpler way since it can be proven (20) that the adjusted directions are nothing but the arithmetic mean of s observed directions (reduced to the reference direction) and the individual residuals for ith set can be calculated from

$$v^i = d^i - \frac{[d^i]^s}{n}, \qquad (3\text{-}66)$$

where d^i is a difference between the correspondingly adjusted direction and its observed value in ith set.

Estimation of a standard deviation of an adjusted observation is calculated from

$$\hat{\sigma}_\delta = \frac{\hat{\sigma}_o}{\sqrt{s}}.$$

Standard errors computed from the spread of the observations should be looked upon with caution since they do not reflect the influence of systematic-error sources, such as lateral refraction. The observer should nevertheless compute them for his or her own benefit, although they may not necessarily be required for further processing of the observation data.

An example of the station adjustment for a group of four directions is given in Table 3-5.

It is noted that the illustrated procedure of station adjustment does not apply if any one of the sets has been incompletely observed.

Table 3-5. Station Adjustment

Set	Sta 1	Sta 2	Sta 3	Sta 4
1	$0^g.0000$	$105^g.4398$	$158^g.0868$	$211^g.6908$
2	.0000	.4400	.0871	.6906
3	.0000	.4396	.0872	.6910
4	.0000	.4404	.0868	.6905
5	.0000	.4401	.0864	.6903
6	.0000	.4395	.0866	.6908
$\bar{\delta} =$ (Mean)	$0^g.000\ 00$	$105^g.439\ 90$	$158^g.086\ 82$	$211^g.690\ 67$

	Differences $\bar{\delta} - \delta^i = d^i$			$\dfrac{[d^i]}{n}$
d^1	0.0^{cc}	1.0^{cc}	0.2^{cc}	-1.3^{cc}	0.0^{cc}
d^2	0.0	-1.0	-2.8	0.7	-0.8
d^3	0.0	3.0	-3.8	-3.3	-1.0
d^4	0.0	-5.0	0.2	1.7	-0.8
d^5	0.0	-2.0	4.2	3.7	1.5
d^6	0.0	4.0	2.2	-1.3	1.2

	Residuals $d^i - [d^i]/n = v^i$			$[v]$
v^1	0.0^{cc}	1.0^{cc}	0.2^{cc}	-1.3^{cc}	-0.1
v^2	0.8	-0.2	-2.0	1.5	0.1
v^3	1.0	4.0	-2.8	-2.3	-0.1
v^4	0.8	-4.2	1.0	2.5	0.1
v^5	-1.5	-3.5	2.7	2.2	-0.1
v^6	-1.2	2.8	1.0	-2.5	0.1

$$[vv] = 108.48 \qquad n = 4 \qquad s = 6$$

$$\hat{\sigma}_o = \frac{\sqrt{108.48}}{15} = 2.7^{cc}$$

$$\hat{\sigma}_\delta = \hat{\sigma}_o/\sqrt{s} = 1.1^{cc}$$

Prorated observations. If the direction observations of a fixed configuration of lines have been distributed among several visits to the station and the means have been calculated separately on each visit, the final station adjustment of all observations can be accomplished by computing, for each direction, its *weighted mean L*, by the formula

$$L = \frac{p_1 l_1 + p_2 l_2 + \cdots}{p_1 + p_2 + \cdots} = \frac{[pl]}{[p]}, \tag{3-67}$$

where l_1, l_2, ... are the individual means and the weights p_1, p_2, ... show the number of sets in each mean. Note that for $p_1 = p_2 = \cdots = 1$, the formula reduces to the calculation of the simple arithmetic mean. The weight of the weighted mean is equal to the sum of the weights $[p]$.

Example 3-5. Compilation of Observed Means.

Number of sets (=weight)	Mean			
6	0g.000 00	105g.439 90	158g.086 82	211g.690 67
3	.000 00	.440 03	.086 63	.689 98
5	.000 00	.439 86	.086 70	.690 33
Weighted mean	0g.000 00	105g.439 91	158g.086 74	211g.690 40

$[p] = 14$.

Subdivided direction groups. Means from subdivided direction groups sets must not be recombined, even if they possess a common line as zero reference. For instance, for 11 directions observed in 2 subgroups of 6 directions each, the means from both subgroups should be entered separately into the net adjustment.

Incomplete sets of observations. Drifting fog, a passing rain shower, or a like disturbance may sometimes obscure a signal when a pointing should be made. Should the signal become visible later on, a missing direction in a previous set can be reconstructed later by observing it together with another direction. In the most usual case of broken sets, only one station is incomplete:

Set	Sta 1	Sta 2	Sta 3	Sta 4
1	0g.0000	105.4398	158.0868	211.6908
2	.0000	.4400	.0871	.6906
3	.0000	.4396	.0872	.6910
4	.0000	.4404	.0868	.6905
5	.0000	.4401	—	.6903
6	.0000	.4395	—	.6908
Mean	0g.0000	105g.439 90	158g.086 98	211g.690 67

		Group means				Number of repetitions (weight)
	Set	Sta 1	Sta 2	Sta 3	Sta 4	
Group I	1 to 4	0.000 00	105.439 95	158.086 98	211.690 72	4
Group II	5 to 6	0.000 00	105.439 80		211.690 55	2
II−I		0.000 00	−0.000 15		−0.000 17	

In the adjustment, this station will receive a reduced weight, and a small correction of its arithmetic mean, which can be computed from the group means as follows:

$$\text{weight of complete directions, } p = s = 6;$$

$$\text{weight of incomplete direction, } p' = \frac{s(s - s')(n - 1)}{s(n - 1) + s'} = 3.6;$$

$$\text{correction to incomplete direction, } \frac{s'[II - I]}{s(n - 1)} = -0.000\,04.$$

Here, n is the number of directions in a complete set and s is the total number of repetitions, of which s' repetitions miss out one station. The station-adjusted mean directions can be treated in further computations as uncorrelated direction observations. Because of their different weights, however, it is often simpler to do without them, and enter instead the group means I and II into the net adjustment separately.

Sta 1	Sta 2	Sta 3	Sta 4
$0^g.000\,00$	$105^g.439\,90$	$158^g.086\,94$	$211^g.690\,67$
weight 6	6	3.6	6

The adjustment may also be done by a method of iterations, as is demonstrated in Example 3-6.

If more than one station has been incompletely observed, station adjustment will produce correlated means that are impractical to deal with in the net adjustment. Such observations, therefore, are best resolved into groups of complete direction sets, the means from which can be entered separately into the net adjustment.

Example 3-6. Station Adjustment of Incomplete Direction Sets

Observed

0.0000	105.4398	158.0868	211.6908
.0000	.4400	.0871	.6906
.0000	.4396	.0872	.6910
.0000	.4404	.0868	.6905
.0000	.4401	—	.6903
.0000	.4395	—	.6908
$0^g.000\,00$	$105^g.439\,90$	$158^g.086\,98$	$211^g.690\,67$

1. Mean—observed

0.0^{cc}	$+1.0^{cc}$	$+1.8^{cc}$	-1.3^{cc}	$+1.5$
0.0	−1.0	−1.2	+0.7	−1.5
0.0	+3.0	−2.2	−3.3	−2.5
0.0	−5.0	+1.8	+1.7	−1.5
0.0	−2.0		+3.7	+1.7
0.0	+4.0		−1.3	+2.7
0.0	0.0	+0.2	+0.2	Sum

2. Subtract mean of each row in 1

−0.4	+0.6	+1.4	−1.7	−0.1
+0.4	−0.6	−0.8	+1.1	+0.1
+0.6	+3.6	−1.6	−2.7	−0.1
+0.4	−4.6	+2.2	+2.1	+0.1
−0.6	−2.6		+3.1	−0.1
−0.9	+3.1		−2.2	0.0
−0.5	−0.5	+1.2	−0.3	Sum

3. Subtract mean of each column in 2

−0.3	+0.7	+1.1	−1.6	−0.1
+0.5	−0.5	−1.1	+1.2	+0.1
+0.7	+3.7	−1.9	−2.6	−0.1
+0.5	−4.5	+1.9	+2.2	+0.1
−0.5	−2.5		+3.2	+0.2
−0.8	+3.2		−2.1	+0.3
+0.1	+0.1	0.0	+0.3	Sum

4. Subtract mean of each row in 3

$$
v = \begin{cases}
-0.3 & +0.7 & +1.1 & -1.6 & -0.1 \\
+0.5 & -0.5 & -1.1 & +1.2 & +0.1 \\
+0.7 & +3.7 & -1.9 & -2.6 & -0.1 \\
+0.5 & -4.5 & +1.9 & +2.2 & +0.1 \\
-0.6 & -2.6 & & +3.1 & -0.1 \\
-0.9 & +3.1 & & -2.2 & -0.0 \\
\end{cases}
$$

−0.1	−0.1	0.0	+0.1	Sum

Station-adjusted observations

400.000 03	105.439 83	158.086 83	211.690 83
399.999 95	.439 95	.087 05	.690 55
.999 93	.439 53	.087 13	.690 93
.999 95	.440 35	.086 75	.690 45
400.000 06	.440 16	—	.690 36
.000 09	.439 59	—	.690 89
$400^g.000\,00$	$105^g.439\,90$	$158^g.086\,94$	$211^g.690\,67$

weight: 6	6	3.6	6

$[vv] = 93$; number of redundant
observations $= 13$

$$\hat{\sigma}_o = \sqrt{\frac{93}{13}} = 2.7^{cc}.$$

REDUCTION OF OBSERVATIONS

Corrections to Station-Adjusted Directions and Angles

Eccentric observations. For the direction correction due to an *eccentric instrument* position, Figure 3-37 gives $\sin \varepsilon = e(\sin \omega)/S$ or, because ε is a small angle,

$$\varepsilon'' = \frac{\rho'' e}{S} \sin \omega, \qquad (3\text{-}68)$$

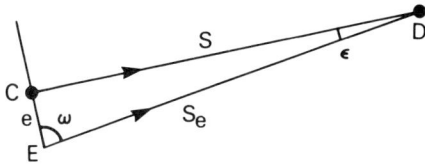

Figure 3-37. Angle ε added to eccentric direction ED gives the centric direction CD.

where $\rho'' = 206\ 265''$. The approximation involved in Eq. (3-68) is negligible ($\leq 0''.05$) unless ε is greater than $40'$. To obtain the proper algebraic sign of the correction, the angle ω of eccentricity should always be determined at E in the clockwise direction from the centric station C to D. If a group of directions is observed at E, each direction will receive an individual correction from Eq. (3-68). The individual angles ω are obtained by reorienting the set of observations, which includes the direction to C, so that line EC becomes the zero direction. Distances S can be scaled from the network layout or computed from preliminary station coordinates.

The centering correction of an angle is equal to the direction correction for its left arm subtracted from the direction correction for its right arm.

Direction corrections due to an *eccentric signal* position are determined in the same way, by calculating the correction ε from Eq. (3-68), where e and ω are elements of the eccentricity of the signal.

In a case of double eccentricity, i.e., when the theodolite and the signal have to be set up eccentrically, the correction to the observed direction is calculated as a sum of the two corrections with their proper algebraic signs,

$$\varepsilon_S'' + \varepsilon_T'' = \frac{\rho'' e_S}{S} \sin \omega_S + \frac{\rho'' e_T}{S} \sin \omega_T \qquad (3\text{-}69)$$

where e_S and ω_S are eccentricity parameters of the signal and e_T and ω_T are the parameters for the theodolite station.

There is no need to be afraid of eccentric measurements if they are based on a proper accuracy preanalysis. By applying the rule of error propagation to Eq. (3-68), we obtain an expression for the variance of ε in the form

$$\sigma_\varepsilon^2 = \left(\frac{\varepsilon''}{e} \sigma_e \right)^2 + \left(\frac{\varepsilon''}{S} \sigma_S \right)^2 + \left(\frac{\varepsilon'' \cot \omega}{\rho''} \sigma_\omega \right)^2. \qquad (3\text{-}70)$$

Example 3-7. Calculation of Eccentricity Correction and Its Standard Deviation. Given:

$$e = 10 \text{ m}, \ S = 1000 \text{ m}, \ \omega = 45°$$

$$\sigma_e = 2 \text{ mm}, \ \sigma_S = 0.2 \text{ m}, \ \sigma_\omega = 60''$$

From Eq. (3-68) we have $\varepsilon = 1458''.5$, and from Eq. (3-70)

$$\sigma_\varepsilon = 1458''.5 \sqrt{4 \times 10^{-8} + 4 \times 10^{-8} + 8 \times 10^{-8}} = 0''.6.$$

Although the computations involved in centerings are very simple, care should be taken to avoid errors; confusing field notes in the recording of e and ω are a common source of error.

Reduction to the ellipsoid and the projection plane. A horizontal direction is reduced to the surface of the reference ellipsoid by applying a correction

$$d\alpha = (\eta \cos \alpha - \xi \sin \alpha)\tan h + 0.11\, H_2 \sin 2\alpha \cos^2 \phi, \qquad (3\text{-}71)$$

where α is the azimuth and h is the elevation angle of the observed line, ξ and η are the components of the deflection of the vertical at the point of observation, ϕ is the latitude, and H_2 is the elevation of the observed point in kilometers above the ellipsoid. In mountainous terrain, the first term may well exceed $3''$, and the second term $0''.3$; in city surveys at high altitudes, the corrections can be significant and should not be ignored. If the deflections of the vertical are not known, large differences in elevation are best avoided between the stations of the first-order net.

Example 3-8. Reduction to the Ellipsoid.

$\xi = +21''.3$	$h = +1° 45'$	$\alpha = 313° 10'$	$\sin \alpha = -0.729\ 37$
			$\cos \alpha = +0.684\ 12$
$\eta = +12''.4$	$H_2 = 1.8$ km	$\phi = 39° 35'$	$\sin 2\alpha = -0.997\ 95$
			$\tan h = +0.030\ 55$
	$d\alpha = +0''.73 - 0''.12 = +0''.6.$		$\cos \phi = 0.770\ 70$

The reduced directions, or angles, can be entered, without further correction, into the net adjustment only if the network is computed in terms of geodetic latitude and longitude, which is seldom the case in urban surveying. For the reduction procedure required in plane coordinate computation, see pp. 32–35 and 36–37.

Corrections to EDM in the First-Order Networks

Curvature correction (long distance). Light and radio microwaves propagate through the atmosphere along paths that curve slightly upward from a straight line. Because of the curvature of the path, which is much smaller than the curvature of the earth, the actual distance measured is neither the *chord length* C' nor the *geodetic slope length* S'; the latter is measured along the line of constant slope α' (Fig. 3-38). A further correction arises from the fact that the average refractive index of air obtained from meteorological readings at the end points of a line and used in the computation of electromagnetic length D is likely to be too small for long lines that pass through much lower, and consequently denser, atmospheric layers than the line of constant slope.

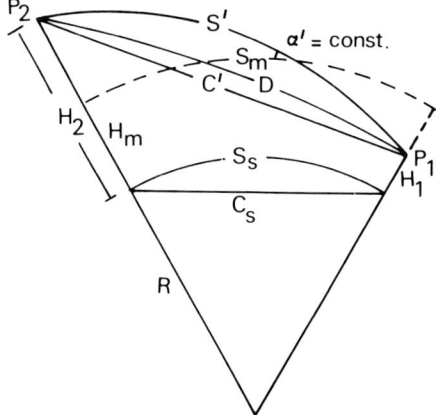

Figure 3-38.   Diagram showing the reduction of EDM distance D to sea level arc S_s.

The general formulas for reduction are

$$S' = D + \frac{(1 - k)^2 D^3}{24R^2} \tag{3-72}$$

$$C' = D - \frac{(2k - k^2)D^3}{24R^2}, \tag{3-73}$$

where R is the curvature radius of the earth and k is the *coefficient of refraction* defined as the ratio of R to the curvature radius of the path. These give the standard corrections (for $R = 6370$ km and $k = 0.20$) which are applicable to all electromagnetic measurements (light and microwave) within the given range.

D (km)	$S' - D$ (ppm)	$C' - D$ (ppm)
15	+0.1	−0.1
20	+0.3	−0.1
25	+0.4	−0.2
30	+0.6	−0.3

The above corrections due to the path curvature are negligibly small in ordinary urban surveys; for horizontal or moderately sloping lines not exceeding 20 km in length, an electromagnetic measurement may be accepted as direct measure of chord length C'.

If the elevation difference of the measured line, $H_2 - H_1$, exceeds 500 m, the standard curvature corrections should be supplemented adding a term

$$\frac{(k_1 - k_2)(H_2 - H_1)D}{12R} = \frac{5}{2}\left[\sin h_{12} - \sin h_{21} - \frac{6(H_2 - H_1)}{5D}\right](H_2 - H_1)$$

$$-\frac{1}{2}(\sin h_{12} - \sin h_{21})^2 D, \qquad (3\text{-}74)$$

which takes into account the rate of change of k with height. This correction can be significant on steep lines, and it frequently shows large daily and seasonal variations. Meade (21) cites an 18-km geodimeter line, with an elevation difference of 1.0 km, on which the correction changed from a value of -2.1 ppm for a late afternoon in May to $+3.4$ ppm at night in September. It was determined by observing simultaneous reciprocal elevation angles h of the Geodimeter light beam in conjunction with each distance measurement. Note that the correction vanishes if

$$H_2 - H_1 = \tfrac{1}{2}(\sin h_{12} - \sin h_{21})D. \qquad (3\text{-}75)$$

This is a well-known formula for trigonometric determination of heights. It follows that substantial corrections to the measured distances can be expected under those atmospheric conditions that also produce large errors in trigonometric heights. These corrections can be determined from reciprocal elevation angles only if the observed height difference ΔH is accurately known (an error of 1 cm in the value of ΔH will cause an error of $\Delta H/D$ cm in the distance correction).

Surface anomalies in vertical refraction generally make theodolite observations from tripod height useless for reduction purposes. The use of reciprocal elevation angles, however, is recommended for improving the accuracy of electrooptical distance measurement between tower stations for lines exceeding 40 km in length or 500 m in elevation difference. The reduction formulas

$$S' = D + 2\left[\sin h_{12} - \sin h_{21} - \frac{5(H_2 - H_1)}{4D}\right](H_2 - H_1)$$

$$-\tfrac{1}{3}(\sin^2 h_{12} + \sin^2 h_{21} - \tfrac{5}{2}\sin^2 h_{12}\sin h_{21})D \qquad (3\text{-}76)$$

$$C' = S' - \frac{D^3}{24R^2} = S'\left(1 - \frac{(S'/R)^2}{24}\right) \qquad (3\text{-}77)$$

combine all previous corrections due to the path curvature. For small, or unknown, elevation difference, a simpler formula,

$$S' = D + \tfrac{1}{24}(\sin h_{12} + \sin h_{21})^2 D, \qquad (3\text{-}78)$$

can be used instead of Eq. (3-76). This is equivalent to the correction given by Eq. (3-72), with

$$k = 1 + \frac{R}{D}(\sin h_{12} + \sin h_{21}). \qquad (3\text{-}79)$$

Reciprocal elevation angles h_{12} and h_{21} of the light path are calculated from the observed values by adding a correction for instrument height, $\rho''(i/D)\sec h$, where i is the height of the horizontal axis of the theodolite above the level of the path. Unless the angles have been observed simultaneously with distance measurement, they should be interpolated from observations made immediately before and after.

An interesting example of the reduction calculation is given below, with numerical data taken from the U.S. transcontinental traverse measurements in 1971.

Example 3-9. Reduction of Geodimeter Measurement with Reciprocal Angles. Given:

$$D = 51.872\ 186 \text{ km} \qquad h_{12} = +1° 30' 50''$$
$$H_1 = 0.050\ 23 \text{ km} \qquad h_{21} = -1° 54' 46''$$
$$H_2 = 1.597\ 49 \text{ km} \qquad \sin h_{12} = +0.0264\ 19$$
$$\sin h_{21} = -0.0333\ 78.$$

Compute: S', C'.

Result: $S' = 51.872\ 399$ km,
$C' = 51.872\ 256$ km.

Note that $D < C'$, indicating atmospheric conditions under which the end-point measurements give too large a value for the mean refractive index.

The path curvature of radio microwaves is generally different from that of visible light. Since the curvature cannot be determined by visual observation, we are compelled to use some nominal value of k (say, $k = 0.25$) in the curvature correction of microwave distance measurements. This can lead to some error on very long or steep lines, which can be avoided if the actual coefficients of refraction are determined on the basis of meteorological observations (22). Unfortunately, the coefficient of refraction involves measurements of vertical temperature and humidity gradients, and this is seldom possible in practical surveying work.

Slope correction. Geodetic slope length S' is reduced to horizontal distance S_m by the formula

$$S_m = \sqrt{S'^2 - (H_2 - H_1)^2}. \qquad (3\text{-}80)$$

This is the horizontal distance at the mean altitude between the end points of the line, or more precisely, at height

$$H_m = \frac{H_1 + H_2}{2} - \frac{(H_2 - H_1)^2}{3R}. \qquad (3\text{-}81)$$

The second term, which amounts to -0.8 m if $H_2 - H_1$ is 4000 m, can usually be ignored.

Conventionally, slope corrections have been computed from the series expansion $S_m - S' = -(H_2 - H_1)^2/(2S') - (H_2 - H_1)^4/(8S'^3) - \cdots$, to avoid the extraction of square root (Eq. 3.80). With a 10-digit electronic calculator, however, the use of the direct formula will be more convenient.

An error x in elevation difference $H_2 - H_1 = \Delta H$ produces an error of $-(\Delta H/S')x$ in horizontal distance S_m (if $\Delta H = 0$, the error is $-\frac{1}{2}x^2/S'$). This means that a relatively large uncertainty in a small elevation difference has little or no effect on the slope correction. Some surveyors mistakenly reverse this rule and in so doing grossly underestimate the importance of an accurate value of ΔH in the reduction of inclined lines.

For trigonometric determination of heights, see pp. 157–159.

Reduction to sea-level arc. Chord length C' is reduced to "sea-level" arc S_s by applying the formulas

$$C_s = \sqrt{\frac{C'^2 - (H_2 - H_1)^2}{(1 + H_1/R_\alpha)(1 + H_2/R_\alpha)}},\tag{3-82}$$

$$S_s = C_s + \frac{C_s^3}{24R_\alpha^2},\tag{3-83}$$

where H_1 and H_2 are the elevations of the end points above sea level and R_α is the curvature radius of the reference ellipsoid, computed from Eq. (2-5) for the mean azimuth and latitude of the observed line.

Because Eq. (3-82) contains the correction for slope, any error in elevation difference $H_2 - H_1$ will have an effect similar to that described in the preceding section.

Alternatively, S_s can be computed from horizontal distance S_m by the formula

$$S_s = \left(\frac{R_\alpha}{R_\alpha + H_m}\right)S_m.\tag{3-84}$$

The value for H_m is determined from Eq. (3-81) for heights above sea level. An error of 1 m in the calculation of H_m will produce an error of 0.16 ppm in S_s.

Example 3-10. Reduction to Sea Level (all distances in kilometers). Given

$$C' = 9.032\,455 \quad \text{or} \quad S' = 9.032\,456$$
$$H_1 = 0.050\,23 \quad\quad \alpha = 169°30'$$
$$H_2 = 1.267\,50 \quad\quad \phi = 35°45'.$$

From Eq. (2-5) we shall obtain $R_\alpha = 6358.0$ using parameters of the Clarke 1866 spheroid.

From Eq. (3-82) is calculated $C_s = 8.949\,128$ and, finally, from Eq. (3-83)

$$S_s = 8.949\,129.$$

The same value of S_s is obtained from Eq. (3-84) by substituting $S_m = 8.950\,057$ and $H_m = 0.6589$ calculated from Eqs. (3-80) and (3-81).

Reduction to the ellipsoid and the projection plane. Sea-level arc S_s is reduced to the surface of the reference ellipsoid by applying a correction

$$S - S_s = (\bar{\xi} \cos \alpha + \bar{\eta} \sin \alpha)(H_2 - H_1) - \left(\frac{\bar{N}S_s}{R_\alpha}\right), \qquad (3\text{-}85)$$

where $\bar{\xi}$ and $\bar{\eta}$ are the average deflection components along the line, expressed in radian measure, and $\bar{N}$ is the average height of the geoid (i.e., sea level) above the surface of the reference ellipsoid. The first term in this correction is analogous to that in direction correction given by Eq. (3-71) and may exceed 0.5 m in mountainous terrain. The second term, which amounts to 1.0 ppm for each 6.4 m of geoid height, can be significant in lowland regions. Failure to apply this correction will result in a scale error of the control net.

Geoid heights $\bar{N}$ can usually be read with sufficient accuracy from a small-scale geoid chart. Note that an error in the vertical datum of heights used in the sea-level reduction will, in effect, produce an error similar to an error in $\bar{N}$.

The reduced ellipsoidal lengths S can be entered without further correction into the net adjustment only if the network is computed in terms of geodetic latitude and longitude. For the reduction procedure required in plane coordinate computation, see pp. 35–36.

Example 3-11. Reduction to the Ellipsoid. Given

$$\begin{aligned}
\xi_1 &= +21''.3 & H_1 &= 627 \text{ m}\\
\eta_1 &= +12''.4 & H_2 &= 1748 \text{ m}\\
\xi_2 &= +23''.7 & \bar{N} &= -16 \text{ m}\\
\eta_2 &= +13''.6 & \alpha_{12} &= 210°12'\\
S_s &= 9374 \text{ m} & R_\alpha &= 6368 \text{ km}
\end{aligned}$$

Computed

$$\bar{\xi} = \tfrac{1}{2}(\xi_1 + \xi_2) = 22''.5 = +0.000\ 1091 \text{ rad.}$$
$$\bar{\eta} = \tfrac{1}{2}(\eta_1 + \eta_2) = 13''.0 = +0.000\ 0630 \text{ rad.}$$

By substituting the above values into Eq. (3-85) we shall obtain the reduction correction

$$S - S_s = -141.2 \text{ mm} + 23.6 \text{ mm} = -118 \text{ mm}.$$

Eccentric distance measurements. A horizontal displacement e of the measuring instrument will change the observed length by an amount obtained from the (approximate) formula

$$D = D_e - e \cos \omega \cos h, \qquad (3\text{-}86)$$

where D is the centric distance
$\quad D_e$ is the eccentric distance
$\quad h$ is the elevation angle of the electromagnetic path
$\quad \omega$ is the horizontal angle of eccentricity (Fig. 3-37).

The error of Eq. (3-86) is ≤ 0.1 ppm if $e{:}D \leq 1{:}2000$.

For lines on which elevation angle h has not been measured,

$$\cos h_{12} = 1 - \tfrac{1}{2}(H_2 - H_1)\left(\frac{H_2 - H_1}{D^2} - \frac{1-k}{R}\right)$$

with $(1-k)/R = 0.000\,13$/km. Slopes of less than $1°$ can be ignored ($\cos h = 1$).

Centerings that involve large eccentricities require more computation. We shall give one method, derived from the triangle in Figure 3-37.

Let H_E, H_C, and H_D be the heights of eccentric point E, center C, and observed point D, respectively. Let V be the angle of eccentricity on the plane of the triangle, with chord lengths $\overline{EC} = e'$, $\overline{ED} = C_e'$, and $\overline{CD} = C'$. Denoting by α the elevation angles of the chords at point E, we have

$$\sin \alpha_C = \frac{(R+H_C)^2 - (R+H_E)^2 - e'^2}{2e'(R+H_E)}$$

$$\sin \alpha_D = \frac{(R+H_D)^2 - (R+H_E)^2 - C_e'^2}{2C_e'(R+H_E)} \tag{3-87}$$

$$\cos V = \cos \alpha_C \cos \alpha_D \cos\omega + \sin \alpha_C \sin \alpha_D$$

$$C' = \sqrt{C_e'^2 + e'^2 - 2C_e'e' \cos V}.$$

These formulas, in which the centering correction is applied to the chord length, are accurate for any value of e'. The computation can be made simpler if the chords are first reduced to sea level.

Example 3-12. Centering of Distance Measurement (all lengths in kilometers).

$C_e' = 12.044\,655$	$\omega = 348°16'48''$
$H_E = 0.029\,17$	$e' = 0.176\,260$
$H_C = 0.050\,23$	$\cos \omega = +0.979\,152$
$H_D = 1.597\,49$	$(R = 6370)$
$\sin \alpha_C = +0.119\,466$	$\cos \alpha_C = 0.992\,838$
$\sin \alpha_D = +0.129\,279$	$\cos \alpha_D = 0.991\,608$

$$\cos V = +0.979\,426$$
$$C' = 11.872\,075.$$

After reduction to sea level,

$$C_e = 11.940\,587 \qquad R_\alpha = 6358.0$$

$$e = 0.174\,996 \qquad \cos V = \left(1 - \frac{C_e^2}{8R^2}\right)\cos \omega$$

$$= +0.979\,152$$
$$C_s = 11.769\,293.$$

Corrections to Distance Measurements in Lower-Order Networks

The corrections listed below are essentially the same as in previous sections, with the exception of the curvature correction of EDM, which is omitted. In these networks, we have $D = S' = C'$; i.e., an electromagnetic distance corrected for the refractive index can be considered directly a measure of chord length C' or slope length S'.

1. Slope correction, Eq. (3-80). In traversing with EDM on sloping lines, particular care should be taken to avoid errors in the determination of slope corrections. A common mistake is to supply height difference $H_2 - H_1$ directly from a vertical angle measurement (see pp. 157–159) and disregard the differences in the instrument, EDM reflector, and theodolite signal heights. Although these differences are usually no larger than a few centimeters, it must be kept in mind that the slope correction of inclined lines requires an accurately determined height difference that must be specifically referred to the terminal heights of the measuring beam.
2. Reduction to sea level, Eq. (3-84). This reduction amounts to -1.57 mm for each 100 m in horizontal distance S_m and in altitude H_m.
3. Reduction to the ellipsoid, Eq. (3-85). It should be noted that the first term in this correction, being independent of the measured distance, can also be significant in the lower-order networks in mountain areas.
4. Corrections for eccentric measurements. These corrections have been explained in the preceding section.
5. Reduction to the projection plane. This reduction is done according to the rules given in Chapter 2.

Checking Procedures

A serious blunder in observations can very possibly escape the scrutiny of an experienced observer, as for instance, a line observed to a wrong signal. The reduction of observations should not be considered complete until the reduced directions and distances have been adequately checked for geometric consistency.

The main checking procedure is calculation of angular and linear misclosures in closed geometrical figures of the network, such as triangles or loop traverses.

The theoretical sum of n inner angles in a closed geometrical figure reduced to a plane is equal to $(n - 2)180°$. Owing to errors of measurement, the actual sum of observed (and reduced) angles differs from the theoretical by the value

$$w = \sum \beta - (n - 2)180°. \tag{3-88}$$

If standard deviations σ_β of observed angles are properly estimated, the angular misclosure should satisfy the condition

$$w \leq 1.96\sigma_\beta\sqrt{n} \tag{3-89}$$

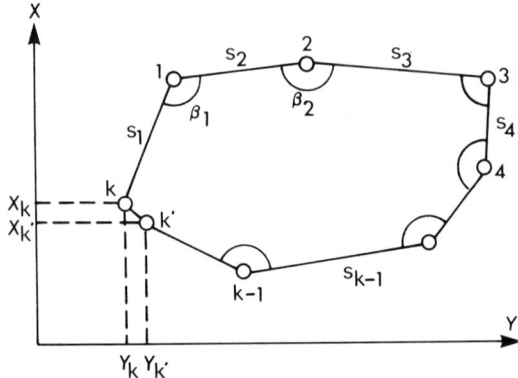

Figure 3-39. Misclosure of a single traverse loop.

at the 95% confidence level. If Eq. (3-89) is not satisfied, we may suspect that blunders or systematic errors have taken place in the measurements.

If angles and distances are measured in the network, single traverse loops may be selected for checking purposes by calculating misclosures of coordinates of point k (Fig. 3-39) in the traverse: $k, 1, 2, \ldots, k - 1, k$.

The coordinates of point k are calculated as in an open traverse: The calculations are started from the fixed direction k to 1 and come to k, which is again treated as a different point k'. Theoretically, the differences between the original coordinates X_k and Y_k and those calculated through the traverse $X_{k'}$ and $Y_{k'}$ should be equal to zero. The eventual misclosures $X_{k'} - X_k$ and $Y_{k'} - Y_k$ should satisfy the conditions

$$X_{k'} - X_k \leq 1.96\sigma_{X_{k'}}$$
$$Y_{k'} - Y_k \leq 1.96\sigma_{Y_{k'}}$$
(3-90)

at the 95% confidence level, where $\sigma_{X_{k'}}$ and $\sigma_{Y_{k'}}$ are estimated (*a priori*) standard deviations of the coordinates $X_{k'}$ and $Y_{k'}$.

The values of $\sigma_{X_{k'}}$ and $\sigma_{Y_{k'}}$ may be calculated (7) from the equations

$$\sigma_{X_{k'}}^2 = \sum_{i=1}^{k-1}(Y_k - Y_i)^2\sigma_{\beta_i}^2 + \sum_{i=1}^{k}\left(\frac{X_i - X_{i-1}}{S_i}\right)^2\sigma_{s_i}^2$$

$$\sigma_{Y_{k'}} = \sum_{i=1}^{k-1}(X_k - X_i)^2\sigma_{\beta_i}^2 + \sum_{i=1}^{k}\left(\frac{Y_i - Y_{i-1}}{S_i}\right)^2\sigma_{s_i}^2,$$
(3-91)

where σ_β and σ_s are anticipated standard deviations of measured angles and distances. It must be pointed out that the linear misclosures of traverses should be calculated with unadjusted (not compensated) angles. Otherwise, Eqs. (3-90) and (3-91) are not valid. This is a very frequent mistake of practicing surveyors who first adjust the angles by distributing the angular misclosure and then try to evaluate the precision of their traverse by calculating the misclosure of coordinates.

The checking procedures of the closed figures may be extended to the traverses that close at both ends to different points of a higher-order control network. In this case, the expected linear misclosure of the traverse will be larger than that of a loop traverse because provision must be made for the relative errors of coordinates of the terminal points of the higher-order control.

NET ADJUSTMENT AND COMPUTATION OF COORDINATES

General Remarks on the Adjustment

To calculate the coordinates of new points (unknowns), the number of observations n must be at least equal to that of the unknowns u. Geodetic networks typical of urban horizontal control usually contain observations far greater than the number of unknowns. An adjustment of the observations, which is necessary before any definitive coordinate values can be computed, should be performed by a rigorous method such as the parametric least-squares method, which allows for a full statistical analysis of the accuracy of the network.

Until the advent of large computers, rigorous methods of adjustment were usually applied only on first-order city networks, and the adjustment was performed for small groups of points without giving the full data for an error analysis. The second- and lower-order densification control nets were, as a rule, adjusted by using approximate methods, with some exceptions—when the "patching" method (pp. 51–52) of densification was used and when the least-squares method was used simultaneously for the adjustment of a very limited number of points.

Results of approximate methods, such as the separate adjustment of azimuths and coordinates of junction points by calculating weighted means, usually agree with the results of the rigorous methods, within the standard deviations of the coordinates, and the differences may be neglected for some practical applications. However, the approximate methods do not provide information on the accuracy of the adjusted coordinates. This is a most serious drawback. Today, with the available computational techniques, the whole control network, including the lowest order, should be adjusted by rigorous methods. The parametric method of least-squares adjustment is recommended since it gives directly, as a by-product, the variance–covariance matrix of the adjusted coordinates.

Ideally, the whole control network for the entire urban area should be adjusted simultaneously (pp. 47–49). Practically, the densification of the control is accomplished in steps through a number of years; the adjustment must also be done in steps, so that portions of the network can be used before total densification has been completed. It is strongly recommended, however, that the whole first-order network, including as many second-order points as possible, should be adjusted simultaneously. Adjustment of the remaining portion of the second- and lower-order nets may be done later in steps whereby large groups of points are simultaneously adjusted as soon as a "block" of the higher-order control is filled up.

In some cases, although it should be avoided, the densification within a block of the higher-order points is spread over a long period of time. In those circumstances, coordinates of partially completed densification control must be calculated for an intermediate use before the densification of the whole block is completed. These coordinates should be treated as temporary, and as soon as the densification surveys within the block are completed, a new adjustment, including all the points, should be performed. This leads, of course, to a change in previous coordinates, an unwelcome occurrence from the practical point of view. However, it is the only way to avoid problems similar to that of the ill-designed control net illustrated in Figure 3-7. One has to get used to the fact that the coordinate system in integrated surveys has to be dynamic. The frequency of readjustment and the resulting change of coordinates depend upon the quality of the network design.

One conclusion is obvious: The design, planning, and computation of the entire control network must be in the hands of one survey institution that should have the power and ability to decide which parts of the network should be adjusted simultaneously. The institution should also decide which co-ordinates should be treated as temporary and when the readjustment should take place. These decisions require long-range policy and very competent management of the control surveys. The execution of control surveys and the above-mentioned decisions should not be the responsibility of practicing individual surveyors.

Preparation of Data for the Adjustment

Adjustment of large control networks does not impose any technical difficulty if a large computer is available to the city survey office. Much time may be wasted if the data are not properly checked and fed into the computer.

The problem of handling the large amount of field observations, and data management in general, is becoming more serious than the adjustment procedure itself. Errors in numbering the survey stations, mistakes in identification of observed lines, and errors in punching the data on computer cards are common and lead to the time-consuming process of running successive adjustments of the network. Therefore, time spent in checking and the proper organization of the input data is never left unrewarded. Data management requires a good knowledge of all phases of adjustment, coordinate computations, and analysis of the computer output.

Some checks on the reduced field data have already been discussed on pp. 119–121. As a basis of all checking procedures and preparation for the adjustment, a large scale plot (1:2000 or larger) of the measured network must be made. The reduced field observations, coordinates of fixed points, and numbers of stations are written on the sketch. The plot of the network, with all the written data, allows an experienced person to find mistakes quickly.

Some computer programs for network adjustment require the renumbering of survey stations in a sequence convenient for computer processing. The renumbering should also be done on the large-scale sketch. Otherwise, chaos is created in the data, and finding of errors is difficult.

Review of the Parametric (Observation-Equation) Method of Least-Squares Adjustment

For each observation l in the network, we can write the equation

$$l_{obs} + v = l_{adj} = l_{approx} + dl, \tag{3-92}$$

where l_{obs} is the field observed quantity reduced to the surface of the coordinate system,

l_{adj} is the value of the observation calculated from adjusted coordinates $X_1, Y_1, \ldots, X_k, Y_k$,

v is the difference between the adjusted and the field-observed values of the observation,

l_{approx} is the value of observation calculated from approximate co-ordinates of the network $X_1{}^0, Y_1{}^0, \ldots, X_k{}^0, Y_k{}^0$, and dl is the correction to the approximate value of the observation.

If we express l_{approx} in Eq. (3-92) as a function of coordinates of the points between which the observation was made and assume that the correction dl is differentially small, Eq. (3-92) may be written in the general form of an *observation equation*,

$$v = \frac{\partial l}{\partial X_1} dX_1 + \frac{\partial l}{\partial Y_1} dY_1 + \cdots + \frac{\partial l}{\partial X_k} dX_k$$

$$+ \frac{\partial l}{\partial Y_k} dY_k + l_{approx} - l_{obs}. \tag{3-93}$$

Values of the partial differentials are calculated for approximate values of the coordinates, which must be known beforehand.

The functional relationship between coordinates and different types of observations, as well as the differentials of those functions, were given in Eqs. (3-20) through (3-27).

Placing the corresponding equations into Eq. (3-93), we obtain the following observation equations for particular types of observations:

1. For *distance observations* between points i and j,

$$v_s = \frac{X_j - X_i}{s} dX_j + \frac{Y_j - Y_i}{s} dY_j - \frac{X_j - X_i}{s} dX_i$$

$$- \frac{Y_j - Y_i}{s} dY_i + s_{approx} - s_{obs}. \tag{3-94}$$

2. For *azimuth observations* between points i and j,

$$v_\alpha = \frac{X_j - X_i}{s^2} dY_j - \frac{Y_j - Y_i}{s^2} dX_j - \frac{X_j - X_i}{s^2} dY_i$$

$$+ \frac{Y_j - Y_i}{s^2} dX_i + \alpha_{approx} - \alpha_{obs}. \tag{3-95}$$

3. For direction observations (Fig. 3-11),

$$v_\delta = \frac{X_j - X_i}{s^2} dY_j - \frac{Y_j - Y_i}{s^2} dX_j - \frac{X_j - X_i}{s^2} dY_i$$

$$+ \frac{Y_j - Y_i}{s^2} dx_i - d\omega + \delta_{approx} - \delta_{obs}. \tag{3-96}$$

4. For angle observations (Fig. 3-12),

$$v_\beta = \frac{X_R - X_C}{s_R{}^2} dY_R - \frac{Y_R - Y_C}{s_R{}^2} dX_R - \frac{X_L - X_C}{s_L{}^2} dY_L + \frac{Y_L - Y_C}{s_L{}^2} dX_L$$

$$+ \left(\frac{X_L - X_C}{s_L{}^2} - \frac{X_R - X_C}{s_R{}^2} \right) dY_C$$

$$- \left(\frac{Y_L - Y_C}{s_L{}^2} - \frac{Y_R - Y_C}{s_R{}^2} \right) dX_C + \beta_{approx} - \beta_{obs}. \tag{3-97}$$

The coefficients of Eqs. (3-94) through (3-97) should be calculated in the same units as those of $l_{approx} - l_{obs}$.

The differentials dX and dY in the above equations are the unknown corrections to the approximate coordinates; they are obtained from the least-squares adjustment.

If direction observations are used in the adjustment, additional unknowns $d\omega$ (one for each direction group) must be taken into account in the adjustment. The adjusted corrections $d\omega$ are usually not needed for further calculations, and therefore, they are called nuisance unknowns. They may be removed from the adjustment process by the inclusion of *sum equations*, a computational device that is treated like ordinary observation equations.

A sum equation is formed by adding together the direction equations containing the same orientation unknown $d\omega_i$, which is omitted, and assigning to the resulting equation a fictitious weight of $p = -1/(n\sigma_i{}^2)$, where n is the number of directions in the sum, each having weight $p = 1/\sigma_i{}^2$.

If an observation includes a point in the network that is treated as fixed (errorless) in the process of adjustment, the corresponding corrections dX and dY to that point equal zero, and the corresponding terms in the observation equation disappear.

Using matrix algebra, a set of observation equations may be written in the short form

$$\mathbf{V} = \mathbf{AX} + \mathbf{L}, \tag{3-98}$$

where $\mathbf{V}$ is a vector matrix of residuals,

A is the design matrix (Eq. 3-17) of the coefficients of the observation equations,

X is a vector matrix of the unknown corrections dX, dY, and if applicable, $d\omega$, and

L is a vector matrix of the differences $l_{approx} - l_{obs}$.

We have as many observation equations as observations used in the adjustment process.

The least-squares solution for finding the vector matrix $\mathbf{X}$ is obtained by applying the condition $\mathbf{V}^T\mathbf{V}$ = minimum. If the observations are not of the same accuracy, we have to introduce the weight matrix $\mathbf{P}$, which was discussed on pp. 60–65, and the condition is (in the matrix form)

$$\mathbf{V}^T\mathbf{PV} = \text{min.} \qquad (3\text{-}99)$$

The above condition, if applied to Eq. (3-98), gives the so-called *normal equations*

$$(\mathbf{A}^T\mathbf{PA})\mathbf{X} + \mathbf{A}^T\mathbf{PL} = 0, \qquad (3\text{-}100)$$

where $\mathbf{A}^T\mathbf{PA} = \mathbf{N}$ is a square and symmetrical matrix.

The number of normal equations is equal to the number u of unknowns. Therefore, a unique solution for the unknowns is now possible in the form

$$\mathbf{X} = -(\mathbf{A}^T\mathbf{PA})^{-1}\mathbf{A}^T\mathbf{PL} = -\mathbf{N}^{-1}\mathbf{A}^T\mathbf{PL}. \qquad (3\text{-}101)$$

In summary, the adjustment is done in the following steps:

1. calculation of approximate coordinates of the new points using some of the field observations and the simplest possible calculation method, e.g., angular intersection, a polar method, or a single traverse;
2. calculation of l_{approx} values for the observations using coordinates from step 1;
3. calculation of elements of the matrix $\mathbf{L}$;
4. calculation of coefficients of observation equations (matrix $\mathbf{A}$) by using approximate coordinates;
5. calculation of weights (matrix $\mathbf{P}$);
6. placing $\mathbf{A}$, $\mathbf{P}$, and $\mathbf{L}$ into Eq. (3-101) and finding the solution for corrections to the approximate coordinates (matrix $\mathbf{X}$).

It should be noted that the above procedure of the adjustment is based on the assumption that the corrections dX and dY are differentially small. If the approximate coordinates, obtained from preliminary calculations, differ considerably from the adjusted values, the result of the adjustment will not be correct. If there is any doubt about the quality of the approximate coordinates, the adjustment should be repeated, with the coordinates obtained from the first adjustment as the approximate ones. If the first adjustment is correct, the values dX and dY from the second adjustment should be negligibly small. Sometimes two or more adjustments (iterations) may be needed. Computer

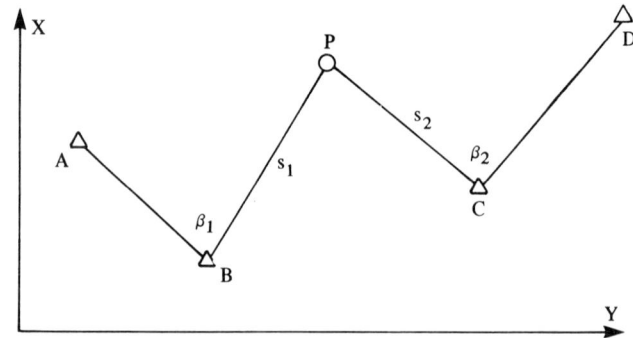

Figure 3-40. Determination of a new point by angular and distance measurements.

programs automatically take care of the updating procedure of the approximate coordinates and stop iterating when the corrections from the adjustment reach negligibly small (preprogramed) values.

A simple example of the least-squares adjustment demonstrates the principle.

Example 3-13. Least-Squares Adjustment [data taken from Ref. (24)]. Two angles and two distances (Fig. 3-40) have been measured from fixed points A, B, C, and D to determine the coordinates of point P. Points A, B, C, and D are taken as errorless.

The following data have been given:

Observations

$$\beta_1 = 90° \ 00' \ 10'' \qquad \sigma_{\beta_1} = 10''$$

$$\beta_2 = 89° \ 59' \ 55'' \qquad \sigma_{\beta_2} = 10''$$

$$s_1 = 707.00 \text{ m} \qquad \sigma_{s_1} = 0.053 \text{ m}$$

$$s_2 = 424.15 \text{ m} \qquad \sigma_{s_2} = 0.041 \text{ m}$$

Coordinates

Pt	X (m)	Y (m)
A	700.00	300.00
B	300.00	700.00
C	500.00	1500.00
D	1000.00	2000.00

Approximate coordinates of P have been calculated by the angular intersection taking rounded-off values of $\beta_1 = 90° \ 00' \ 00''$ and $\beta_2 = 90° \ 00' \ 00''$, which gives $X_P{}^0 = 800.00$ m and $Y_P{}^0 = 1200.00$ m.

Approximate values of observations, calculated from the given coordinates, are

$$\beta_1 = 90° \ 00' \ 00'' \qquad s_1 = 707.11 \text{ m}$$

$$\beta_2 = 90° \ 00' \ 00'' \qquad s_2 = 424.27 \text{ m}.$$

Matrix **L** is therefore

$$\mathbf{L} = \begin{pmatrix} -10'' \\ +5'' \\ +0.11 \\ +0.12 \end{pmatrix}$$

The *weight matrix* has been calculated from the given standard deviations, with $p_0 = 1$ for $\sigma_0 = 1$

$$\mathbf{P} = \begin{pmatrix} \dfrac{1}{\sigma_{\beta_1}^2} & & & \\ & \dfrac{1}{\sigma_{\beta_2}^2} & & \\ & & \dfrac{1}{\sigma_{s_1}^2} & \\ & & & \dfrac{1}{\sigma_{s_2}^2} \end{pmatrix} = \begin{pmatrix} 0.01 & & & \\ & 0.01 & & \\ & & 356 & \\ & & & 595 \end{pmatrix}$$

Matrix **A** has been calculated using Eq. (3-94) for distances and Eq. (3-97) for angles. The latter was multiplied by $\rho'' = 206\ 265''$ since values of $\beta_{approx} - \beta_{obs}$ were taken in seconds. We have, therefore,

$$\mathbf{A} = \begin{array}{c} dX_P \qquad dY_P \\ \begin{pmatrix} -206 & 206 \\ -344 & -344 \\ 0.707 & 0.707 \\ 0.707 & -0.707 \end{pmatrix} \begin{array}{l} \beta_1 \\ \beta_2 \\ s_1 \\ s_2 \end{array} \end{array}$$

Placing **A**, **P**, and **L** matrices into Eq. (3-100), we obtain two *normal equations* for the unknowns

$$2083\ dY_P + 639\ dX_P + 81.57 = 0$$
$$639\ dX_P + 2083\ dY_P - 60.59 = 0,$$

since

$$\mathbf{N} = \mathbf{A}^T\mathbf{PA} = \begin{pmatrix} 2083 & 639 \\ 639 & 2083 \end{pmatrix} \qquad \text{and} \qquad \mathbf{A}^T\mathbf{PL} = \begin{pmatrix} 81.57 \\ -60.59 \end{pmatrix}.$$

The inverse $\mathbf{N}^{-1}$ is equal to

$$(\mathbf{A}^T\mathbf{PA})^{-1} = \begin{pmatrix} 0.000\ 530 & -0.000\ 163 \\ -0.000\ 163 & 0.000\ 530 \end{pmatrix},$$

and the final solution is

$$\mathbf{X} = \begin{pmatrix} dX_P \\ dY_P \end{pmatrix} = -\mathbf{N}^{-1}(\mathbf{A}^T\mathbf{PL}) = \begin{pmatrix} -0.053 \\ 0.045 \end{pmatrix}.$$

The adjusted coordinates, therefore, are equal to

$$X_P = X_P{}^0 + dX_P = 799.947 \qquad Y_P = Y_P{}^0 + dY_P = 1200.045.$$

Error Analysis of the Adjusted Network

Estimated variance factor. A basic procedure in error analysis is finding the variance factor $\hat{\sigma}_0{}^2$ as derived from the adjustment that allows calculation of the standard deviation of an observation for which $p = 1$.

The variance factor is calculated from

$$\hat{\sigma}_0{}^2 = \frac{[vpv]_1^n}{n - u}. \qquad (3\text{-}102)$$

If the mathematical model of the network and the *a priori* estimation of the accuracy of observations has been correct, the calculated $\hat{\sigma}_0{}^2$ value should agree with the *a priori* accepted $\sigma_0{}^2$ value within the confidence interval at a specified probability level. Chi-square (χ^2) distribution is used for testing the agreement of the value $\hat{\sigma}_0{}^2$ with $\sigma_0{}^2$ by using the confidence interval

$$\left[\frac{(n - u)\hat{\sigma}_0{}^2}{\chi_{P_2}{}^2} \le \sigma_0{}^2 \le \frac{(n - u)\hat{\sigma}_0{}^2}{\chi_{P_1}{}^2} \right], \qquad (3\text{-}103)$$

with $\chi_{P_2}{}^2$ and $\chi_{P_1}{}^2$ found from statistical tables for the chi-square distribution at the required probability level.

Table 3-6 lists some values of $\chi_{P_1}{}^2$ and $\chi_{P_2}{}^2$ for 95 and 99 % probability level as a function of degrees of freedom (difference $n - u$) in the network. If the number of degrees of freedom is very small, as in Example 3-13, the certainty

Table 3-6. Chi-square Distribution

$n - u$	Probability 95 %		Probability 99 %	
	$\chi_{P_2}{}^2$	$\chi_{P_1}{}^2$	$\chi_{P_2}{}^2$	$\chi_{P_1}{}^2$
1	5.02	0.0010	7.88	0.0000
2	7.38	0.0506	10.6	0.010
4	11.1	0.484	14.9	0.207
6	14.4	1.24	18.5	0.676
8	17.5	2.18	22.0	1.34
10	20.5	3.25	25.2	2.16
20	34.2	9.59	40.0	7.43
30	47.0	16.8	53.7	13.8
40	59.3	24.4	66.8	20.7
50	71.4	32.4	79.5	28.0
80	106.6	57.2	116.3	51.2
100	129.6	74.2	140.2	67.3

attached to estimated $\hat{\sigma}_0^2$ is, of course, very small. This is reflected by a very large confidence interval for the *a priori* value of σ_0^2.

Example 3-14. Calculation and Testing of $\hat{\sigma}_0^2$.

Taking the data from the adjustment calculation given in Example 3-13, we obtain the following values for the residuals when we replace the calculated corrections dX and dY in the observation equations:

$$v_{\beta_1} = 10'', \qquad v_{\beta_2} = 8'', \quad v_{s_1} = 0.10, \quad \text{and } v_{s_2} = 0.05.$$

The estimated variance factor is thus equal to

$$\hat{\sigma}_0^2 = \frac{[vpv]}{n-u} = \frac{7.07}{2} = 3.53$$

The *a priori* variance factor is $\sigma_0^2 = 1$.

We shall test whether the *a priori* value agrees with the estimated one within a confidence interval of 95%.

We have, from Table 3-6, for $n - u = 2$,

$$\chi_{P_2}^2 = 7.38 \qquad \text{and } \chi_{P_1}^2 = 0.0506$$

and, from Eq. (3-103),

$$\frac{2 \times 3.53}{7.38} \leq 1 \leq \frac{2 \times 3.53}{0.0506},$$

which gives

$$0.96 \leq 1 \leq 140.$$

The test shows there is a 95% probability that our mathematical model and estimation of the accuracy of measurements are correct. Because of the very small number of redundant observations in the above example, the certainty attached to the estimated $\hat{\sigma}_0^2$ is very small and as we can see, our *a priori* estimated accuracy could be as much as $\sqrt{140}$ larger and the test would still pass the results of the adjustment. Therefore, the application of statistical tests to the network analysis is not very meaningful if the number of redundant observations is small. In this case, the error analysis of the network must be based on a very good (*a priori*) knowledge of the accuracy of our measurements. Practically, the calculated $\hat{\sigma}_0^2$ from the adjustment may be accepted as a basis for further error analysis of the network if the number of degrees of freedom is comparatively large, say more than 30, and, of course, if the χ^2 test is passed. If the χ^2 test fails, the reason must be found before a further error analysis is made.

A frequent reason for the failure of the test is the presence of a blunder or systematic errors in the observations. A quick glance at the list of residuals obtained from the adjustment may help to find it. The residuals should not exceed about twice the expected standard deviations of the corresponding measurements, and their sum should tend toward zero.

Use of the variance–covariance matrix. The solution of the least-squares adjustment, as given by Eq. (3-101), includes the inverse matrix $(\mathbf{A}^T \mathbf{PA})^{-1}$, which should be identical with the inverse matrix $\mathbf{Q}$ obtained from preanalysis (Eq. 3-16) if the shape and type of observations in the adjusted network are the same as in the design. The only question in calculating the variance–covariance matrix after the adjustment is whether the matrix $\mathbf{Q}$ should be multiplied by the *a priori* σ_0^2 or by $\hat{\sigma}_0^2$ calculated from the adjustment. Different authors give different answers, but it seems clear that if the comparison of σ_0^2 and $\hat{\sigma}_0^2$ passes the χ^2 test, any value of σ_0^2 that is within the obtained confidence is equally good statistically, and therefore, as a rule, the *a priori* value of σ_0^2 may be used.

The variance–covariance matrix allows the calculation of the following accuracy parameters of the adjusted network: standard deviations of coordinates, which in the general form will be represented by the standard-error curves (Eq. 3-12), absolute and relative standard-error ellipses (Eqs. 3-3 through 3-11), and standard deviations of functions of adjusted coordinates, e.g., standard deviations of a distance or an azimuth calculated from the coordinates of any two points in the network. A numerical example of a calculation of the standard deviations of coordinates and standard-error ellipse was given in Example 3-1 on p. 63.

Calculation of standard deviations of the functions of adjusted coordinates is extremely important to the users of the control network in legal and engineering surveys since practicing surveyors should be able to estimate the accuracy of observations calculated from given coordinates.

The calculation of the errors of the functions follows the general rule of error propagation given by Eq. (3-13),

$$\Sigma_U = \mathbf{B} \Sigma_Z \mathbf{B}^T.$$

For instance, if we have an observation l, which is calculated from coordinates of points i and j, the standard deviation of l will be calculated from

$$\sigma_1^2 = \left(\frac{\partial l}{\partial X_i}, \frac{\partial l}{\partial Y_i}, \frac{\partial l}{\partial X_j}, \frac{\partial l}{\partial Y_j} \right) \begin{pmatrix} \sigma_{X_i}^2 & \sigma_{X_iY_i} & \sigma_{X_iX_j} & \sigma_{X_iY_j} \\ \sigma_{Y_iX_i} & \sigma_{Y_i}^2 & \sigma_{Y_iX_j} & \sigma_{Y_iY_j} \\ \sigma_{X_jX_i} & \sigma_{X_jY_i} & \sigma_{X_j}^2 & \sigma_{X_jY_j} \\ \sigma_{Y_jX_i} & \sigma_{Y_jY_i} & \sigma_{Y_jX_j} & \sigma_{Y_j}^2 \end{pmatrix} \begin{vmatrix} \dfrac{\partial l}{\partial X_i} \\ \dfrac{\partial l}{\partial Y_i} \\ \dfrac{\partial l}{\partial X_j} \\ \dfrac{\partial l}{\partial Y_j} \end{vmatrix}.$$

$$(3\text{-}104)$$

The partial differentials are calculated at the estimated values of the coordinates.

Example 3-15. Standard Deviation of a Distance Calculated from Coordinates.

The coordinates of two points are

$$X_1 = 300.00 \text{ m} \qquad X_2 = 1000.00 \text{ m}$$
$$Y_1 = 500.00 \text{ m} \qquad Y_2 = 1500.00 \text{ m},$$

and their variance–covariance matrix in square meters is

$$\Sigma_{x_{1,2}} = \begin{matrix} & X_1 & Y_1 & X_2 & Y_2 & \\ & \begin{pmatrix} 0.36 & 0.26 & 0.30 & 0.20 \\ 0.26 & 0.49 & 0.10 & 0.20 \\ 0.30 & 0.10 & 0.25 & 0.20 \\ 0.20 & 0.20 & 0.20 & 0.25 \end{pmatrix} & \begin{matrix} X_1 \\ Y_1 \\ X_2 \\ Y_2. \end{matrix} \end{matrix}$$

Calculated distance is $s_{12} = 1220.66$ m.

The partial derivatives of the distance equation may be taken from Eq. (3-23) by placing $j = 1$ and $i = 2$. We have, therefore, the matrix:

$$\mathbf{B} = (-0.57 \ - \ 0.82 \ + \ 0.57 \ + \ 0.82);$$

and from the product $\mathbf{B}\Sigma_{x_{1,2}}\mathbf{B}^T$ we shall obtain

$$\sigma_s^2 = 0.378 \qquad \text{and} \qquad \sigma_s = 0.61 \text{ m}.$$

Special Problems in Net Adjustment

Crossing of zonal boundaries. Net adjustment in plane coordinates by no means implies that all stations in the adjustment should be expressed in the same rectangular coordinate system. In very large networks extending over several projection zones, each station can be treated in its own projection zone if attention is paid to proper forming of observation equations for those lines that cross a zonal boundary.

Let $P_A P_B$ be such a line, with initial point P_A located in the western zone and end station P_B in the eastern zone. Transformation of the approximate coordinates of P_B from the eastern to the western zone allows the observation equations of the line to be formed in the coordinate system of the western zone. Before these can be used in the adjustment, however, it will be necessary to transform coordinate corrections dX_B and dY_B into their respective values dX_B' and dY_B' in the coordinate system of the eastern zone in which station P_B is adjusted. The transformation formulas are

$$dX_B = dX_B' \cos \alpha + dY_B' \sin \alpha$$
$$dY_B = dY_B' \cos \alpha - dX_B' \sin \alpha, \tag{3-105}$$

where α is the angle of tilt between the two coordinate systems ($\alpha = L \sin \phi$, where L is the width of projection zone and ϕ is the latitude).

Similarly, the transformation formulas

$$dX_A' = dX_A \cos \alpha - dY_A \sin \alpha$$
$$dY_A' = dY_A \cos \alpha + dX_A \sin \alpha$$

(3-106)

must be applied to the observation equations of line $P_B P_A$ formed in the co-ordinate system of the eastern zone.

Observation equations for measured distances can be formed in either direction. It is only important to ensure that each distance measurement has been reduced to the same plane on which it becomes adjusted.

Example 3-16. Observation Equation for Measured Distance in UTM Zones 75°W and 81°W (Clarke 1866 Spheroid).

Given: Approximate coordinates of terminal points:

A: $N =$ 5 097 641.916   B: $N =$ 5 086 192.327
$\quad E = 81°716\,814.112$ $E = 75°273\,638.073$
$\quad \phi \sim 46°00'$ $\phi \sim 45°54'$
$\quad \lambda \sim 78°12'$ $\lambda \sim 77°55'.$

Measured length of AB, reduced to the ellipsoid:

$$S = 24\,856.616.$$

Compute: Observation equation for line AB in (a) Zone 81°W and (b) Zone 75°W.

1. Coordinate transformations into the adjacent zone, as in Example 2-8:
   A: $N = 5\,098\,809.555$ B: $N = 5\,086\,672.661$
   $\quad E = 75°252\,213.475$ $E = 81°739\,125.970.$
2. UTM scale factors of line AB, according to the description on pp. 35–36:
   Zone 75°W, $\bar{m} = 1.000\,291\,46$ Zone 81°W, $\bar{m} = 1.000\,239\,44$
3. Coefficients and constant terms of the observation equations:

	Zone 75°		Zone 81°	
	N^0	E^0	N^0	E^0
A	5 098 809.555	252 213.475	5 097 641.916	716 814.112
B	5 086 192.327	273 638.073	5 086 672.661	739 125.970
B − A	− 12 617.228	+21 424.598	− 10 969.255	+22 311.858
ctg t_{AB}		−0.588 913		−0.491 633
cos $t_{AB} = c_{AB}$		−0.5075		−0.4412
sin $t_{AB} = d_{AB}$		+0.8617		+0.8974
s^0		24 853.786		24 862.493
$\bar{m}S$		.861		.568
l_{AB}		−0.075		−0.075

4. Approximate tilt angle α:

$$\alpha = L \sin \phi = 6° \times \sin 45°57' = 4°.31$$
$$\sin \alpha = 0.0752$$
$$\cos \alpha = 0.9972.$$

Result: Substitution of (a) $dX_B = 0.9972\, dX_B' + 0.0752\, dY_B'$ and $dY_B = 0.9972\, dY_B' - 0.0752\, dX_B'$ into observation equation

$$-0.4412(dX_B - dX_A) + 0.8974(dY_B - dY_A) - 0.075 = v_{AB} \text{ (Zone 81°)}$$

or (b) $dX_A' = 0.9972\, dX_A - 0.0752\, dY_A$ and $dY_A' = 0.9972\, dY_A + 0.0752\, dX_A$ into observation equation

$$-0.5075(dX_B' - dX_A') + 0.8617(dY_B' - dY_A') = v_{AB} \quad \text{(Zone 75°)}$$

gives the final observation equation in the form

$$0.4412\, dX_A - 0.8974\, dY_A - 0.5075\, dX_B' + 0.8617\, dY_B' - 0.075 = v_{AB.}$$

The numerical result obtained in Example 3-16 demonstrates that coordinate transformations in Eqs. (3-105) and (3-106) can be effected by computing the coefficients of the observation equations in the proper projection zone for each unknown.

Tying of the city control net to the national geodetic network. As mentioned previously, the city control network should be tied to the geodetic network in order to maintain an integrated survey system at the national level.

As a rule, survey ties should always be made to all geodetic control points in the vicinity of the urban area. The inclusion of geodetic points in the adjustment of the urban network, however, should be treated with caution. The following cases may be considered:

1. Accuracy of the geodetic network is better than the required accuracy of the first-order city network.
2. Accuracy of the geodetic network is unknown, and there is a suspicion that it is inferior to the city network.
3. Accuracy of the geodetic network is the same or worse than the required accuracy.

Each case will require a different approach in fitting the urban network into the geodetic control net. Some discussion of this topic is given in Ref. (23).

In the first case, the coordinates of the geodetic points may be held fixed (errorless) in the process of adjustment of the urban network.

In the second case, the urban network should be adjusted as an independent network, including the geodetic points as new points. The adjustment is performed with minimum constraints, only one direction and one point of the network being held as fixed, with the assumption that some distances in the network were measured. New coordinates will be obtained for the geodetic points included in the adjustment. A transformation is performed by using translation and rotation of the new network in order to fit the new coordinate system into the old in the best possible way. The scale of the new network should not be changed in the process of the transformation, since this could distort good observations in the city network. To keep information on the accuracy of the city network, the variance–covariance matrix, obtained from the independent adjustment, should also be transformed to fit the new orientation of the network after its rotation.

In the third case, when the geodetic network has the same or lower accuracy than the urban network, two alternatives are possible: Either the adjustment procedure is performed as in the second case or the coordinates of the geodetic points are included in the adjustment as quasi-observables with proper weights ($\mathbf{P}_X$ matrix from Eq. 3-28) calculated from their known variances and covariances. The second alternative is recommended since it results in a rigorous fitting together of the two networks.

In both cases, when using the transformation or the $\mathbf{P}_X$ matrix approach, two sets of coordinates for the common geodetic control points will result: the old coordinates in the national system and the new ones, slightly different, in the urban system.

MAINTENANCE AND RECORD-KEEPING

As emphasized in the introduction, the control network should be treated as a public utility. Therefore, the responsibility for the establishment, updating, and record-keeping of the control network should be in hands of one central survey office under the direct jurisdiction of the city authorities. Field surveys and computations may be subcontracted to different agencies or private survey companies but must be performed according to the design and under the supervision of the city office.

The maintenance of the control network includes:

1. A frequent check, at least once a year, of the state of the survey markers and protection of the markers from corrosion and other possible damage. Missing markers should be immediately replaced and remeasured. The new survey should be adjusted to the existing points by using the $\mathbf{P}_X$ matrix approach;
2. Periodic remeasure of the parts of the network that are suspected of being affected by ground movements, caused, for instance, by construction excavations, changes in water level, or mining and tunneling activities. This should be continued until ground stability is back to normal. Updating of coordinates of the affected points may be necessary.

The city survey office is responsible for keeping the record of up-to-date coordinates of the control points and allowing recognized users easy access to the record. The record should include:

1. A large-scale map of the urban area with marked control points, including the numbering and order of the network to which the points belong as well as intervisibility lines to other points;
2. A catalog of control points (it can be computerized) which shows: the number of each point, the order of the survey, the address of its location and the file number of the topographical descriptions, and coordinates updated (if applicable) to the dates of computations;
3. A file of topographical descriptions (it can be combined with the catalog of control points);
4. The variance–covariance matrix of the network (usually stored in the computer).

The integrated survey system can be successful only if control points are readily available in all parts of the city and the up-to-date information on coordinates and accuracy of the network is easily accessible to lawful users.

NEW DEVELOPMENTS IN CONTROL SURVEYS

As we have seen, a control network that meets the stringent accuracies demanded by its users is not easy to accomplish. The major obstacle arises from the adverse influence of the atmosphere in the measurements, which, at the present time, can be taken care of only by time-consuming repetition of observations during different times of the day and on different days. Practicing surveyors who deal with the most accurate work should be kept informed of new developments in the elimination of refraction from the measurements, which is theoretically possible by utilizing the dispersion of different wavelengths in the electromagnetic spectrum (25–27). As to EDM, measurements have already been made with blue and red lasers, which provided a continuous output of distance to within ± 0.1 ppm (28). It is likely that two-color EDM instruments will soon become commercially available (29). Similarly, encouraging progress has been reported (30) in the energetic research that has long been going on toward the elimination of refraction from angular measurements.

Another innovation, even more spectacular, is the inertial surveying, or navigation, systems that are capable of accurate sensing of motion in three-dimensional space (31). These systems, which are being developed mainly in the United States, have recently become operational in second-order densification of national geodetic nets (32). They measure and record all required geodetic parameters (latitude, longitude, height, gravity, deflection components) with highly complex and sophisticated equipment installed in a helicopter or a land vehicle. It may well be that inertial navigation can eventually solve most of our current problems in establishing horizontal control, although its accuracy is not yet sufficient and the cost of the hardware, about half a million dollars, is beyond the means of smaller surveying organizations.

References

1. Vanicek, P., and Merry, C. Determination of the geoid from deflections of the vertical using a least squares surface fitting technique, *Bulletin Géodésique*, No. 109, 1973.

2. *Specifications and Recommendations for Control Surveys and Survey Markers*, Surveys and Mapping Branch, Department of Energy, Mines and Resources, Ottawa, 1973.

3. Strobel, A. Die elektronische Entfernungsmessung bei der Festpunktbestimmung in Baden-Württemberg, *Zeitschrift für Vermessungswesen*, 93:12, 1968.

4. Chrzanowski, A., and Steeves, P. Control networks with wall monumentation as a basis for integrated survey systems in urban areas, *The Canadian Surveyor*, September 1977.

5. Rainesalo, A., and Saastamoinen, J. Les Graphiques pour la Détermination des Visées Géodésiques, *Bulletin Géodésique*, No. 29, 1953.

6. Adler, R. K., and Papo, B. H. *Lateral refraction in proximity of structures*, American Society of Civil Engineering, Vol. 95, 1969.

7. Chrzanowski, A. Design and Error Analysis of Surveying Projects, Lecture Notes No. 47, University of New Brunswick, 1977.

8. Steeves, R. R. *Computer Programme for Pre-Analysis and Adjustment of Plane Co-ordinates of Horizontal Control Networks*, Dept. of Surveying Engineering, University of New Brunswick, 1976.

9. Bossler, J. Grafarend, E., and Rainer, K. Optimal design of geodetic nets, 2, *Journal of Geophisical Research*, September, 1973, Vol. 78, No. 26.

10. Grafarend, E., and Harland, P. Optimales Design Geodätischer Netze I, Deutsche Geodätische Kommission, Reihe A, *Höhere Geodäsie*, Heft 74, Bayerische Akademie der Wissenschaften, Munich, 1973.

11. Grafarend, E. Genauigkeitsmasse geodätischer Netze, Deutsche Geodätische Kommission, Reihe A, *Theoretische Geodäsie*, Heft 73, Munich, 1972.

12. Bramorski, K., Gomoliszewski, J., and Lipinski, M. *Geodesja Miejska*, Państwowe Przedsiębiorstwo Wydawnictw Kartograficznych, Warsaw, 1973.

13. Chrzanowski, A. Investigations on control survey requirements for integrated survey systems for urban areas, technical report submitted to EMR, University of New Brunswick, 1974.

14. Chrzanowski, A., and Wilson, P. Underground measurement with the Tellurometer, *The Canadian Surveyor*, June 1966.

15. Bomford, G. *Geodesy*, 3rd ed., Oxford University Press, 1971.

16. Chrzanowski, A., and Konecny, G. Theoretical comparison of triangulation, trilateration and traversing, *The Canadian Surveyor*, December 1965.

17. Cooper, M. A. R. *Modern Theodolites and Levels*, Crosby Lockwood & Son Ltd., London, 1971.

18. Deumlich, F. *Instrumentenkunde der Vermessungstechnik*, Verlag für Bauwesen, Berlin, 1972.

19. Heuvelink, H. J. Bestimmung des regelmässigen und des mittleren zufälligen Durchmesser-Teilungsfehlers bei Kreisen von Theodoliten und Universalinstrumenten, *Zeitschrift für Vermessungswesen*, No. 17, 1913.

20. Szpunar, W. *Geodezja Wyzsza i Astronomia Geodezyjna*, Państwowe Wydawnictwa Naukowe, Warsaw, 1964.

21. Meade, B. K. High precision distance measurements in the United States, *Proceedings of the International Symposium on Terrestrial Electromagnetic Distance Measurements and Atmospheric Effects on Angular Measurements*, Vol. 1, No. 20, Stockholm, Sweden, 1974.

22. Saastamoinen, J. On the path curvature of electromagnetic waves, *Bulletin Géodésique*, No. 78, 1965.
23. Chrzanowski, A., and Canellopoulos, N. Problems arising from redefinition in densification surveys, *The Canadian Surveyor*, December 1974.
24. Hausbrandt, S. *Rachunki Geodezyjne*, Państwowe Przedsiębiorstwo Wydawnictw Kartograficznych, Warsaw, 1953.
25. Owens, J. C., and Bender P. L. Multiple wavelength optical distance measurements, *Electromagnetic Distance Measurement*, Hilger & Watts Ltd., London, 1967.
26. Prilepin, M. T. Some problems in the theory of determining geodetic refraction by dispersion method, *Bulletin Géodésique*, No. 108, 1973.
27. Thompson, M. C., and Wood, L. E. The use of atmospheric dispersion for the refractive index correction of optical distance measurements, *Electromagnetic Distance Measurement*, Hilger & Watts Ltd., London, 1967.
28. Hugget, G. R., and Slater L. E. Electromagnetic distance-measuring instrument accurate to 1×10^{-7} without meteorological corrections, *Proceedings of the International Symposium on Terrestrial Electromagnetic Distance Measurements and Atmospheric Effects on Angular Measurements*, Vol. 1, No. 11, Stockholm, Sweden, 1974.
29. Shipley, G. Georan I, a new two-colour laser ranger, *Proceedings of the International Symposium on Terrestrial Electromagnetic Distance Measurements and Atmospheric Effects on Angular Measurements*, Vol. 1, No. 12, Stockholm, Sweden, 1974.
30. Tengström, E. Elimination of refraction at vertical angle measurements using lasers of different wavelengths, *Österreichische Zeitschrift für Vermessungswesen*, Sonderheft 25, 1967.
31. Litton Systems Inc. *Inertial Positioning System Test Data Summary Report* ETL-0028, U.S. Army Engineer Topographic Laboratories, Fort Belvoir, Va., September 1975.
32. O'Brien, L. S. Investigations of an inertial survey system by the Geodetic Survey of Canada, *The Canadian Surveyor*, December 1976.

Additional Readings

Adler, R., Papo, H., and Perlmutter, A. Establishment of Standards for Lower Order Horizontal Control, *Proceedings of XV International Congress of Surveyors*, Stockholm, 1977.
Angus-Leppan, P. V. Practical Application of Accuracy Standards in Traversing, *The Australian Surveyor*, Vol. 25, No. 1. March 1973.
Brown, D. C. Densification of urban geodetic nets, *Proceedings of the Semi-Annual Convention of the American Congress on Surveying and Mapping*, September 1976.
Burnside, C. D. *Electromagnetic Distance Measurement*, Crosby Lockwood & Son Ltd., 1971.
Chrzanowski, A. Pre-analysis and design of surveying projects, *Northpoint* (journal of the Survey Technicians and Technologists of Ontario), Vol. 11, 1974; Vol. 12, 1975; Vol. 13, 1976; Vol. 14, 1977.
Hirvonen, R. A. *Adjustment by Least Squares in Geodesy and Photogrammetry*, Frederick Ungar Publishing Co., New York, 1971.
Johnson, P. C. *A Measure of the Economic Impact of Urban Horizontal Geodetic Surveys*, U.S. National Geodetic Survey Publication, August 1975.
Laurila, S. H. *Electronic Surveying and Navigation*, John Wiley & Sons, New York–Toronto, 1976.

Mikhail, E. M., and Ackermann, F. *Observations and Least Squares*, IEP—A Dun-Donnelley Publisher, New York, 1976.

Richardus, P., and Allman, J. *Project Surveying*, North-Holland Publishing Co., Amsterdam, 1966.

Saastamoinen, J. J., ed. *Surveyor's Guide to Electromagnetic Distance Measurement*, University of Toronto Press, 1967.

Vanicek, P., and Krakiwsky, E. *The Concepts of Geodesy*, North-Holland Publishing Co., to be published in 1979.

Wells, P., and Krakiwsky, E. The method of least squares, lecture notes No. 18, University of New Brunswick, Fredericton, 1971.

Chapter 4

Vertical Control

GENERAL CHARACTERISTICS OF VERTICAL CONTROL NETS IN URBAN AREAS

Urban planners, public works departments, engineering companies, and those involved in physical changes in a city area must work on a three-dimensional model of the city. Therefore, a knowledge of vertical distances between various points in urban areas is as important as the knowledge of their horizontal position.

A vertical control net of well-marked and easily identifiable points (bench marks) at known elevations above a reference horizontal surface serves as a basis for the determination of the heights (Z-coordinates) of the three-dimensional model.

Vertical control nets should be tied to the first-order national geodetic network. If the geodetic bench marks are too distant or their quality is doubtful, the city vertical net may be established as an independent net, with heights referred to an arbitrarily chosen horizontal surface. The vertical control is, or should be, a part of an integrated survey system for an entire urban area. The network should be of uniform quality, and the heights should be referred to the same reference level.

To assure uniform quality, all surveys and computations, as well as maintenance of the network, should be carried out to the same specifications and supervised by one central office for the entire area.

Vertical control nets in urban areas are characterized by:

High accuracy requirements, particularly in flat areas, mainly owing to requirements of sewer systems.

High density of bench marks because of the frequent changes (expansion and reconstruction works) in urban areas.

Difficulties in measurement due to city traffic; and

Frequent changes of heights of bench marks due to ground movements caused by construction and excavation works, changes in hydrological conditions, motor-vehicle traffic, and subsidence of buildings.

These characteristics require careful design of the monumentation of bench marks, survey procedures, and maintenance of the control network.

It is convenient to have bench marks spaced at about 200-m intervals in the core of the city and at about 500-m intervals in suburban areas. It is usually impossible to cover the whole city simultaneously by such a dense network of points. Vertical control, therefore is, established in two or three steps, and two or three orders of networks are created.

The first-order network usually consists of control points spaced from 2 to 4 km apart and is densified by the second-order network, with bench marks spaced from 0.5 to 1 km apart. The third-order network gives the final densification, with spacing of bench marks from 0.1 to 0.3 km.

Measurement of the vertical control net is done by the well-known method of differential leveling using spirit or automatic (self-compensating) levels. If difficulties arise in connecting networks across rivers, a reciprocal simultaneous leveling is used.

ACCURACY SPECIFICATIONS

Accuracy of leveling is influenced by random and systematic errors of measurements (pp. 146–149). The standard error of leveling may be expressed as a function of the length L of the leveling line in the form

$$\sigma^2 = \varepsilon^2 L + \delta^2 L^2, \tag{4-1}$$

where ε is the random component per unit distance and δ is the systematic component per unit distance.

Influence of systematic errors and their propagation have been the subject of controversy for a number of years. Different authors propose different formulas for including systematic errors in calculations of the accuracy of leveling nets. Bomford (1) reviews some theories and empirical results. Some systematic errors accumulate proportionally to the length L; others are a function of differences in elevation or of time. Generally, the influence of systematic errors is much smaller than that of random errors if leveling lines do not exceed a few kilometers, as is the case in urban networks, and proper procedures are employed in field measurements. In such cases, the accuracy of leveling may be characterized by a standard deviation σ_0 per unit distance with a propagation over the distance L expressed by

$$\sigma = \sigma_0 \sqrt{L}. \tag{4-2}$$

Table 4-1. Maximum Permissible Discrepancies Between Independent Levelings

Canada	United States	Australia
Special order 3 mm $\sqrt{L}$	First order, class 1, 3 mm $\sqrt{L}$	First order 4 mm $\sqrt{L}$
First order 4 mm $\sqrt{L}$	First order, class 2, 4 mm $\sqrt{L}$	Second order 8.4 mm $\sqrt{L}$
Second order 8 mm $\sqrt{L}$	Second order, class 1, 6 mm $\sqrt{L}$	Third order 12 mm $\sqrt{L}$
Third order 24 mm $\sqrt{L}$	Second order, class 2, 8 mm $\sqrt{L}$	
	Third order 12 mm $\sqrt{L}$	

L is in kilometers.

For example, specifications for geodetic leveling in Poland (2) are based on the following values of σ_0:

first-order nets $\qquad \sigma_0 = 1$ mm/km

second-order nets $\qquad \sigma_0 = 2$ mm/km

third-order nets $\qquad \sigma_0 = 4$ mm/km.

As a general rule, leveling of each section of a network is done twice; therefore, σ_0 characterizes the accuracy of double leveling.

Specifications for field procedures in leveling are usually based on a maximum permissible discrepancy between two independent levelings of the same line. If the maximum error of a double leveling is accepted as equal to $2\sigma_0\sqrt{L}$ (at 95% confidence level), the maximum discrepancy Δ between two levelings is equal to

$$\Delta = 4\sigma_0\sqrt{L}. \qquad (4\text{-}3)$$

Table 4-1 gives examples of Δ from national specifications of Canada (3), the United States, and Australia (4).

Accuracy of leveling nets in urban areas is dictated mainly by requirements of street construction works, particularly in the grading of sewers. In flat areas, the average grade of main sewers is of the order of $1^\circ/_{oo}$ and it may be necessary in some cases to keep it as low as $0.2^\circ/_{oo}$, i.e., 2 cm per 100 m. In such cases, an error of a few millimeters in height determination of aligning points of the sewer may be critical. Therefore, an error of 5 mm in the difference in elevation between the lowest-order bench marks is usually accepted as the maximum allowable error. Since the average distance between third-order bench marks is about 200 m, the allowable error per kilometer for third-order differences in elevation is $5\sqrt{5}$ mm/km. This error is composed of leveling errors M_{III} of the third-order network and errors of the higher-order nets to which the third order is adjusted. Thus

$$(5\sqrt{5})^2 = M_{III}^2 + M_{II}^2 + M_{I}^2. \qquad (4\text{-}4)$$

If the maximum errors are taken as equal to 2.5σ at the 99% confidence level, Eq. (4-4) may be expressed in terms of standard deviations as

$$\left(\frac{5\sqrt{5}}{2.5}\right)^2 = \sigma_{o_{III}}^2 + \sigma_{o_{II}}^2 + \sigma_{o_{I}}^2. \tag{4-5}$$

Accuracies of higher-order networks are usually designed at least twice as high as that of the lower-order networks:

$$\sigma_{o_{II}} = \frac{\sigma_{o_{III}}}{2} \quad \text{and} \quad \sigma_{o_{I}} = \frac{\sigma_{o_{II}}}{2}. \tag{4-6}$$

From Eqs. (4-5) and (4-6), the accuracies of the vertical control in urban areas are as follows:

first order $\qquad \sigma_o \cong 1$ mm/km
second order $\qquad \sigma_o \cong 2$ mm/km
third order $\qquad \sigma_o \cong 4$ mm/km.

Allowable discrepancies between two levelings of the same line, calculated from Eq. (4-3) with the above values of σ_o, are as follows:

first order $\qquad \Delta = 4\sqrt{L}$ mm
second order $\qquad \Delta = 8\sqrt{L}$ mm
third order $\qquad \Delta = 16\sqrt{L}$ mm,

where L is in kilometers.

By comparing the above values with the specifications listed in Table 4-1, it can be concluded that accuracy specifications for vertical control in urban areas are the same as those for national geodetic leveling. Thus the accuracy of the primary network in urban areas should be the same as that of the first-order national net.

DESIGN OF VERTICAL CONTROL

A leveling network is designed in two stages: (1) preliminary design and (2) working design. The preliminary design is preceded by ascertaining the location and quality of the national and other bench marks in the area. The quality determination may require control leveling between some bench marks, particularly if the elevations of the bench marks have not been checked within the last few years. The accepted existing bench marks are plotted on a map of the city on a scale of 1:10 000 or larger, and the map serves as a basis for the preliminary design of a new net. The design includes:

The description of existing bench marks.

Proposed areas (approximately) for placing new bench marks according to density requirements.

Preliminary selection of routes of leveling (with alternatives), including connecting surveys to existing bench marks and a subdivision of the proposed network into orders of accuracy.

Geological and hydrological information on the selected areas.

The approximate cost of the new survey.

The preliminary design is followed by a field reconnaissance to select the type and exact location of bench marks in the proposed areas. Proposed routes of leveling are checked for grade, traffic, possible obstacles, and type of the terrain surface.

The final working design includes:

Final routes of leveling shown on the map with a description of their lengths, grades, and proposed accuracy and method of measurement.

Location of existing bench marks.

Location of proposed bench marks with a description of their monumentation.

Proposed method of adjustment.

Explanation of differences between the preliminary and final design.

The detailed cost of the whole project.

Proper monumentation of bench marks is the most important part of the project. The location of a bench mark should be easily accessible, easy to find, and, more importantly, should ensure stability of the bench mark for a long period of time. Therefore, the design of the type and location of the bench mark is more important than the design of the routes of leveling. However, in designing the routes, the geometrical shape of the network, which influences the error propagation, should be carefully considered. For example, if two routes of leveling are designed as shown in Figure 4-1, it is obvious that a large error in the difference in elevation between bench marks A and B may be expected

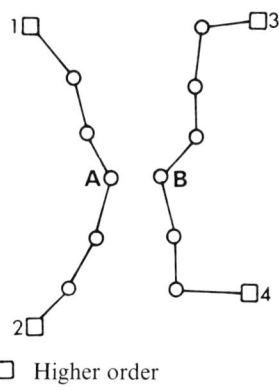

☐ Higher order
○ Lower order

Figure 4-1. An ill-designed leveling net.

since heights have been determined from two separate adjustments between higher-order bench marks 1 and 2 and 3 and 4. In this particular case, bench marks A and B should be directly connected by leveling, and a simultaneous adjustment should be applied to the whole net: 1, A, 2, 3, B, 4. An accuracy preanalysis of the proposed network is desirable before accepting the final design and starting the field surveys.

MONUMENTATION

Bench marks monumented in walls or foundations of large buildings, old churches, or bridge embankments are the most popular leveling markers in urban areas. They are inexpensive, easy to install, and comparatively stable. Figure 4-2 shows typical wall markers, usually made of bronze or brass.

Electric drills of a rotary-hammer type facilitate the preparation, in a few minutes, of holes for cementing in the wall markers, with portable gasoline generators used as the power supply.

Figure 4-3 shows an older type of wall marker used in primary leveling networks in Europe (9, 10), which requires special adaptors for leveling rods. The British Ordnance Survey (6) uses quite a complicated wall marker in the form of a flush bracket (Fig. 4-4). A movable attachment is needed that can be hung from the bracket to provide a staff support at exactly the same height as the permanent reference surface.

Wall markers should not be placed in newly constructed buildings, because even small or medium size buildings usually show a subsidence of several centimeters in the first year after construction and sometimes require a number of years for complete settlement.

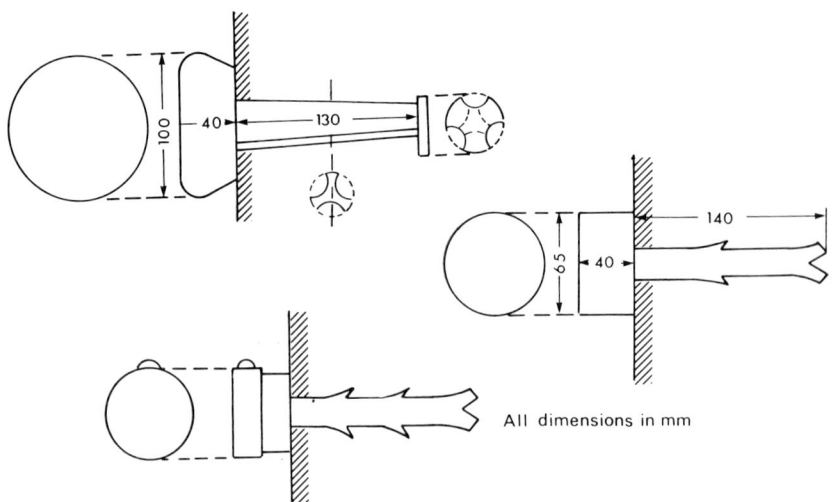

Figure 4-2. Typical wall markers, with dimensions in millimeters (2).

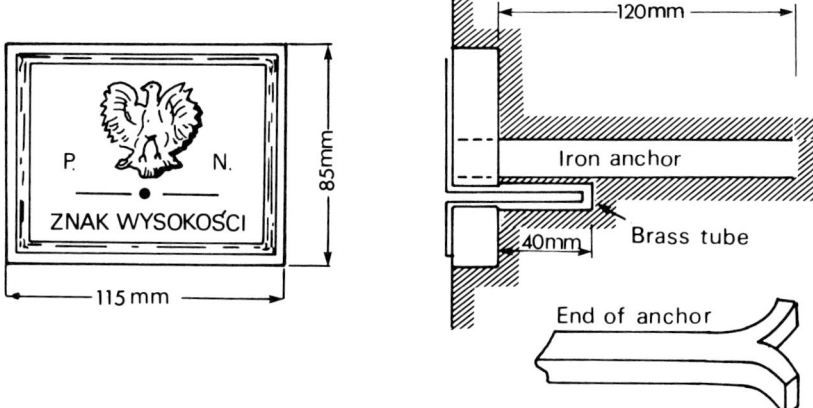

Figure 4-3. Older type of wall marker (primary network) (9).

Specifications of the Surveys and Mapping Branch in Canada (3) give a detailed description of different types of ground bench marks used in various types of soil and terrain conditions.

Ground monumentation requires careful topographical reconnaissance and study of the geological, hydrological, and soil conditions. Bench marks of the higher-order networks should be placed as distant as possible from areas designated for major excavation and construction work and areas that may be flooded. Excavation works may cause great changes in underground water levels and can result in ground movement several hundred meters away.

Stability of bench marks should be regularly checked every few years by releveling portions of the network and checking that differences in height are within the allowable limits discussed on pp. 140–142.

Since most wall and ground bench marks may be subjected sooner or later to vertical movements, each city should have a few so-called fundamental bench marks that are monumented directly on bedrock. This sometimes requires very deep markers, to 20 m or more. One type of fundamental bench

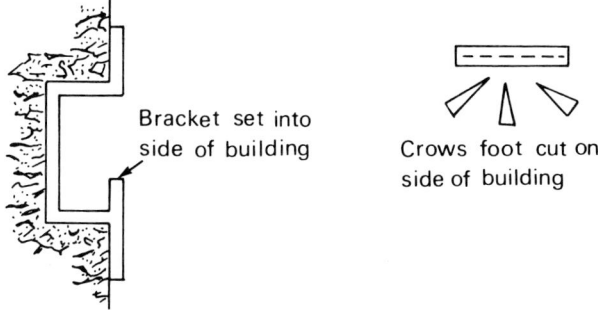

Figure 4-4. Wall bench marks of the Ordnance Survey in England (6).

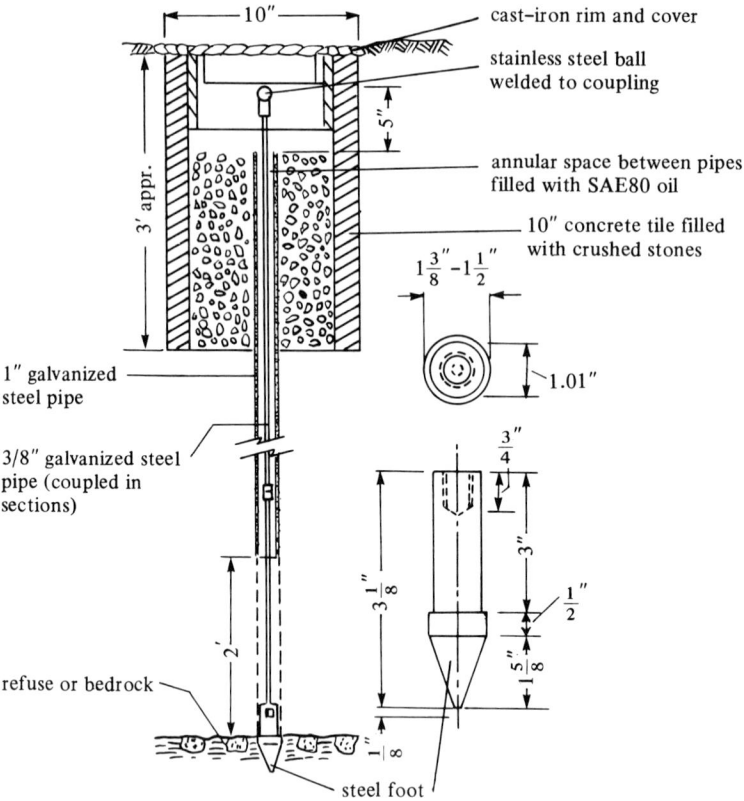

Figure 4-5. A fundamental bench mark (dimensions in feet and inches).

mark, designed at the National Research Council of Canada (3), is shown in Figure 4-5.

Fundamental bench marks serve as a basis for checking the stability of the leveling network. For example, the vertical control of the city of Warsaw is based on 14 fundamental bench marks (2).

INSTRUMENTS AND FIELD PROCEDURES

Basically the same instruments and the same field procedures are used in leveling in urban surveying as in geodetic or engineering surveys with the equivalent accuracy requirements.

A brief discussion of choice of instruments and recommendations for field procedures is given, with the assumption that the reader is familiar with the basic principles of handling leveling instruments and performing differential leveling (7–9). An excellent review of leveling instruments and instrumental errors is given in Ref. (10). Any general manual on surveying gives adequate information on field procedures and the basic adjustment of instruments.

Table 4-2 gives the main characteristics of leveling instruments used in vertical control surveys in urban areas. They are classified in three groups according to the accuracies achievable with the instruments if they are properly adjusted and if proper field procedures are applied. In each group, either a spirit-level type or an automatic (self-compensating) instrument can be selected.

Generally, the self-compensating instruments are preferable in all orders of leveling; in the near future, they probably will completely replace spirit levels. The main advantage of the automatic level is that the measurements are considerably faster and less strenuous for the observer. Also, automatic levels are less affected by heat radiation and do not need to be shaded from the sun's rays unless the highest precision is required.

Some types of automatic instruments, particularly the older models, may be difficult to work with in windy weather or in heavy street traffic because of vibrations of the line of sight. Most new models show a significant improvement in damping the vibrations of the self-compensating systems and do not give less accurate results than spirit levels in adverse conditions. Test leveling in heavy street traffic with different models of automatic levels is recommended before selecting and purchasing an instrument for urban surveys.

A review and summary of results of investigations on sources of errors and measuring techniques with different models of automatic levels is given in Ref. (7).

All sources of errors are the same for automatic and for spirit leveling except, of course, the error of setting the line of sight horizontal.

The compensating systems of automatic levels may produce systematic errors in leveling because of mechanical hysteresis of the compensator and tilt of the instrument. To reduce the error, the following precautions should be observed in the field use of automatic levels in first- and second-order leveling:

1. The spherical level, which is used for a preleveling of the instrument, should be periodically checked and adjusted; it should always be carefully centered during field observations.
2. The sequence of the forward (F) and backward (B) readings should follow the pattern BFFB, FBBF, BFFB, etc.

Both automatic and spirit levels should be frequently checked and adjusted for the collimation error (error of nonhorizontality of the axis of sight) by the well-known "two-peg" method.

To reduce errors of leveling to allowable quantities, the following instruments and field procedures should be used in urban areas:

Instruments with a minimum magnification of 40, 30, and 20× should be used in first-, second-, and third-order leveling, respectively.

Instruments with parallel glass plate micrometers and invar rods with double scales should be used in first- and second-order surveys.

Table 4-2. Characteristics of Some Leveling Instruments

		Telescope		Spirit level ($''/2$ mm)	Weight (kg)	σ_o (mm/km)
Model	Producer	$M\times$ Diameter (mm)	Length (mm)			
High-precision levels with parallel-plate micrometers $\dot\sigma_o \leq 1$ mm/km						
Ni-004	Zeiss–Jena	44 56	375	$10''$	6.2	± 0.4
Ni-Al	MOM-Budapest	40 65	314	10	4.2	± 0.3
N-3	Wild, Heerbrugg	42 50	297	10	3.5	± 0.2
Nl	USSR	49 60	420	10	6.9	± 0.5
Ni-1	C. Zeiss, Oberkochen	50 50	305	Auto.	5.2	± 0.2
Ni-2	C. Zeiss, Oberkochen	32 40	270	Auto.	2.1	± 0.3
Ni-002	Zeiss–Jena	40 55	370	Auto.	6.5	± 0.2
Ni-A3	MOM-Budapest	50 67	210	Auto.	3.0	± 0.2
Ni-007	Zeiss–Jena	31.5 40	"Periscope"	Auto.	3.9	± 0.5
Precision engineering levels $\sigma_o \leq 2$ mm/km						
GK23	Kern, Aarau	30 45	170	18	1.5	$\pm 2\ (0.5)^*$
5169	Filotecnica Salmoiraghi	30 45	160	20	3.4	$\pm 1.3\ (0.3)^*$
N2	Wild, Heerbrugg	26 40	196	30	2.8	± 2
Ni-030	Zeiss–Jena	25 35	195	30	1.8	$\pm 2\ (0.8)^*$
N2	USSR	40 52	390	10	6.0	± 1
NA2	Wild Heerbrugg	30 45	250	Auto.	2.7	$\pm 1.5\ (0.4)^*$
SNA2	SLOM, Paris	25 40	220	Auto.	3.2	± 1.5
FNA2	O. Fennel, Kassel	32 40	250	Auto.	3.0	± 1
S77	Vickers Ltd.	32 39	267	Auto.	2.7	± 1.5
Medium-accuracy levels $\sigma_o \leq 5$ mm/km						
N10	Wild Heerbrugg	20 32	155	60	1.7	± 3
Ni3	C. Zeiss, Oberkochen	19 25	135	30	1.1	± 3
N-3	USSR	30 40	175	15	1.8	± 4
GK1	Kern, Aarau	22 30	118	45	0.9	± 4
GK1A	Kern, Aarau	25 45	125	Auto.	1.6	± 2.5
FNA1	O. Fennel, Kassel	25 30	200	Auto.	2.2	± 2
Ni-025	Zeiss–Jena	20 30	195	Auto.	1.7	± 3
NAK1	Wild, Heerbrugg	25 36	201	Auto.	2.2	± 2

* Parallel-plate micrometer optional.

Wooden (unfolded) rods with 1-cm divisions and visual estimation of the readout are permitted in third-order leveling.

Minimum sensitivity of the level vials (or equivalent accuracy of the compensators in automatic levels) should be $10''$, $20''$, and $40''/2$ mm for first-, second-, and third-order leveling, respectively.

Leveling rods in first- and second-order leveling should be equipped with self-supporting stands. Freehand holding is permitted in the third order. All rods should be equipped with adjusted circular levels. Heavy foot plates or, preferably, steel pins driven into the ground should be used as turning points. Ball bearings may be used as the turning point on asphalt.

Lines of sight should not exceed 40, 80, and 120 m with differences between backsight and foresight distances at each setup not exceeding 2, 10, and 20 m, for first-, second-, and third-order leveling, respectively.

Lines of sight in first- and second-order should not be lower than 0.5 m above the ground.

Lines of sight in first- and second-order should be shortened to 30 m when leveling on a steady slope, such as along railroads or highways, even if the grade permits longer sights.

Leveling rods should be compared for a zero difference. If two rods are used, each rod should be used alternatively for backward and forward sightings, always with an even number of instrument setups between the bench marks.

When leveling on soft ground or on soft asphalt, the tripod legs should not be "dug-in" deeply; supporting plates should be placed under the tips of the legs. Large-diameter coins may serve the purpose.

If sinking or upheaval of the instrument is suspected, the time interval between the backward and forward readings should be as short as possible.

Leveling of the first- and second-order bench marks should be done twice in opposite directions and, if possible, in different atmospheric conditions. A double leveling is also recommended for third-order bench marks.

If a river crossing is necessary, the well-known reciprocal and simultaneous leveling is performed with instruments set up on each side of the river. The river crossing in the first- and second-order surveys should be repeated twice on different days to decrease the refraction influence. The use of laser levels and self-centering detectors seems to be a promising method for this type of survey (11).

ADJUSTMENT AND COMPUTATIONS OF HEIGHTS

Systems of Heights

Heights of terrain points may be defined as vertical distances to a reference horizontal surface for which the height is arbitrarily chosen as zero.

A horizontal surface is a geopotential surface that may be defined as an equipotential surface of the actual gravity field.

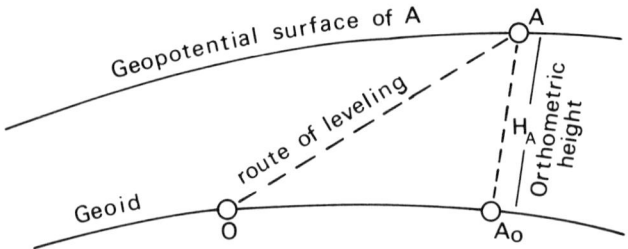

Figure 4-6. Orthometric height.

Usually, the *geoid*, or the geopotential surface of the mean sea level, is accepted as the reference surface for height systems.

The system of heights in which a height H_A of a terrain point A, (Fig. 4-6) is determined as the distance between point A and the geoid surface (point A_o), measured along the true plumb line of A, is called the *true orthometric system*.

Owing to the ellipsoidal shape of the earth and irregularities in mass distribution, the equipotential surfaces at different distances from the geoid are not parallel to each other (Fig. 4-6). Since the orientation of the axis of sight of leveling instruments is dependent on the direction of local gravity, it becomes obvious that a result of the differential leveling between the reference geopotential surface and terrain point A is dependent on the route of leveling. For example, the field result of the leveling from O to A (Fig. 4-6) gives a height of A above the geoid that is different from the vertical distance (orthometric height) $A_o - A$. In some cases, the misclosure may be of the order of several millimeters per kilometer of the distance between the bench marks, particularly in mountainous areas with large differences in elevation. Even at low elevations and in flat areas, the misclosures may reach values of the order of 1 mm/km. Therefore, practicing surveyors should be aware of this problem when establishing the first- and second-order leveling networks for a city.

To free the results from the above effect, field observations must be corrected. The orthometric correction requires a knowledge of gravity values along the leveling route, as well as along the plumb lines of the bench marks. The direct observation of gravity along the plumb lines is usually impossible. A number of different theories have been developed for the approximate estimation of a mean gravity value along the plumb lines, which result in approximate orthometric heights that may differ quite significantly from the true values and from each other. In some countries, orthometric corrections are determined on a basis of so-called normal gravity values, which are calculated with the assumption that the earth is an ellipsoid with an homogeneous mass distribution. This, of course, leads to a different system of orthometric heights.

Two points with the same orthometric heights are not necessarily on the same equipotential surface, which may lead to confusion in precision-engineering construction. This is overcome if dynamic heights (or geopotential numbers), instead of orthometric heights, are used for expressing the vertical coordinates of bench marks.

A detailed discussion of height systems and gravity corrections is beyond the scope of this manual. Reference should be made to up-to-date manuals on physical geodesy, including a good review and clarification of different height systems and calculations of gravity corrections by Nassar and Vanicek (12).

The problems presented by different height systems should not be underestimated by a surveyor in charge of establishing vertical control or connecting the city network to the national geodetic bench marks. A consultation with the national geodetic office is necessary to determine the corrections that should be applied to field surveys.

It is again emphasized that computations of heights on a basis of field leveling only may lead to misclosures much larger than those permitted by accuracy specifications.

Preliminary Computations and Choice of Adjustment Method

During the net adjustment, the net points counted are only those bench marks that already have fixed values of heights and those new points (junction points) from which at least three leveling lines lead to other net points. If there are no fixed points, one net point must be selected as the initial point with an arbitrarily fixed height. New bench marks from which only one or two lines have been leveled are not included in the net adjustment but are connected or interpolated later.

A leveling network consists of a number of leveling lines (traverses) that are run between the junction bench marks and the fixed points of the higher-order network. In preparatory computations for the network adjustment, each leveling line is treated separately as a unit of computation. First, a difference between two levelings of each line is checked against the allowable discrepancy (see p. 142). If the discrepancy is larger than is permitted, the leveling of the line must be repeated.

Mean differences h_i of elevations from repeated and accepted levelings are calculated for each line, and gravity corrections are added if necessary. The means are tabulated and written on a large-scale sketch of the network; arrows are used to indicate directions of the computations.

The sketch should also show numbering of the leveling lines, lengths S_i of the lines and heights Z_i of the higher-order bench marks. An example (taken from Ref. 13) of a network with four junction points is given in Figure 4-7. Mean differences in elevations of 14 lines are shown in meters, distances (in parentheses) in kilometers, and heights of the four fixed points in meters.

As the next step in preliminary computations, approximate heights of the junction points are calculated on several different routes from the fixed points. As a rule, one should use as many routes of calculations as leveling lines joining the junction point. The spread of the results and a comparison of the means with

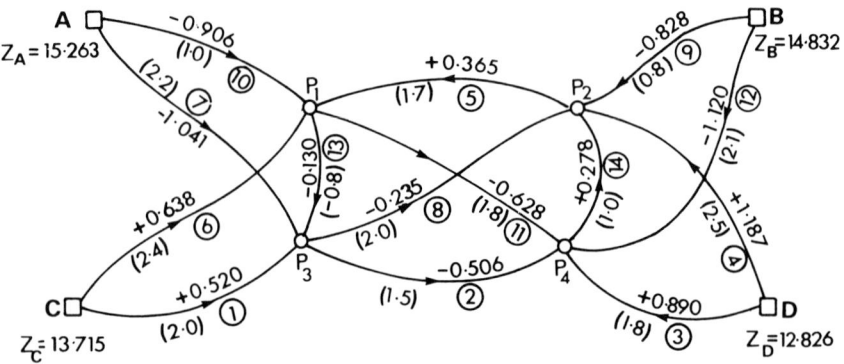

Figure 4-7. Example of a leveling network.

individual approximate heights serve as a further check on the quality of the measurements and indicate whether the heights of the fixed points of the higher-order network have changed.

If the deviations from the mean are within $2\sigma_o\sqrt{S}$, the field results and the heights of the fixed points may be accepted for the final computation of the coordinates of the junction points. Table 4-3 shows check computations for the network shown in Figure 4-7. In the example, the desired accuracy of leveling was $\sigma_o \leq 4$ mm/km. All deviations in Table 4-3 are within the value of $8\sqrt{S}$.

The final computation of heights of the junction points may be done by either the least-squares or another approximate adjustment.

The approximate method of adjustment follows the procedure shown in Table 4-3, except that the adjusted heights of junction points are calculated not as the arithmetic means but as weighted means with weights equal to $P_i = 1/S_i$, where S_i is the length of the calculation route. Heights of other bench marks along each individual level line are then calculated by an interpolation of the misclosure between the adjusted coordinates of the junction points and the field results of the differences in elevation.

The rigorous least-squares adjustment may be done by either a parametric method (observation-equations method) or a conditional method. The first- and second-order leveling nets are usually adjusted by the least-squares method, although differences in heights adjusted approximately or rigorously are within the accuracy of the measurements. The least-squares adjustment allows a better statistical evaluation of the accuracy of the network. The least-squares adjustment of the leveling networks is much simpler than the adjustment of horizontal networks. The linear observation equations (in the parametric method) or linear condition equations (in the conditional method) can be formed directly from the observed quantities.

Only the parametric method is reviewed here since examples of the conditional method of adjustment of leveling nets may be found in most manuals on advanced surveying and geodetic computations.

Table 4-3. Check Computations

Pt	Route of calculations	Height (m)	Deviation (mm)
P_1	A + 10	14.357	+1
	B + 9 + 5	14.369	−11
	C + 6	14.353	+5
	D + 3 − 11	14.344	+14
	C + 1 − 13	14.365	−7
		$z_1 = 14.358$	
P_2	A + 10 − 5	13.992	+9
	B + 9	14.004	−3
	C + 1 + 8	14.000	+1
	D + 4	14.013	−12
	D + 3 + 14	13.994	+7
		$z_2 = 14.001$	
P_3	A + 7	14.222	+7
	A + 10 + 13	14.227	+2
	B + 9 − 8	14.239	−10
	C − 1	14.235	−6
	D + 3 − 2	14.222	+7
		$z_3 = 14.229$	
P_4	A + 10 + 11	13.729	−7
	B + 12	13.712	+10
	B + 9 − 14	13.726	−4
	C + 1 + 2	13.729	−7
	D + 3	13.716	+6
		$z_4 = 13.722$	

Least-Squares Adjustment (Parametric Method)

Each leveling line of a leveling net between points i and j supplies one observation equation of the form

$$-dz_i + dz_j + (h_{ap} - h_{ob}) = v, \qquad (4\text{-}7)$$

where dz_i and dz_j are unknown corrections (parameters) to the approximate coordinates (heights) of the new points,

h_{ap} is a difference in elevation between points i and j calculated from the approximate heights of the new points and given heights of the fixed points,

h_{ob} is the measured difference in elevation,

v is a residual which is a correction to h_{ob} after the adjustment.

If the leveling line begins or ends on a fixed point, the value of dz_i or dz_j equals zero.

For example, if we take the leveling net shown in Figure 4-7 and the approximate coordinates of the four new points calculated in Table 4-3 (the mean values from several routes of calculations), the observation equations for leveling lines are

$$dz_3 \qquad - 6 = v_1 \qquad \text{for line 1,}$$
$$-dz_3 + dz_4 - 1 = v_2 \qquad \text{for line 2, etc.}$$

A set of observation equations may be written in the matrix form as

$$\mathbf{AZ} + \mathbf{L} = \mathbf{V}, \qquad (4\text{-}8)$$

where $\mathbf{A}$ is the matrix of the coefficients of the unknown parameters in the observation equations,

$\mathbf{Z}$ is a column matrix of the unknown parameters dz_i and dz_j,
$\mathbf{L}$ is a column matrix of the differences $(h_{ap} - h_{ob})$, and
$\mathbf{V}$ is a column matrix of the residuals.

Applying the condition that $\mathbf{V}^T\mathbf{PV} = \min$, one obtains the solution for the unknown parameters

$$\mathbf{Z} = -(\mathbf{A}^T\mathbf{PA})^{-1}(\mathbf{A}^T\mathbf{PL}) = -\mathbf{Q}(\mathbf{A}^T\mathbf{PL}), \qquad (4\text{-}9)$$

where

$$\mathbf{Q} = (\mathbf{A}^T\mathbf{PA})^{-1} \qquad (4\text{-}10)$$

and $\mathbf{P}$ is a diagonal matrix of the weights P_i of the observations. The weights of observations for each line are usually calculated as

$$P_i = \frac{1}{S_i}. \qquad (4\text{-}11)$$

Table 4-4 gives calculations of weights (diagonal elements of the $\mathbf{P}$ matrix) and elements of the matrix $\mathbf{L}$ for the numerical example (Fig. 4-7). The approximate heights of the new points are accepted from the preliminary calculations (Table 4-3) as $z_1 = 14.358$ m, $z_2 = 14.001$ m, $z_3 = 14.229$ m, and $z_4 = 13.722$ m.

Matrix $\mathbf{A}$ is given in Table 4-5 for 14 observed lines and 4 unknown parameters.

Introducing matrices $\mathbf{A}$, $\mathbf{L}$, and $\mathbf{P}$ into Eq. (4-9), the solution for the corrections (unknowns) is

$$\mathbf{Z} = \begin{pmatrix} dz_1 \\ dz_2 \\ dz_3 \\ dz_4 \end{pmatrix} = \begin{pmatrix} -0.55 \\ -0.53 \\ -0.08 \\ -0.90 \end{pmatrix}$$

Table 4-4. Calculation of Weights and Matrix **L**

No.	Line lengths S [km]	Weights $P = 1/S$	Diff in elev (m) h_{appr}	h_{obs}	$L_i = h_{ap} - h_{ob}$
1	1	0.50	0.514	0.520	-6
2	1.8	0.67	-0.507	-0.506	-1
3	1.8	0.56	0.896	0.890	6
4	2.5	0.40	1.175	1.187	-12
5	1.7	0.59	0.357	0.365	-8
6	2.4	0.42	0.643	0.638	5
7	2.2	0.45	-1.034	-1.041	7
8	2.0	0.50	-0.228	-0.235	7
9	0.8	1.25	-0.831	-0.828	-3
10	1.0	1.00	-0.905	-0.906	1
11	1.8	0.56	-0.636	-0.628	-8
12	2.1	9.48	-1.110	-1.120	10
13	0.8	1.25	-0.129	-0.130	1
14	1.0	1.00	0.279	0.278	1

Table 4-5. Coefficients of Observation Equations (Matrix **A**)

No.	dz_1	dz_2	dz_3	dz_4
1			1	
2			-1	1
3				1
4		1		
5	1	-1		
6	1			
7			1	
8		1	-1	
9		1		
10	1			
11	-1			1
12				1
13	-1		1	
14		1		-1

The solution is in millimeters because elements of matrix $\mathbf{L}$ are in millimeters. The adjusted coordinates (heights) of the new points are therefore

$$
\begin{aligned}
z_1 &= 14.358 - 0.000\ 55 = 14.357 \text{ m,} \\
z_2 &= 14.001 - 0.000\ 53 = 14.000 \text{ m,} \\
z_3 &= 14.229 - 0.000\ 08 = 14.229 \text{ m,} \\
z_4 &= 13.722 - 0.000\ 90 = 13.721 \text{ m.}
\end{aligned}
$$

In this particular example, the least-squares adjustment did not introduce any significant change in the approximate coordinates.

By placing the solution-matrix $\mathbf{Z}$ into Eq. (4-8), the corrections v_i to the observed h_{ob} differences in elevations are obtained. The values of matrix $\mathbf{V}$ are used in the calculations of an estimated standard deviation $\hat{\sigma}_o$ of an observation with the unit weight

$$
\hat{\sigma}_o = \sqrt{\frac{(\mathbf{V}^T \mathbf{PV})}{n - u}}, \tag{4-12}
$$

$\mathbf{V}^T \mathbf{PV} = 285.54$, and $n - u = 14 - 4 = 10$. This gives

$$
\hat{\sigma}_o = \sqrt{\frac{285.54}{10}} = 5.3 \text{ mm.}
$$

Since the weights were calculated from Eq. (4-11) with S in kilometers, the obtained value of $\hat{\sigma}_o$ represents an estimated standard deviation of leveling per kilometer. The result shows that the actual accuracy of observations is slightly lower than the aforementioned expected accuracy of 4 mm/km. However, the chi-square (χ^2) test (see Chapter 3, p. 128) at 95% probability level for $df = 10$ gives 3.7 mm $\leq \sigma_o \leq$ 9.4 mm; therefore, the survey may be accepted.

To calculate estimated standard deviations $\hat{\sigma}_{z_i}$ of the adjusted heights, the corresponding diagonal elements q_{ii} of the $\mathbf{Q}$ matrix are taken from Eq. (4-10) and placed in the equation

$$
\hat{\sigma}_{z_i} = \sigma_o \sqrt{q_{ii}}. \tag{4-13}
$$

Matrix $\mathbf{Q}$ in the numerical example gave

$$
\mathbf{Q} = (\mathbf{A}^T \mathbf{PA})^{-1} = \begin{pmatrix} 0.357 & 0.115 & 0.176 & 0.132 \\ 0.115 & 0.343 & 0.123 & 0.150 \\ 0.176 & 0.123 & 0.410 & 0.152 \\ 0.132 & 0.150 & 0.152 & 0.407 \end{pmatrix},
$$

from which the accuracy of the adjusted heights is

$$
\hat{\sigma}_{z_1} = 5.3\sqrt{0.357} = 3.2 \text{ mm} \qquad \hat{\sigma}_{z_3} = 5.3\sqrt{0.410} = 3.4 \text{ mm}
$$

$$
\hat{\sigma}_{z_2} = 5.3\sqrt{0.343} = 3.1 \text{ mm} \qquad \hat{\sigma}_{z_4} = 5.3\sqrt{0.407} = 3.4 \text{ mm.}
$$

TRIGONOMETRIC DETERMINATION OF HEIGHTS

The use of EDM in the horizontal-control network requires height information for the calculation of slope corrections that may not be readily available from the leveling network of the city. For instance, the Z-coordinates of rooftop monuments cannot be determined practically by the leveling method.

This kind of height information can best be provided by trigonometric determinations from vertical-angle measurements, often called trigonometric leveling. The basic formula is

$$\Delta H_{AB} = S'_{AB} \sin h_{AB} + \left(\frac{1-k}{2R}\right) S'^{2}_{AB} + i_A - m_B, \qquad (4\text{-}14)$$

where ΔH_{AB} denotes height difference $H_B - H_A$,
   S'_{AB} is the slope length of AB,
   h_{AB} is the vertical angle (elevation angle) measured at A,
   k is the coefficient of refraction,
   R is the radius of the earth $(=6370$ km$)$,
   i_A is the height of the horizontal axis of the theodolite above point A, and
   m_B is the height of the signal above point B.

The second term in Eq. (4-14), which represents the combined effect of the curvature of the earth and of atmospheric refraction, has the following orders of magnitude:

S' (m)	Earth curvature and refraction effect
100	0.1 cm
400	1.0 cm
1 200	0.1 m
4 000	1.0 m
12 500	10.0 m

The value of $k = 0.15$ is applicable on lines running at least 20 meters above the ground. At tripod heights, however, k varies greatly and the trigonometric determination of heights should be based upon simultaneous reciprocal observations of the vertical angle. The height difference may then be calculated from equation

$$2\Delta H_{AB} = S'_{AB} (\sin h_{AB} - \sin h_{BA}) + (i_A - i_B) - (m_B - m_A) \qquad (4\text{-}15)$$

without knowledge of the refraction coefficient. The latter can be determined from Eq. (4-14) by substituting ΔH_{AB} from Eq. (4-15) and solving for k.

Example 4-1. Calculation of Height Difference from Reciprocal Vertical Angles (all lengths in meters).

	A → B	B → A
h	$-0^g.0168$	$-0^g.0040$
$\sin h$	$-0.000\ 264$	$-0.000\ 063$
S'	2274.5	2274.5
$S' \sin h$	-0.600	-0.143
i	1.550	1.475
$-m$	-1.475	-1.550
$S' \sin h + i - m$	-0.525	-0.218

Height difference $\Delta H_{AB} = -0.154$.

The curvature and refraction term is 0.371 and corresponds to coefficient of refraction $k = 0.09$.

For best results, the reciprocal angles should be observed simultaneously at both ends of a line, but this procedure cannot be recommended in practical work. In triangulation networks, vertical angles are most conveniently measured in conjunction with the observation of horizontal directions; the most favorable time is during the afternoon hours preceding observations of horizontal directions. It is good practice to observe the vertical angle for every station in the network schedule. Two readings in both telescope positions should be taken for each point.

Inclusion of vertical angle measurements in the observation period immediately after sunrise (pp. 103–104) is a waste of time; such observations are generally useless for the computation of heights.

In lower-order traversing on tripods, station heights are ordinarily determined by third-order leveling. If vertical angles are used, they should be observed reciprocally even on short sights, since this will improve the reliability of the results. The working schedule should be arranged for the shortest possible time lapse between the two measurements; i.e., the observations at each traverse station should start with the vertical angle observed backward and end with the vertical angle observed forward.

A network of trigonometric heights can be adjusted by the junction-point method explained on pp. 151–153, but with weights proportional to $1/S_i^2$ for each height difference computed from a reciprocal observation. Observations that show exceptionally large deviations owing to erratic refraction should be discarded.

The accuracy of trigonometric height determination depends almost entirely on prevailing conditions of atmospheric refraction, which are frequently un-

favorable in the urban environment. In the outskirts of a city, however, trigonometric heights may be considered as an alternative to third-order leveling.

RECORD-KEEPING AND MAINTENANCE

Records of the location, quality, and heights of all bench marks should be kept
in two files: (1) according to the sequential number of the bench marks and
(2) according to street location in alphabetical order of the streets. The record of
each bench mark should contain: sequential number of the bench mark; order
of the network; address of the location, with a topographical sketch and the
number of the property registry; type of marker used; dates of the surveys and
computations of heights; number of the field book and computational data file;
height with a specification of the system of heights; and remarks on the quality
and maintenance of the bench mark.

An up-to-date approximate location of existing bench marks with their
sequential numbers should be shown on a medium-scale map of the city.

At least every 3 years, bench marks should be checked for corrosion or other
damage and, if necessary, covered with an anticorrosive paint. At least once
every 5 years, the heights of bench marks should be checked and, if necessary,
readjusted and recomputed. This requires a well-organized city survey office
and a staff member permanently responsible for keeping the records up to date
and scheduling the maintenance surveys. A separate catalog of heights and their
changes (with dates) of the bench marks is most useful, not only for the surveying
needs of the city office, but also for planners and construction engineers who
may draw conclusions on the stability of buildings and the surrounding ground.

References

1. Bomford, G. *Geodesy*, 3rd ed., Oxford University Press, 1971.
2. Bramorski, K., Gomoliszewski, J., and Lipinski, M. *Geodezja Miejska*, Państwowe
   Przedsiębiorstwo Wydawnictw Kartograficznych, Warsaw, 1973.
3. Surveys and Mapping Branch, Canada. *Specifications and Recommendations for Control Surveys and Survey Markers*, Ottawa, 1973.
4. National Mapping Council of Australia. *Standard Specifications and Recommended
   Practices for Horizontal and Vertical Control Surveys*, 1970.
5. Chrzanowski, A. Pre-analysis and design of surveying projects, *Northpoint* (journal of
   the Survey Technicians and Technologists of Ontario), Vol. 11, 1974; Vol. 12, 1975;
   Vol. 13, 1976; Vol. 14, 1977.
6. Clark, D. *Plane and Geodetic Surveying*, 6th ed., Constable and Co., London, 1971.
7. Cooper, M. A. R. *Modern Theodolites and Levels*, Crosby Lockwood & Son Ltd.,
   London, 1971.
8. Jordan/Eggert/Kneissl. *Handbuch der Vermessungskunde*, Vol. 3, Stuttgart, 1956.
9. Kowalczyk, Z. *Geodesy; Vol. 2; Leveling*, U.S. Dept. of Commerce, 1968.
10. Deumlich, F. *Instrumentenkunde der Vermessungstechnik*, Berlin, 1972.

11. Chrzanowski, A., and Janssen, D. Use of laser in precision leveling, *The Canadian Surveyor*, Vol. 26, No. 4, 1972.
12. Nassar, M., and Vanicek, P. *Levelling and Gravity*, Technical Report No. 33, University of New Brunswick, 1975.
13. Hausbrandt, S. *Rachunek Wyrownawczy i Obliczenia Geodezyjne* (Adjustment Calculus and Surveying Computations, in Polish), Państwowe Przedsiębiorstwo Wydawnictw Kartograficznych, Warsaw, 1971.

Chapter 5

Ground Surveying

INTRODUCTORY REMARKS

The purpose of surveying is to provide precise quantitative and, occasionally, qualitative information on the earth's surface, topography, and natural and man-made features. Surveying must be based on a suitable net of control points since only a survey capable of providing the correct relative positions of various terrain details is of lasting value. We assume that such a control net has been established and that control points of the lowest order are available at almost every street intersection. The control points are either monumented or referenced to permanent features, so that their positions can be easily re-established with an accuracy of at least 1 cm (pp. 69–76).

Previously, the survey of detail in the field often took the form of graphical mapping utilizing the traditional plane table technique. In this approach, the dimensions of the mapped objects could be supplemented by direct ground measurements if the dimensions derived from the map proved unsatisfactory. When the plane table technique is used, the coordinates of any one point can be derived only by reading them from the actual map.

When direct measurements in the field are used, the situation is reversed. From the ground survey the coordinates of a measured point can be directly computed, and the map is derived from the numerically recorded data. The main difference between these two approaches is in accuracy. Mapping that is carried out in the field by graphical means can provide *graphical accuracy* only, whereas surveying that is based on angular and distance measurement provides a higher *numerical accuracy*, commensurate with the accuracy of the measurements. In the latter case, the map is also of higher accuracy than a map produced in the field by graphical methods.

Detailed surveying can also be accomplished by *photogrammetric* means. However, even when photogrammetry is used as the primary technique for basic city surveying and mapping, there will always be a need for direct field-survey determinations. An outline of field procedures encountered in city surveys is presented, in the hope that it will help to introduce a certain uniformity in surveying operations in various countries.

BRIEF REVIEW OF SURVEYING METHODS

Terrestrial surveying methods can be divided into seven basic groups. The elements used in field measurements are the lengths or the angles, or both. Depending upon the underlying geometry and the elements considered, the following methods can be distinguished:

1. intersection of points by distance measurements.
2. the extension method.
3. the orthogonal method.
4. traverse.
5. the polar method.
6. intersection of points by angle measurements.
7. resection of a point.

To illustrate the various methods, we assume that the foundation of a building, *ABCD*, which protrudes slightly over the ground, must be surveyed in the field in a manner that permits tying the building into the control net, represented by points 1, 2, 3 and 4 (Figs. 5-1 to 5-7).

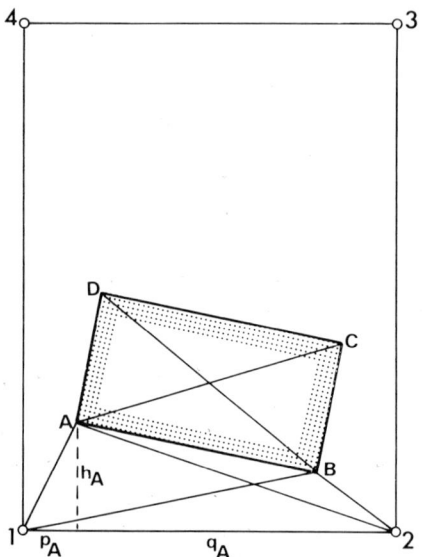

Figure 5-1. Intersection of points by distances only.

Intersection of Points by Distance Measurements

Starting from control points 1 and 2, the positions of foundation corners A and B are determined by measuring distances $\overline{1A}$, $\overline{2A}$, $\overline{1B}$, $\overline{2B}$, or any other distances between suitable points on the side 1–2 and respective corner points A and B. By measuring distances $\overline{AD}$, $\overline{BD}$ and $\overline{AC}$, $\overline{BC}$, the positions of points D and C are related to points A and B and also to control points 1 and 2. The surveyed points can be graphically plotted to their correct relative positions by using a compass, and the coordinates (see p. 12) can be computed as follows, taking point A as an example:

$$Y_A = Y_1 + \left(\frac{Y_2 - Y_1}{\overline{12}}\right)p_A - \left(\frac{X_2 - X_1}{\overline{12}}\right)h_A$$

$$= Y_2 - \left(\frac{Y_2 - Y_1}{\overline{12}}\right)q_A - \left(\frac{X_2 - X_1}{\overline{12}}\right)h_A,$$

$$X_A = X_1 + \left(\frac{X_2 - X_1}{\overline{12}}\right)p_A + \left(\frac{Y_2 - Y_1}{\overline{12}}\right)h_A \qquad (5\text{-}1)$$

$$= X_2 - \left(\frac{X_2 - X_1}{\overline{12}}\right)q_A + \left(\frac{Y_2 - Y_1}{\overline{12}}\right)h_A,$$

where

$$p_A = \frac{\overline{12}^2 + \overline{1A}^2 - \overline{2A}^2}{2 \times \overline{12}}, \qquad q_A = \frac{\overline{12}^2 + \overline{2A}^2 - \overline{1A}^2}{2 \times \overline{12}}$$

and

$$h_A = \frac{1}{2 \times \overline{12}} \times$$
$$\sqrt{\overline{12}^2(\overline{1A}^2 + \overline{2A}^2 - \overline{12}^2) + \overline{1A}^2(\overline{12}^2 + \overline{2A}^2 - \overline{1A}^2) + \overline{2A}^2(\overline{12}^2 + \overline{1A}^2 - \overline{2A}^2)}.$$

This method is frequently used and can provide precise results. It is particularly convenient if the distances to the points to be determined are short, of the order of a few meters. The method is often used, for instance, to reference control points to natural features or distinct points on buildings and for checking other measurements. Naturally, it is important that the geometry of the intersecting elements is suitable; i.e., the intersecting angle at the point to be determined should be close to 100^g (90°).

The Extension Method

This method also enables an exact mathematical derivation of the locations of points A to D in reference to the given points 1 to 4. Lines A–B and C–D are extended until they intersect lines 1–4 and 2–3, and the distances $\overline{aA}$, $\overline{AB}$,

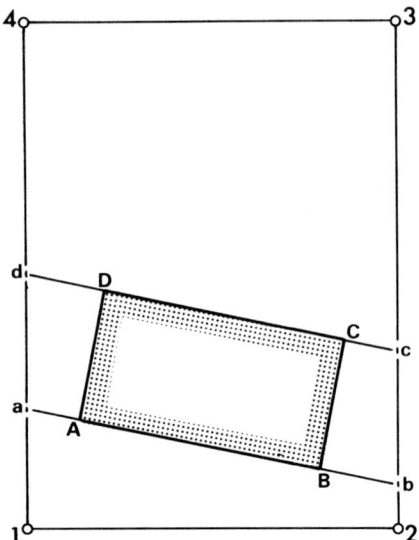

Figure 5-2. The extension method.

$\overline{Bb}$, $\overline{cC}$, $\overline{CD}$, and $\overline{Dd}$, as well as $\overline{1a}$, $\overline{ad}$, $\overline{2b}$, and $\overline{bc}$, are measured. The method is schematically shown in Figure 5-2. It is clear from this figure that the survey results of the extension method are rigidly fitted into the control net represented by points 1 to 4.

The computation of coordinates is somewhat awkward. First, points a and b are computed as

$$Y_a = Y_1 + \left(\frac{Y_4 - Y_1}{14}\right)\overline{1a}, \qquad X_a = X_1 + \left(\frac{X_4 - X_1}{14}\right)\overline{1a}$$

$$Y_b = Y_2 + \left(\frac{Y_3 - Y_2}{23}\right)\overline{2b}, \qquad X_b = X_2 + \left(\frac{X_3 - X_2}{23}\right)\overline{2b},$$

and points c and d from points 2 and 1 are computed similarly. Then points A to D are computed using points a to d, e.g.,

$$Y_A = Y_a + \left(\frac{Y_b - Y_a}{ab}\right)\overline{aA}, \qquad X_A = X_a + \left(\frac{X_b - X_a}{ab}\right)\overline{aA}. \tag{5-2}$$

This method is quite often used as a basic method of surveying boundaries, single standing buildings, and other linear features. The lines a–b and d–c can be considered as secondary reference lines for surveying further details close at hand. Graphical plotting of the results requires only a linear scale and a pencil once the control points have been plotted according to their coordinates.

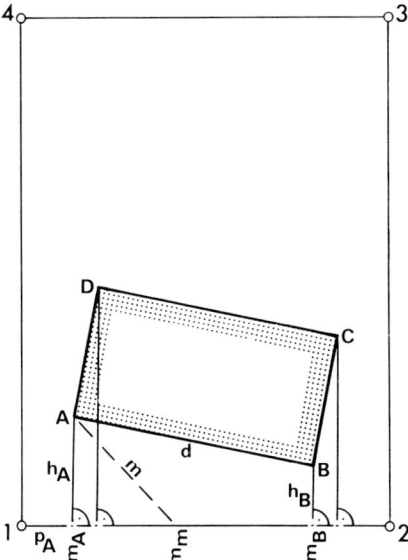

Figure 5-3. The orthogonal method.

The Orthogonal Method

When determining a point by the method described on p. 163 (intersection of points by distance measurements), the auxiliary values p, q, and h were computed first, and the coordinates of the points were then calculated. Assuming that a right angle can be easily set up on a survey line, i.e., a line connecting two known points, the values p (or q) and h can be measured directly in the field. This is the orthogonal method, which is schematically shown in Figure 5-3.

Point A can be computed from points 1 and 2 as follows:

$$Y_A = Y_1 + \left(\frac{Y_2 - Y_1}{12}\right) p_A - \left(\frac{X_2 - X_1}{12}\right) h_A,$$

$$X_A = X_1 + \left(\frac{X_2 - X_1}{12}\right) p_A + \left(\frac{Y_2 - Y_1}{12}\right) h_A.$$

(5-3)

This method is widely used, particularly in city surveys, to measure a large number of points along the traverse lines or auxiliary survey lines. The field equipment is simple, inexpensive, and consists of range poles, a prism for setting up right angles, and a measuring tape. By additional measurements, such as AB in our example, which are usually required, or by providing supplementary measurements m, an immediate check of the correctness of the survey is available by applying the Pythagorean theorem,

$$\sqrt{(m_B - m_A)^2 + (h_A - h_B)^2} = d$$

$$\sqrt{m^2 - (m_m - m_A)^2} = h_A.$$

The plotting of the survey results is also most convenient. In its simplest form, a linear scale and a triangle with a right angle can suffice as plotting tools. In larger projects, however, a small rectangular coordinatograph (see Chapter 9, City Maps) should be considered. It combines high plotting speed with superior accuracy.

Traverse

A traverse can be fully determined if an instrument for the measurement of angles (theodolite) and a distance-measuring device, such as a tape or an optical distance-measuring instrument, are available.

The method is shown schematically in Figure 5-4. Points A through D can be related to the known points 1 and 4 by measuring the distances $\overline{1A}$, $\overline{AB}$, $\overline{BC}$, and $\overline{CD}$ as well as the angles $(41A)$, $(1AB)$, (ABC), and so on. The computation is shown for point A only.

First the azimuth $(1 \rightarrow A)$ is derived as

$$(1 \rightarrow A) = (1 \rightarrow 4) + (41A)$$

and the coordinates of point A are determined as

$$
\begin{aligned}
Y_A &= Y_1 + \overline{1A} \sin(1 \rightarrow A) \\
X_A &= X_1 + \overline{1A} \cos(1 \rightarrow A),
\end{aligned}
\tag{5-4}
$$

with azimuths counted clockwise from the direction of the X-axis; e.g., the Y-axis is at the direction $+90°$.

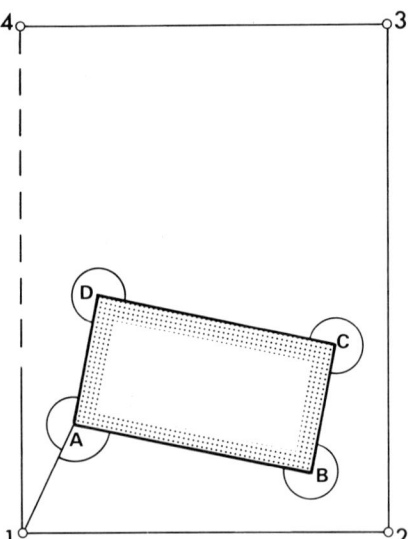

Figure 5-4. The traverse method.

The traverse method is mainly used to determine additional control points. In city areas, it is needed frequently—for instance, to survey backyards with an occasional access through narrow lanes and gates.

In some countries, this method is used in legal surveys to determine the size and shape of individual lots, with boundary corners employed as traverse points. This procedure requires a careful accuracy preanalysis (Chapter 3) because of the danger of propagating errors in traverses. Plotting of a traverse can be accomplished by using a protractor and a linear scale. For more accurate work, a polar coordinatograph (Chapter 9) is recommended.

The Polar Method

The polar method consists of determining the direction and the distance of each new point from a known station. The method is shown in Figure 5-5, where 1 is assumed to be the station, (1 → 4) the reference direction, and A to D new points. Unlike the traverse method, in which each newly determined point becomes the station for the next point, the polar method determines several points from one station. The computation of point A is as given in the preceding section.

Distances in the polar method are usually measured by optical or electromagnetic instruments; thus there is relatively little interference with city traffic. This method is also particularly convenient in areas of significant elevation differences or areas otherwise difficult for taping.

A certain drawback to this method is the difficulty in providing simple field-check measurements unless the same points are determined twice, from

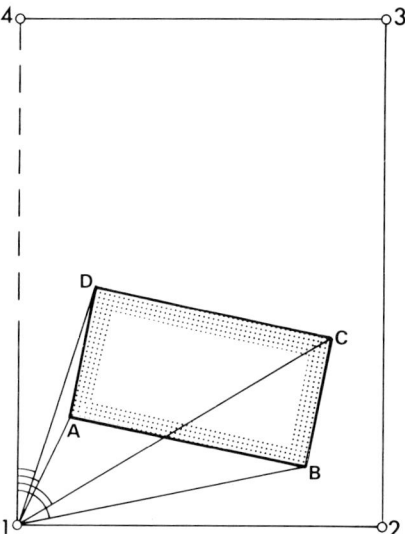

Figure 5-5. The polar method.

two different stations. Another check can be provided by direct measurement of distances between the points determined by the polar method, which are compared with the distances computed from their coordinates. These computations are rather involved and can be performed best in the office. Nevertheless, the polar method is an important technique, and the recent development of electromagnetic instruments with an automatic recording feature has significantly increased application of this technique.

Graphical plotting of the results is simple and efficient. Since several detail points are usually determined from each station, there is an additional justification for the use of a polar coordinatograph. In less demanding projects, the plotting can be made with a simple protractor and a linear scale.

Intersection of Points by Angle Measurements

The position of new points can be established by measuring their directions from a number of known stations. The minimum number of stations is two. By using points 1 and 2 in Figure 5-6 as known stations, the coordinates of points A, B, C, and D can be determined by measuring angles $(A12), (B12), (C12), (D12)$ and $(12A), (12B), (12C), (12D)$. Any known reference direction, such as $(1 \rightarrow 3)$, is acceptable and sufficient. The same applies to station 2. It is only essential that directions $(1 \rightarrow A), (2 \rightarrow A)$, and so on, can be derived.

Assuming that directions $(1 \rightarrow 2)$ and $(2 \rightarrow 1)$ are the reference directions for stations 1 and 2, respectively, the required angles $(A12)$ and $(12A)$ can be measured directly. By using these angles and considering that

$$(2A1) = 180° - (A12) - (12A),$$

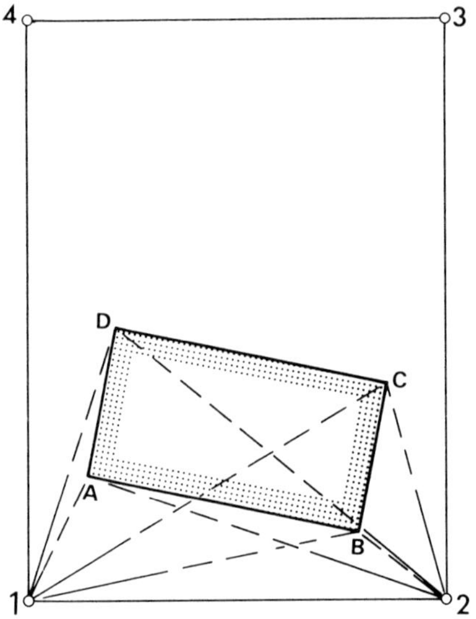

Figure 5-6. Intersection of points by angle measurements.

the sides $\overline{1A}$ and $\overline{2A}$ can be derived, and these, together with the azimuths

$$(1 \rightarrow A) = (1 \rightarrow 2) - (A12)$$

and

$$(2 \rightarrow A) = (2 \rightarrow 1) + (12A),$$

the coordinates of point A,

$$Y_A = Y_1 + \overline{1A} \sin(1 \rightarrow A)$$
$$X_A = X_1 + \overline{1A} \cos(1 \rightarrow A),$$

(5-5)

can be determined in a manner similar to the traverse and polar methods discussed previously. Point A could, of course, also be computed from point 2,

$$Y_A = Y_2 + \overline{2A} \sin(2 \rightarrow A)$$
$$X_A = X_2 + \overline{2A} \cos(2 \rightarrow A).$$

Point A can also be computed by using only the coordinate differences of the known points and the tangents of the two directions to the new point $A, (1 \rightarrow A)$ and $(2 \rightarrow A)$. Hence

$$X_A = X_1 + \frac{(Y_2 - Y_1) - (X_2 - X_1)\tan(2 \rightarrow A)}{\tan(1 \rightarrow A) - \tan(2 \rightarrow A)}$$

$$= X_2 + \frac{(Y_2 - Y_1) - (X_2 - X_1)\tan(1 \rightarrow A)}{\tan(1 \rightarrow A) - \tan(2 \rightarrow A)}$$

(5-6)

$$Y_A = Y_2 + (X_A - X_2)\tan(2 \rightarrow A)$$

$$= Y_1 + (X_A - X_1)\tan(1 \rightarrow A).$$

The intersection method requires the occupation of several stations, so it is used mainly to determine the positions of points that are not easily accessible.

Resection of a Point

Resection permits the determination of the position of a point from angular measurements taken from the point to at least three known points (Fig. 5-7). Several solutions exist for the computation of the resected point. Only the most frequently used solutions, which are known by the names of their authors—Kaestner, Collins, and Cassini—are presented.

Assuming that the measured angles are $(1A4)$ and $(2A1)$, *Kaestner's* solution (Fig. 5-8) requires the computation of angles $(12A)$ and $(A41)$. First, distances $\overline{12}$ and $\overline{14}$ between the given points 1, 2, and 4 are computed, as well as the directions $(1 \rightarrow 4)$ and $(1 \rightarrow 2)$ from the known coordinates of points 1, 2, and 4,

$$\overline{12} = \sqrt{(Y_2 - Y_1)^2 + (X_2 - X_1)^2}$$
$$\overline{14} = \sqrt{(Y_4 - Y_1)^2 + (X_4 - X_1)^2}$$

$$\tan(1 \rightarrow 2) = \frac{Y_2 - Y_1}{X_2 - X_1} ; \qquad \tan(1 \rightarrow 4) = \frac{Y_4 - Y_1}{X_4 - X_1}.$$

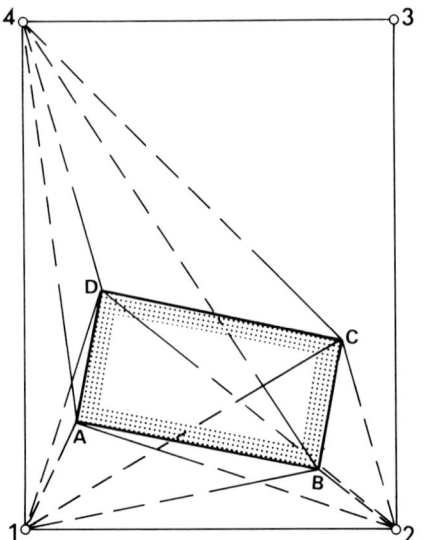

Figure 5-7. Resection method.

Now, from quadrangle 412A,

$$\frac{(12A) + (A41)}{2} = 180° - \frac{(2A1) + (1A4) + (1 \to 2) - (1 \to 4)}{2}$$

and

$$\tan \frac{(12A) - (A41)}{2} = \tan \frac{(12A) + (A41)}{2} \times \text{ctg}(45° + \mu),$$

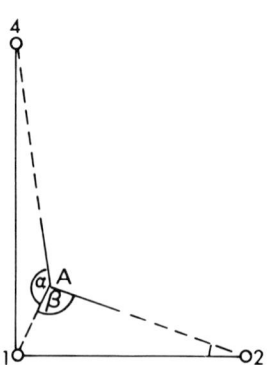

Figure 5-8. Resection solution by Kaestner.

where μ is an auxiliary angle determined by

$$\operatorname{ctg} \mu = \frac{\sin(12A)}{\sin(A41)} = \frac{\overline{14}\sin(2A1)}{\overline{12}\sin(1A4)}.$$

Then

$$(12A) = \frac{(12A) + (A41)}{2} + \frac{(12A) - (A41)}{2}$$

$$(A41) = \frac{(12A) + (A41)}{2} - \frac{(12A) - (A41)}{2}.$$

The required directions $(4 \to A)$ and $(2 \to A)$ are

$$(4 \to A) = (4 \to 1) - (A41)$$

$$(2 \to A) = (2 \to 1) + (12A),$$

(5-7)

with

$$\overline{4A} = \overline{14} \times \frac{\sin[(A41) + (1A4)]}{\sin(1A4)}$$

and

(5-8)

$$\overline{2A} = \overline{1A} \times \frac{\sin[(12A) + (2A1)]}{\sin(2A1)}.$$

The elements $(2 \to A), \overline{2A}$ and $(4 \to A), \overline{4A}$ enable computation of the coordinates of point A from point 2 and point 4, respectively.

Collins' solution (Fig. 5-9) reduces the resection problem to two intersections. First, the auxiliary point Q is intersected from 2 and 4, then A is intersected from

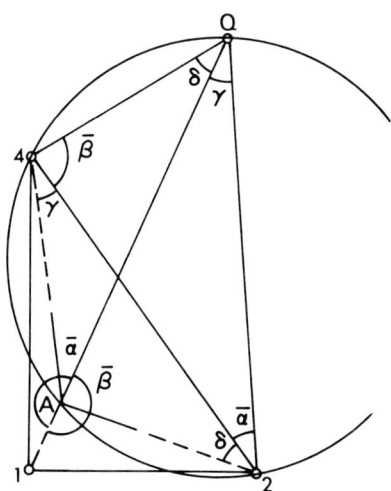

Figure 5-9. Resection solution by Collins.

2 and 4 by using the two angles $(2QA)$ and $(AQ4)$. If a circle is placed through the three points 2, 4, and A and the auxiliary point Q is defined as the point of intersection between this circle and the extension of the line connecting points 1 and A,

$$(4AQ) = (42Q) = \alpha$$

$$(QA2) = (Q42) = \beta,$$

where α and β are angles measured on point A.

Consequently, point Q in triangle $(2Q4)$ can be intersected by using the formulas given in the previous section. Once the coordinates of point Q are known, the angles γ and δ are computed, and since

$$(24A) = (2QA) = \gamma$$

$$(A24) = (AQ4) = \delta,$$

point A can be intersected from points 2 and 4.

Cassini's solution (Fig. 5-10) utilizes two auxiliary points, S and T. Point S is the point of intersection of the circle determined by points 1, 2, and A and the diameter of this circle through point 1. Point T is similarly defined by the intersection of the circle through points 1, 4, and A and the diameter of this circle through 1. Points S, T, and A are located on a straight line, which is perpendicular to $\overline{1a}$. Point A is to be derived as a point of intersection of the two straight lines $\overline{ST}$ and $\overline{1A}$, which are perpendicular to each other. Although this solution of the resection may appear complicated, it offers the most con-

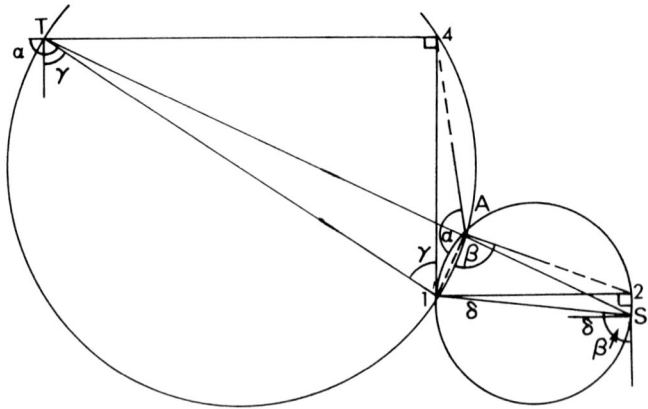

Figure 5-10. Resection solution by Cassini.

venient method for numerical computation. The solution is given by the following formulas:

$$a = Y_4 - Y_1 \qquad\qquad A = a - b\,\text{ctg}(1A4)$$
$$b = X_4 - X_1 \qquad\qquad B = b + a\,\text{ctg}(1A4)$$
$$c = Y_2 - Y_1 \qquad\qquad C = c + d\,\text{ctg}(2A1)$$
$$d = X_2 - X_1 \qquad\qquad D = d - c\,\text{ctg}(2A1)$$
$$E = \frac{A - C}{D - B}; \quad \text{or} \quad F = \frac{D - B}{A - C}$$
$$W = 1 + E^2; \qquad\qquad V = 1 + F^2$$

$$
\left\{
\begin{aligned}
Y_A &= Y_1 + \frac{A + EB}{W} \\[6pt]
&= Y_1 + \frac{C + ED}{W}; \\[6pt]
X_A &= X_1 + E(Y_A - Y_1);
\end{aligned}
\right.
\qquad
\left\{
\begin{aligned}
X_A &= X_1 + \frac{FA + B}{V} \\[6pt]
&= X_1 + \frac{FC + D}{V} \\[6pt]
Y_A &= Y_1 + F(X_A - X_1).
\end{aligned}
\right.
\qquad (5\text{-}9)
$$

Determination of a point by resection is often convenient because the field work is reduced to the measurement of angles at one point only.

If point A, which is to be determined, lies on the circle through known points 1, 2, and 3, called dangerous circle, the solution is undetermined, since it is evident that all points on the circle have the same resection angles (Fig. 5-11).

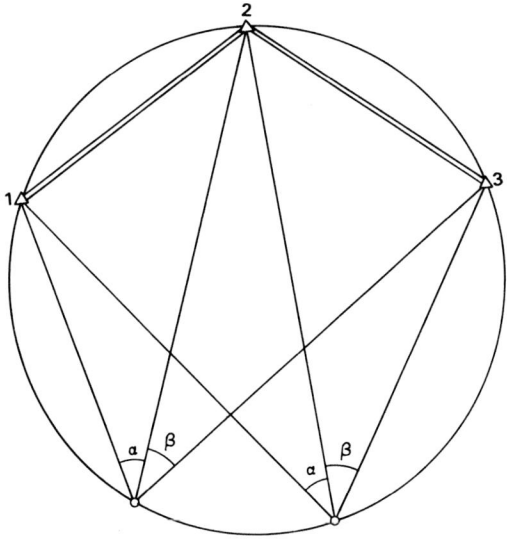

Figure 5-11. Dangerous circle.

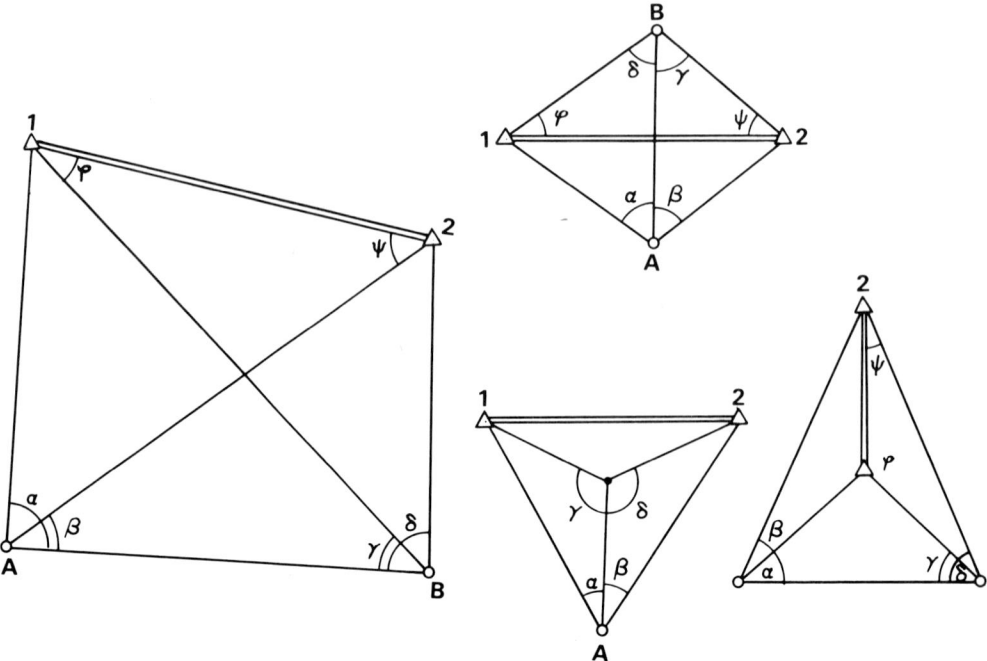

Figure 5-12. Hansen's problem.

Because of a finite accuracy in measurements, points located close to the dangerous circle are also ill-defined and must be avoided. The determination is most precise if point A is located in the center of the dangerous circle. The determination is also well defined if point A is located *inside* triangle 123.

A variant of resection is the problem of *Hansen*, which occurs when only *two points* of known coordinates are available, neither of which can be occupied. In such a case, two new points are resected simultaneously, and the problem consists of a solution of a quadrangle (Fig. 5-12), in which auxiliary angles ϕ, ψ, and $\mu = (\sin \psi/\sin \phi)$ are first computed. Subsequently, the angles are used to solve individual triangles for computing the coordinates of points A and B. The general configuration in Hansen's solution may assume a variety of forms, as indicated in Figure 5-12. Further details on computational procedures can be found in the literature on the subject (e.g. Jordan et al. 1950).

MEASUREMENT OF DISTANCES

Introduction

In surveying, the distance between two points generally refers to horizontal distance. Horizontal distance can be measured directly, or if the two points have different elevations, the inclined or slope distance can be measured and subsequently reduced to its horizontal value. Distances can be determined by mechanical length measurements, optical length measurements, and photo-grammetric, electromagnetic, and geodetic-astronomical methods. Photo-grammetric and electromagnetic determinations are explained in Chapters 8 and 3, respectively. The geodetic-astronomical methods are not used for distance determinations in terrestrial detail surveys and will not be discussed.

Mechanical Length Measurements

A mechanical length measurement consists of comparing the length to be measured with a given length element, which in turn can be a length standard or a multiple of a length standard. Important features of mechanical length measurement are the simplicity and low cost of the equipment used. In several surveying operations, direct length measurement using a tape represents the most efficient solution.

Many tools have been developed and used for mechanical length measurements that lack the necessary stability or are otherwise inconvenient for field use. With the development of band steel, steel tapes became practically the only measuring tools used in surveying, with the exception of invar wires or tapes, which are used occasionally in precise distance measurement.

For shorter distances, particularly for perpendiculars staked out to the measured points, the lighter steel tapes in boxes can be used. For the very short distances that are frequently measured in detailed surveys, a semirigid pocket tape, such as that used by carpenters, is handy.

In addition to the tapes, the following equipment is needed for mechanical length measurement: range poles, steel pins, plumb bobs, colored crayon, field notebook, axe, and material for marking semipermanent points, such as wooden stakes and tags.

For more precise distance measurements with a surveying tape, the following equipment may also be needed: tension poles, spring balance, thermometer; clinometer; target equipment (often mounted on tripods, so that the tape can be read against an index), and anemometer (to measure wind velocity). Unless distances are measured using the tape stretched in a horizontal position, the measured slant distances must be reduced to their horizontal projection by using the formula

$$L' = L \cos \alpha, \tag{5-10}$$

where α is the slope angle and L is the distance measured along the slope. The slope angle can be measured directly in the field (with a clinometer or theodolite) or determined from the elevation difference of the end points of the measured distance.

Preferably, longer distances are staked out before they are measured. This procedure usually consists of setting range poles at each end and in between, depending upon local conditions. The number of intermittent range poles is determined by a simple requirement that, from any point of the measured line,

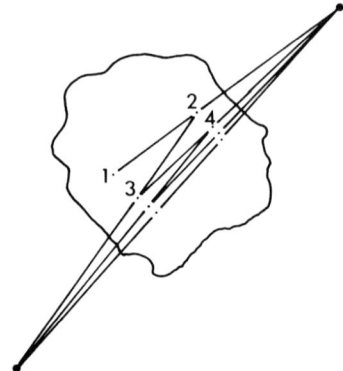

Figure 5-13. Staking out a line over a hill.

at least two range poles, separated by a suitable distance, are visible when looking in either direction of the measured line. If the line is staked out according to this requirement, the two helpers actually measuring the distance can align themselves. When staking out a line over a hill, an iterative procedure must be used, as shown in Figure 5-13, which is self-explanatory.

An identical procedure can be used if the surveyor cannot position him- or herself at the extension of the line because of the nature of the end points (corners of buildings, for instance).

Often, the actual line cannot be staked out or cannot be measured directly because of obstacles. In these cases, either an offset line must be measured or the length must be calculated from auxiliary triangles. The procedures to be used and the respective computations are shown in Figure 5-14.

Sources of error in mechanical length measurements. There are many causes of both systematic and random (accidental) errors in mechanical length measurements that can be virtually eliminated by taking suitable precautions. Because of the high accuracy requirements in city areas, the surveyor must be familiar with correct procedures and, if necessary, should study the pertinent literature. Only the more important sources of error and, where applicable, their mathematical expressions are listed.

Incorrect tape length, the most important error, has a systematic effect ΔL on the measured distance L, either negative or positive,

$$\Delta L = L \cdot \frac{\Delta l_D}{l_o}, \tag{5-11}$$

Δl_D is the length error of the tape and l_o is the nominal length of the tape. If the tape is too short, the correction to the measured length is negative.

Poor alignment always has a systematic effect and makes the measured distance appear longer.

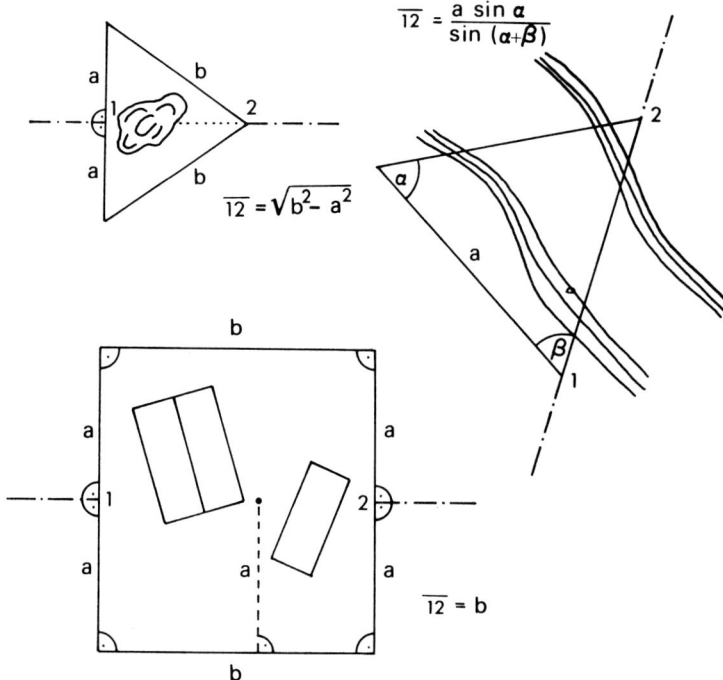

$$\overline{12} = \frac{a \sin \alpha}{\sin (\alpha + \beta)}$$

$$\overline{12} = \sqrt{b^2 - a^2}$$

$$\overline{12} = b$$

Figure 5-14. Measuring procedures for the case in which the line cannot be measured directly.

Thermal changes depend upon the coefficient α of thermal expansion and contraction of the material of which the measuring device is made. The typical coefficients per $1°$ Celsius are:

$$\text{wood } \alpha = 0.000\,0067$$
$$\text{steel } \alpha = 0.000\,0115$$
$$\text{invar } \alpha = 0.000\,0016.$$

The correction dL of a measured length L is

$$-dL = \alpha L(T_s - T), \qquad (5\text{-}12)$$

where T_s is the reference temperature at which the device has been calibrated and T is the temperature during the measuring operation. Manufacturers supply the thermal expansion coefficients and the calibration report of the tape. Tapes used in city surveys should be calibrated periodically. Length measurement errors due to temperature at given atmospheric conditions are of a systematic character.

Incorrect slope angle. The length error dL' of a horizontal distance L' caused by an incorrectly determined slope amounts to

$$dL' = -L \sin \alpha \, d\alpha = -\frac{h}{L} dh, \qquad (5\text{-}13)$$

where L is the measured slant distance, α is the slope angle, $d\alpha$ is the error in assumed slope, h is the elevation difference at end points, and dh is the error in elevation difference. If the horizontal length of a slant line is measured, then $h = dh$, and

$$dL' = -\frac{dh^2}{L}.$$

Sag. A tape not supported along its entire length sags in the form of a catenary. If C is the length of the chord, P is the pull on the tape, and W is the weight of the tape per length unit, the correction due to the sagging of the tape is

$$\Delta L_1 = -\frac{W^2 C^3}{24 P^2}. \qquad (5\text{-}14\text{a})$$

If the chord has a slant position with the slope angle α, a further correction is necessary:

$$\Delta L_2 = \frac{W^2 C^3}{12 P^2} \sin^2 \alpha.$$

Wind may cause a lateral displacement of the tape and increase the sag, calling for a further correction:

$$\Delta L_3 = -\frac{W^2 C^3}{24 P^2} \tan^2 \phi. \qquad (5\text{-}14\text{b})$$

An approximate knowledge of displacement angle ϕ is sufficient. The total correction for sag is

$$dL = \Delta L_1 + \Delta L_2 + \Delta L_3 = -\frac{W^2 C^3}{24 P^2}(1 + \tan^2 \phi)(1 - 2\sin^2 \alpha).$$

$$(5\text{-}14\text{c})$$

This correction yields the right chord length, which must be reduced to a horizontal chord. The need to use the more precise formulas, Eqs. (5-14a and b), arises only in very precise measurements, such as those carried out with invar wires. As a rule, in detailed surveys in urban areas, the tape rests on the ground, and the reading is reduced only to the horizontal distance by applying the slope correction where needed. Paved surfaces of uniform slope suggest this approach.

Sagging of the tape has a systematic effect on the result of distance determination; errors of partial determination of the length accumulate throughout the measurement process and result in the total distance being either too short or too long.

The remaining sources of error have an accidental effect, except under specific conditions when they acquire a systematic character.

Improper plumbing. Plumbing, using a plumb bob or a range pole (held between two fingers to permit achievement of vertical position), is always affected by small errors which may become systematic if there is a strong wind, for instance.

Inaccurate marking. Marking of the partial measurement (end of the tape) by steel pins or colored crayon, for instance, is affected by small random errors.

Elongation of the tape caused by pull. If the prescribed pull P_s is applied, as during tape calibration, via a spring balance, the resulting errors, if any, are of systematic character. If an estimated pull force is applied, differences in pull force will result in variable elongation, of random nature, of the tape. The effect of incorrect pull can be computed from the formula

$$dL = (P_s - P) \cdot \frac{L}{qE}, \tag{5-15}$$

where P is the pull actually applied, q is the cross-sectional area of the measuring device, and E is the modulus of elasticity, which, for steel, is about $2 \cdot 10^6$ kg/cm^2.

Gross errors. In mechanical length measurements, particular care must be taken to avoid gross errors, which are easy to commit by miscount in the number of full tape lengths used, transposition of figures when either reading or recording the taped distances, use of the wrong index as the zero index, etc. Gross errors are avoided by strictly observing standard field procedures, measuring lines twice, and applying other field checks.

Final remark. The development of electromagnetic distance-measuring devices has resulted in a decreasing interest in mechanical distance-measuring procedures for more demanding and precise operations. Polygon sides, other more important reference lines, and longer distances should be measured by electromagnetic devices.

Optical Length Measurements

Use of mechanical length measurements in city areas interferes with traffic and may be inconvenient in some survey operations. Also, the use of a tape is not practical on broken terrain with significant slopes or in areas with open canals, creeks, or other obstacles difficult to cross. Moreover, tape measurements are most efficient when limited to short distances not exceeding the length of a tape. In surveying operations which require a large number of measurements over longer distances, particularly when the measurements originate from the same point (polar method), electromagnetic or optical methods are preferred. Electromagnetic instruments provide superior accuracy, and it is to be expected that they will soon dominate distance-measuring operations entirely. However, many countries have instruments for optical length measurements available, and in the absence of more modern means, there is no reason why these instruments should not be put into good service.

Compared to mechanical length measurement, optical length measurement has several advantages which reduce the risk of gross errors and the dependence on terrain features and facilitate the carrying out of a survey. These advantages are as follows:

A short distance can often be measured directly, i.e., the distance does not have to be broken down into tape lengths;

Length can be measured even along lines that are impassable due to bodies of water, fences, or traffic;

The damage to vegetation is greatly reduced because the distance does not need to be traveled directly;

Elevation differences do not represent a serious obstacle; and

Differences in height are determined simultaneously with distances.

Optical distance-measurement devices can provide accuracies from 1/100 to about 1/10 000. The instrument should be selected according to the accuracy required.

Optical distance-measurement instruments also possess a capability for angular measurement with compatible accuracy. Optical distance measurement is therefore performed jointly with angular measurements, which again offers operational advantages.

The distance to be measured optically is derived from a triangle (Fig. 5-15) in which usually two elements are known and a third is measured. This gives

$$L = \frac{b \sin \sigma}{\sin \gamma},$$ (5-16a)

or

$$L = b \operatorname{ctg} \gamma$$ (5-16b)

if ε is a right angle, which usually is the case.

The problem can, from an instrument-designer's point of view, be solved in many ways. For instance, the base b can be located at the observer or at the target; it can be vertical or horizontal; and it can be of fixed length or be measured.

The instruments for optical distance determination are classified according to use: topographical surveys—tacheometer, self-reducing tacheometer; traverse surveying, detail surveys—precision self-reducing tacheometer, traverse surveying—parallactic determination using theodolite and precise stadia. Typical solutions only will be outlined, without entering into technical details that can be found in the specialized literature.

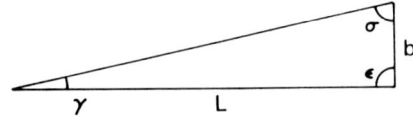

Figure 5-15. Parallactic triangle.

Tacheometers. Tacheometers (Gk. *tachys*, swift) are used for rapid determination of positions of points in a polar coordinate system. The equipment consists of a special theodolite and the usual leveling staffs. The telescope, in addition to the usual cross hairs, has four stadia lines located symmetrically to the central cross in the field of view.

From the staff reading (upper minus lower reticle line), the length of base b is found. The observer must also read the horizontal angle (from a reference direction) and the vertical angle. The general geometry is given in Figure 5-16. The formulas used for the computation of distance D and height difference ΔH are

$$D = (kb + c)\cos^2 \alpha \qquad (5\text{-}17)$$

and

$$\Delta H = D \tan \alpha + i - h, \qquad (5\text{-}18)$$

where k is a multiplication constant, as a rule equal to 100, and c (in modern instruments equal to 0) is the instrument constant, i is the height of the tacheometer, and h is the reading of the center reticle line on the staff. Consequently, the formula for the distance usually has a simple form,

$$D = 100\, b \times \cos^2 \alpha$$

quite convenient for computation, particularly when modern electronic pocket calculators are used. The height difference ΔH is equally simple to calculate.

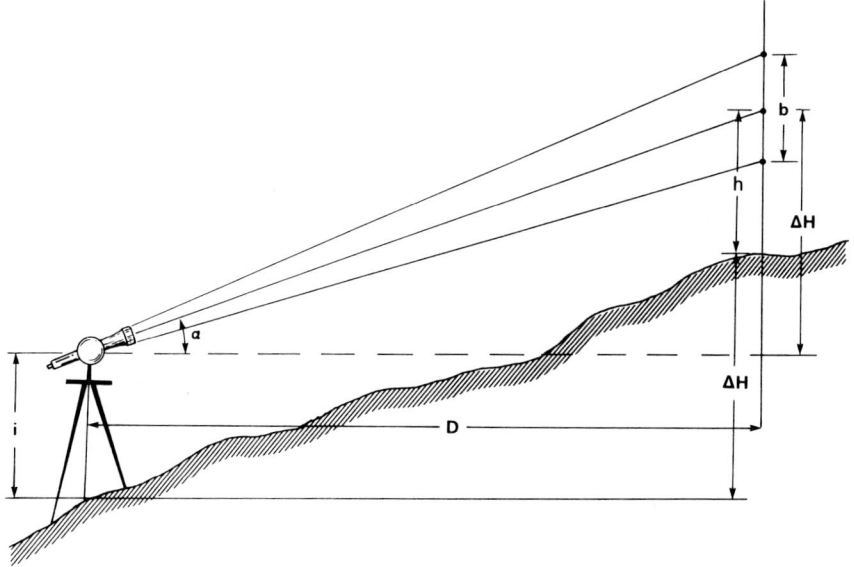

Figure 5-16. Tacheometric measurements.

Conventional theodolites and leveling instruments are usually equipped with stadia reticles so that they can be used as tacheometers. More practical, however, are the so-called *auto-reducing tacheometers*, which permit direct determination of *horizontal distances* and *differences* in elevation. Those sections of the staff that are enclosed between the specific curves visible in the telescope define the horizontal distance and the difference in height. The multiplication constant for determining the distance is 100. For height differences, a variable multiplication constant is usually used, depending upon the inclination angle of the telescope. These constants are engraved on the respective curve sections in such a way that the correct multiplication constant is visible in the field of view (Fig. 5-17).

Auto-reducing tacheometers are quick to use, and supplementary computational work in the office is not needed. A typical field party includes the observer and two helpers with staffs. The accuracy, however, which is of the order of 20 to 30 cm on a distance of 100 m, is somewhat limited; consequently, ordinary tacheometers are suitable for surveying the topography of the terrain but are not adequate for planimetric work in urban areas.

Precision reduction tacheometer. For detailed surveys in which a higher degree of accuracy is expected, precision reduction tacheometers can be considered. These are mainly used with a horizontal staff from which the horizontal distance and the difference in elevation between the instrument and the staff can

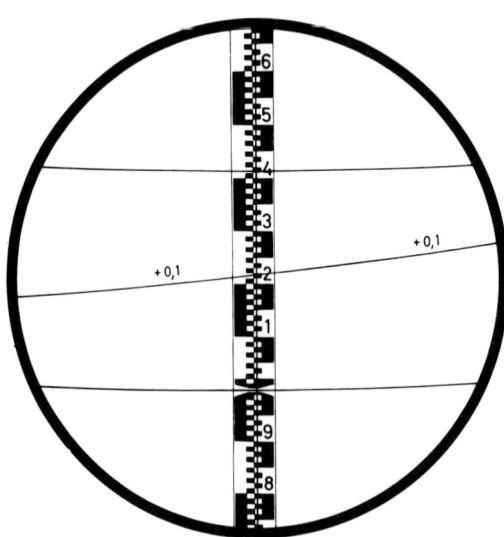

Figure 5-17. Example of a distance and a height difference reading in an auto-reducing tacheometer Wild RDS (courtesy of Wild Heerbrugg Ltd.). Distance 41.3 m. Height $+0.1 \times 21.7 = 2.17$ m.

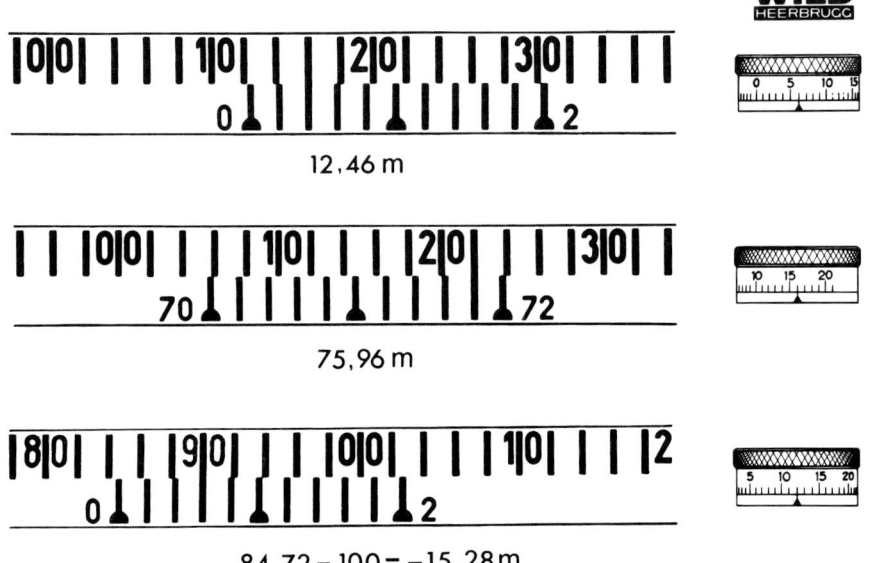

Figure 5-18. Example of staff readings (courtesy of Wild Heerbrugg Ltd.).

be read directly (Fig. 5-18). The last digit of the distance or height difference is obtained from a coincidence drum. The solutions are based on an ingenious auto-reducing optical system combined with the vernier-type arrangement. A knob permits switching from horizontal to vertical determinations and vice versa. Similarly, when reading the distance, the telescope can be changed from single-image operation, required in angular measurements, to double-image operation.

The accuracy quoted by the manufacturers is of the order of 1 to 2 cm in distance determination and 3 to 5 cm in height determination per 100 m. The normal range is up to 150 m if atmospheric conditions are stable. In less favorable conditions, the range must be reduced accordingly if the quoted accuracy is to be maintained.

The main drawback of precision reducing tacheometers with a horizontal staff is the difficulty in setting up the staff over the point to be measured; if this cannot be accomplished, the staff must be used in an offset position, an approach that requires reduction of the recorded data.

Precision invar subtense bar. The invar subtense bar, in conjunction with a precision theodolite, is another device occasionally used for optical determination of distances (Fig. 5-19). The bar, which can be folded to facilitate transport, provides two well-defined targets at a fixed mutual distance, usually 2 m.

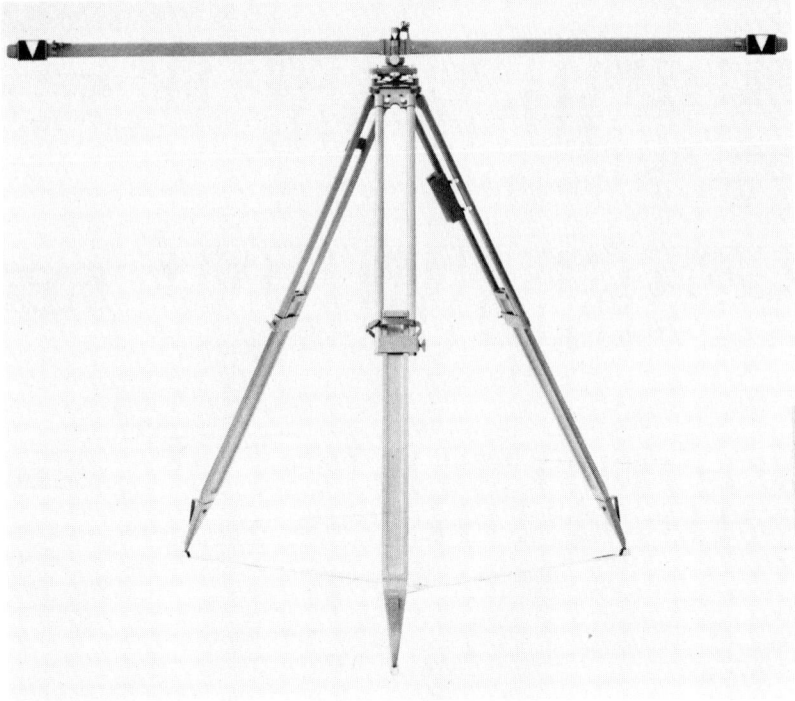

Figure 5-19. Wild precision invar subtense bar (courtesy of Wild Heerbrugg Ltd.).

The bar is set on one end point of the distance to be measured, and the theodolite is set on the other end point. By precise measurement of the angle between the directions to both targets (the bar must be perpendicular to the measured line), the actual distance can be computed or derived directly from a table supplied with the subtense bar. The derived distance is always the horizontal distance; consequently, no further reduction computations are involved. With a 2-m subtense bar and careful measurements of angles, accuracies of the order of 2 to 3 cm at a 100-m distance can be achieved. If the vertical angle is also measured, the difference in elevation can be computed as

$$\Delta H = D \tan \alpha + i - t,$$

where t is the height of the targets.

This method was used for traverse measurement in some countries before electromagnetic instruments became more widely spread. The mean value of the directions to both targets defines the direction of the measured line (traverse side). Longer bars with targets were developed in some countries to increase the permissible range of distances to be measured.

Sources of error in optical length measurements. The accuracy quoted by manufacturers is obtainable only under favorable atmospheric conditions, by a qualified operator using carefully calibrated equipment. Any departure from the nominal values of instrument elements that determine the measured distance, or from the assumed basic geometry, will cause errors in the final results. Therefore, not only must the instruments used be carefully calibrated, but proper care must be taken in setting up the instruments and staff in the prescribed manner during measuring operations. For instance, should the staff deviate from the vertical position or from the position perpendicular to the aiming direction, measuring errors will result. Similarly, any error in judgment by the operator in aiming or establishing coincidence of various division elements will affect the results. It is obvious that adverse atmospheric conditions, such as strong wind or "air vibration" due to excessive insolation, will make the task of the instrument operator more difficult and will affect the results accordingly. Thus it is good practice to cease surveying operations at noon, when insolation is particularly intense.

If auto-reducing tacheometers are used, the surveyor must ensure that slant distances are correctly reduced to horizontal distances and height differences. Therefore, staking out a test range on which the overall performance of the optical distance measuring process can be checked is recommended. The range would consist of a number of distances (e.g., 20, 40, 60, 80, 100 m), with the distances and height differences precisely established by another, more accurate measuring technique. If permanently monumented, such a simple test range is of great practical value since it permits rapid testing of the equipment used.

MEASUREMENT OF ANGLES

Definitions

The following frequently encountered terms are of interest.

A *horizontal angle* is the angular difference between two vertical planes of arbitrary orientation.

The *direction* of a vertical plane is its angle from an established vertical plane of reference. This plane can be erected along the grid reference axis or a grid line parallel to this axis. It can also be at the 0 (zero) reading of a horizontal circle. Consequently, a reading taken from a theodolite circle is a direction.

Azimuth. A distinction is made between true and magnetic azimuths. A *true azimuth* is a direction referenced to geographic north or south. It is determined by geodetic-astronomical methods—e.g., the techniques used for the orientation of triangulation networks. A *magnetic azimuth* is referenced to magnetic north and as such is closely related to the earth's magnetic field, which is characterized by lines of force that, except for small variations, remain fairly constant in direction.

Bearing. If the angular value is kept smaller than 90° with reference to a certain line, we speak of a *bearing angle*. The *bearing* of a line is the bearing angle with the notation of the directional quadrant in which the line lies, e.g., N 34°38′ W

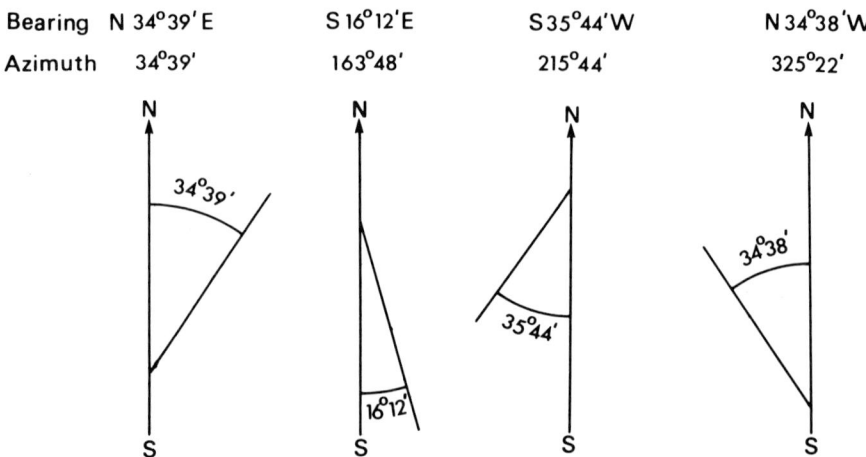

Bearing	N 34°39'E	S 16°12'E	S 35°44'W	N 34°38'W
Azimuth	34°39'	163°48'	215°44'	325°22'

Figure 5-20. Bearing angles in each of the four quadrants and corresponding azimuths from north.

(Fig. 5-20). Bearings may be referenced to any direction, but it is important to indicate clearly the reference direction used. Bearings have been and are still used in legal surveying, particularly in Anglo-American countries.

Devices Defining Fixed Angles

In surveying, the need to determine angles of 100^g and 200^g is frequent. Simple devices, such as angle drums, angle mirrors, and prisms, can be used for this purpose. The first two devices are of historical interest only.

In contrast, prisms are very handy and useful devices. They will be described in more detail since they are not too well known outside of European countries.

Prisms. The path of the rays through prisms can be determined by the laws of refraction, reflection, and total reflection. Three types of prisms are considered here: the three-sided, or Bauernfeind, prism; the five-sided, or pentagon, prism after Goulier; and the five-sided prism after Wollaston.

The three-sided prism has an isosceles triangle as cross section. As seen in Figures 5-21a and b, a ray may pass through the prism in two different ways.

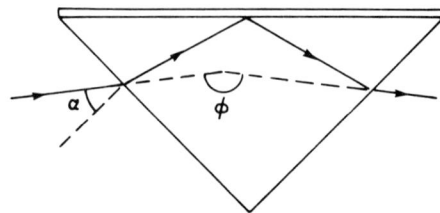

Figure 5-21a. Unstable image in a three-sided prism.

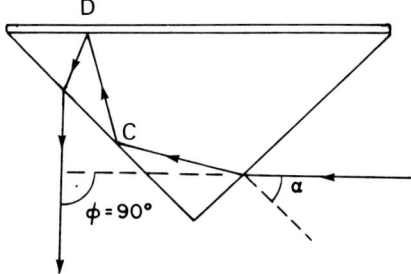

Figure 5-21b. Stable image in a three-sided prism.

The first path (Fig. 5-21a) results in an unstable image because angle ϕ changes if angle α changes. The second case provides a stable image and, with the angles as given, permits laying out a right (100^g) angle. The ray undergoes total reflection at point C and partial reflection at point D. To keep the loss of light intensity to a minimum, the long side of the isosceles triangle is coated as a mirror.

The five-sided prism after Goulier was originally intended to be a four-sided prism (Fig. 5-22); one corner is cut off to give the prism a more compact shape. Since the two reflections inside the prism are only partial, the two reflecting surfaces are coated as mirrors.

The five-sided prism after Wollaston (Fig. 5-23) was also originally a four-sided prism. The two reflections inside the prism are total, so that no surface requires mirror coating.

Each of the prisms discussed is useful for laying out a right angle.

If, in the orthogonal method, point E in relation to the house corner D and the survey line $A-C$ is to be determined by a single observer, a second point B of the line is required in the direction of A so that the observer can

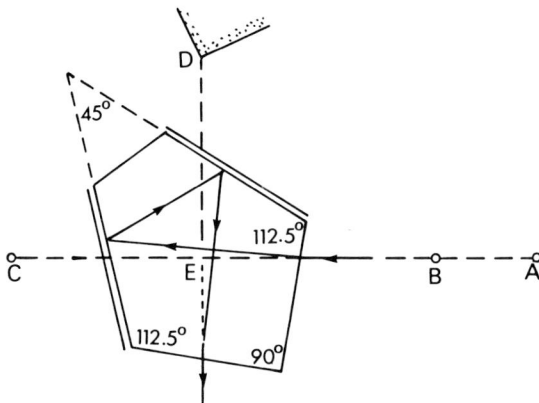

Figure 5-22. Five-sided prism after Goulier.

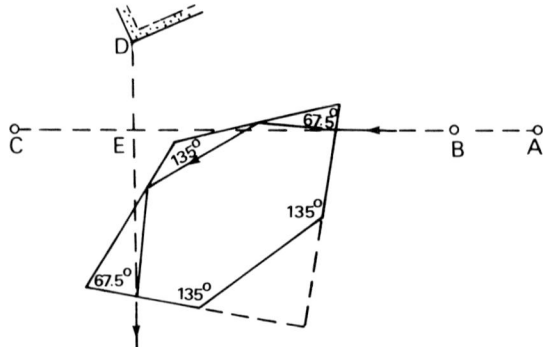

Figure 5-23. Five-sided prism after Wollaston.

see whether he or she is on the line. Use of an extension of *AB* for this purpose usually leads to a loss of accuracy and is not practical. For this reason, double prisms were built and have become popular (Figs. 5-24 and 5-25).

The use of prisms. Prisms are used for two purposes: to lay out a point in a perpendicular direction (or at some other fixed angle) from a certain point to a reference direction and to determine the point of intersection between a given reference line and a line originating from a given point and intersecting the reference line at a fixed angle, usually 100^g.

Double prisms enable an observer to position him- or herself along a given survey line, e.g., line *AC* in Figures 5-24 and 5-25. In the first application, the observer is positioned at point *E*, determined by means of a plumb bob, and a helper is directed onto the line *ED* to be laid out. In the second application, the observer moves along the line *AC* until the range poles placed in *A* and *C* and the edge *D* of the house are seen in one line, in which case he or she has arrived at *E*.

A 100^g angle can be determined by using a double prism with an accuracy of $\pm 2^c$. This corresponds to a lateral shift of ± 1 cm for a point at a 35 m distance.

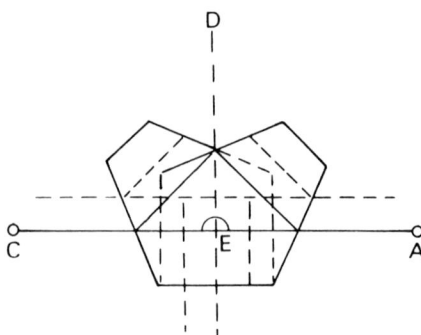

Figure 5-24. Double prism using two prisms after Goulier.

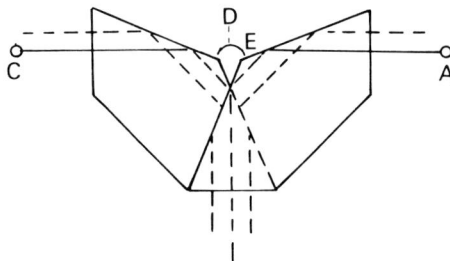

Figure 5-25. Double prism using two prisms after Wollaston.

Since in the orthogonal method, the abscissas do not generally exceed 35 m, this accuracy is sufficient for legal purposes in most countries.

Sources of Error in Fixed-Angle Instruments

Because of the simplicity of the instruments, only one error needs to be considered: deviation from the intended angle. Since prisms can be produced with angular accuracies better than 10^{cc}, they can be considered error free if so proven in the factory.

Theodolites

For measurement of arbitrary horizontal and vertical angles, a *theodolite* is used (Fig. 5-26).

The movable part of the instrument (the alidade) with two alidade supports carrying the telescope rotates about the vertical axis of the instrument. This axis passes through the center of the upper plate (the alidade proper) and the lower plate (or limb). The amount of rotation can be read from the horizontal circle. Depending upon the type of theodolite, a variety of arrangements are

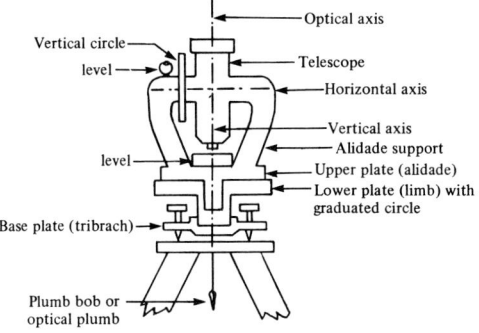

Figure 5-26. Schematic drawing of a theodolite.

Overview

Each terrestrial survey can be carried out.in either a local coordinate system or a countrywide system. The seven basic field-survey procedures were introduced on p. 162. Of these, three—the extension, orthogonal, and polar methods—are particularly suited to the determination of an array of terrain details, such as boundaries, buildings, fences, walls, and trees. These three methods differ considerably in technical and organizational requirements.

In the extension method, a dense net of linear partial systems is created. All measurements are length measurements taken in reference to lines defined by the control net (Fig. 5-27a).

The orthogonal method requires, in addition to the length-measuring device, an instrument for *rapid determination* of points in which the ordinates are perpendicular to the reference axis of the elementary systems. Prisms are generally used for this purpose. Since the use of prisms provides only a somewhat limited accuracy, the ordinates should be rather short. Therefore, the area covered by an elementary system is approximately rectangular (Fig. 5-27b).

The polar method requires the use of an angle-measurement device in addition to the distance-measurement device. It creates elementary systems that are circular or square in shape (Fig. 5-27c).

The relationship between orthogonal and polar coordinates is as follows:

$$Y_i = s_i \sin t_i \qquad s_i = \sqrt{X_i^2 + Y_i^2}$$

$$X_i = s_i \cos t_i \qquad t_i = \tan^{-1} \frac{Y_i}{X_i}. \tag{5-19}$$

The reference axis in both cases is the straight line through points P_o and P_E. If the distance $P_o P_E$ is already known from the coordinates of the end points, the scale of the elementary system is imposed by that of the control net to which points P_o and P_E belong.

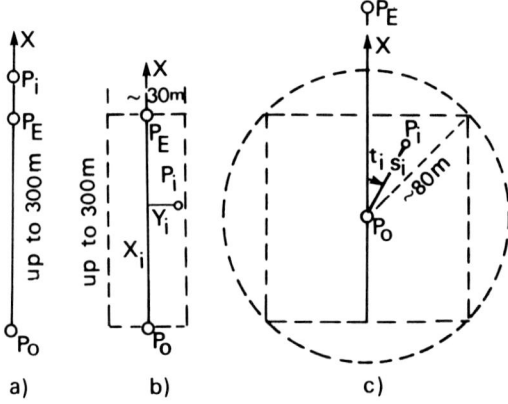

a)   b)   c)

Figure 5-27 Three most frequent approaches used in detailed field surveys: (a) extension method, (b) orthogonal method, (c) polar method. According to Ref. (2).

The extension method (Fig. 5-27a) is obviously a special case of the orthogonal system for $Y_i = 0$ or the polar system for $t_i = 0$.

A number of factors should be considered in the selection of the most practical and efficient field-survey method, or combination of methods. First, the instrumentation for the selected method must be available. In areas with high labor costs and adverse climatic conditions, the most rapid method requiring a minimum of field personnel should be used. This also entails good preparation of the field work. Available source material must be searched out and evaluated beforehand. The survey lines and stations must be placed in optimum locations. If the work consists of small survey jobs, several should be planned for one field mission. If the use of electronic processing of survey data is intended, the method of numbering of each surveyed point must be decided upon.

The survey should be accurate enough to satisfy the most important requirements. The use of the survey results for the derivation of point coordinates, which may then define further elementary systems, requires a homogeneous control net of certain accuracy. In this type of survey, gross errors must be watched for particularly carefully. In cadastral survey, for instance, several countries require the taking of sufficient control measurements to detect gross errors during the survey operation in the field.

It is obviously much easier to control orthogonal measurements than polar measurements.

Field Records

Depending upon the surveying method used, various types of field notes are produced. These can be divided into graphical and tabular field records. Graphical field records are basically sketches depicting relative positions of surveyed points and terrain details, complemented by numerical values of measured quantities (distances and angles), names, numbers, and other information, in descriptive and symbolized form, in order to render the sketch a complete surveying document. In the office, sketches are used for plotting final maps and the elaboration of surveying documents.

Field measurements should be systematically recorded in tabular form suitable for the further processing of survey data. Examples are tabular records produced in surveying traverses and tabular records in tacheometric work.

All final records must be produced with the utmost conscientiousness because they are permanent records which, in the case of real-property surveys, may also have legal significance.

Graphical field records. Depending upon filing arrangements for field records and their later use, the standard size of the form for graphical field records may vary widely.

Graphical field records or field sketches should display details roughly to scale. In West Germany, the recommended scale ranges are as follows: 1:150 to 1:300 in densely built-up areas, 1:300 to 1:500 in sparsely built-up areas, and

1:500 to 1:1000 in open areas. These scales should permit a clear and readable presentation of all surveyed terrain details together with pertinent numerical data and symbols.

Each field sketch should present a given area, e.g., a city block. A certain administrative area should be divided prior to field surveys into suitable territorial units, which in turn are numbered to create a unique reference to the surveyed area.

Graphical field records should be suitable for copying and should be able to withstand field conditions. Climate-resistant, transparent material permits copying by dyeline and photostatic printers. Notes on this transparent material should be made in water-resistant ink or in pencil. Existing survey lines should be drawn onto the record beforehand in order to facilitate field work. Each measurement must be clearly written into the field sketch to permit recognition of its relationship to other measurements.

The graphical field record usually contains: all survey lines and the numbers of end points of survey lines, names or numbers for administrative units, street names and house numbers, information on usage of buildings, all monumented boundary points, all boundaries, classification of streets, a north direction indicator and names or numbers of neighboring field records, for orientation purposes; and numbers of all points that will later be subjected to automatic data processing.

Notation of length measurement. Figure 5-28 shows three different notations of length measurement along the wall of a building: (a) *running measurement* in which lengths are measured from one corner point, with the figures written looking in the direction of measurement; (b) three *separate measurements* of the same wall; and (c) *overall measurement* made from end to end in addition to the separate measurements. Owing to the accumulation of errors, the total line length derived from the separate measurements is likely to have a ($\sqrt{3}$ times) larger error than the running measurement, for which reason the latter is always preferable.

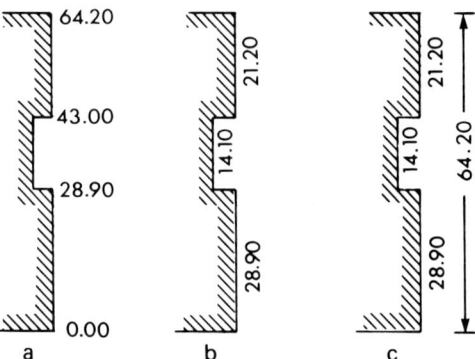

Figure 5-28. Notations of length measurement.

The numbering of points. All points defining survey lines should be shown with their numbers, so that coordinates for the location of these points on maps may be found easily in the further processing of the measurements.

In addition, each new point that is to be given coordinates must be numbered as well. This number can be assigned directly in the field or later in the office. The number will also serve to correlate measurements, some of which are recorded on a graphical field record and the rest of which are recorded in a numerical field record. If point numbers are to be assigned in the field, suitable preparation will usually be required.

The question of point numbering has become increasingly important with the introduction of automatic data processing (p. 211).

Symbols for the presentation of terrain features and measurements in graphical field records. The following suggested symbols (3) are by no means complete. Cities with well-established survey procedures may prefer to continue to adhere to their own systems. The philosophy underlying these symbols is, however, generally valid.

Presentation of points. Points can be monumented in many different ways, or may be unmonumented (Fig. 5-29). Each point that is subject to measurement

Monumented points

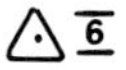

Control point, part of a well-monumented and referenced adjusted net determined by geodetic procedures.

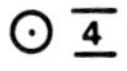

Control point, part of a well-monumented and referenced, possibly also adjusted, net determined by a breakdown of geodetic control to a spacing practical for surveying by field methods, primarily polar and orthogonal methods.

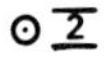

Unreferenced points of a survey line used either to mark major survey lines in the extension method or to break down long survey lines and determined by theodolite ranging.

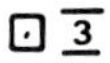

Point of an administrative boundary.

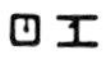

Point of a parcel boundary.

Chisel mark.

Letter to indicate the material used for monumentation, e.g., B for bolt and N for nail.

Subsurface boundary monument only.

Unmonumented points

Free-standing dot.

Figure 5-29. Examples of different survey point symbols. Numerals give suggested dimensions in millimeters.

should be clearly marked on the sketch by a dot. Depending upon the monumentation, the dot may be surrounded by a line feature.

Presentation of lines. Lines connect various classes of points. Therefore, different ways of presenting connections are required; distinctions can be made both in line thickness and line character. Suggested relative line thicknesses are indicated by an *L* and a number.

A connection may be broken into line and point elements. Figure 5-30 gives some lines and their application.

Symbols for linear topographical features. Linear features will always have significance if they have been erected on or beside a boundary. The suggested symbols will therefore include a parcel boundary (Fig. 5-31).

Recording of measurements. To standardize field records, certain conventions for the notation of measurements are desirable. Figure 5-32 represents a section of a field sketch with generally accepted notations of measurements; the numbers represent the following:

1. *Start of a survey line.* It is marked by 0.0 or not at all.
2. *End of a survey line.* It is written like an abscissa value and underlined twice.
3. *Abscissa.* It is written perpendicular to the survey line opposite the ordinate. The orientation of numericals as in running measurements.
4. *Ordinates.* They are written on the ordinate with the foot of the numerals toward the drawn ordinate, or on the extension of the ordinate.
5. *Measured right angle.* It is indicated by two arcs.

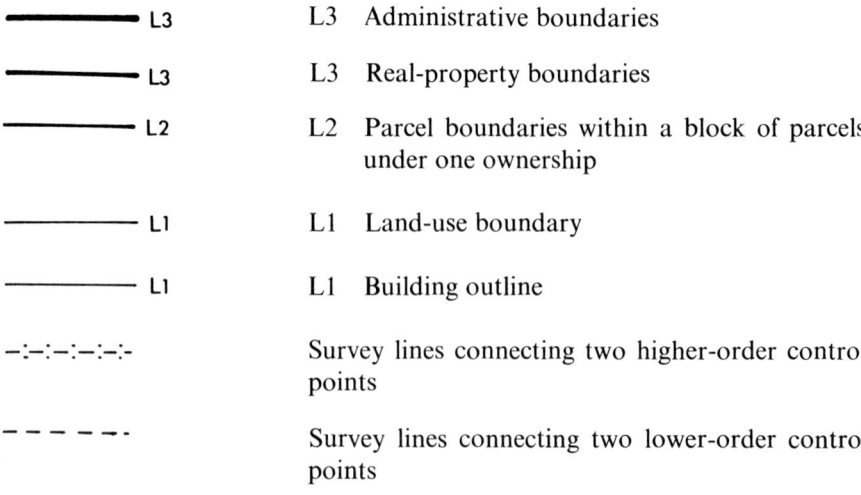

L3	L3	Administrative boundaries
L3	L3	Real-property boundaries
L2	L2	Parcel boundaries within a block of parcels under one ownership
L1	L1	Land-use boundary
L1	L1	Building outline
		Survey lines connecting two higher-order control points
		Survey lines connecting two lower-order control points

Figure 5-30. Examples of different line symbols.

One-sided   Shared

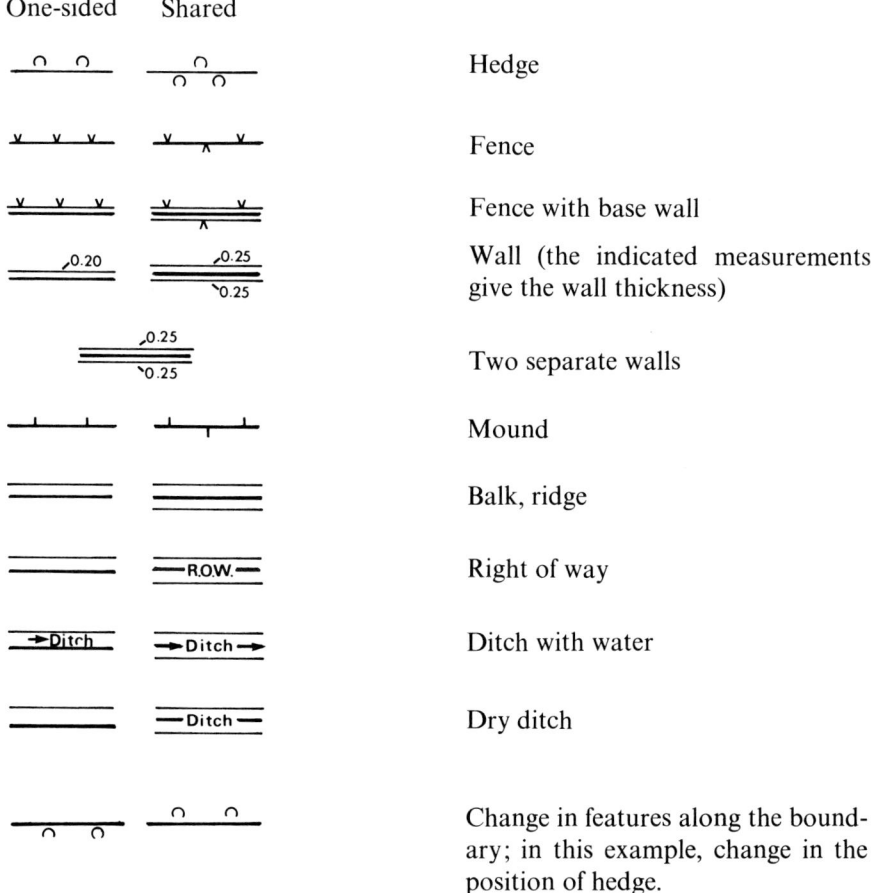

Hedge

Fence

Fence with base wall

Wall (the indicated measurements give the wall thickness)

Two separate walls

Mound

Balk, ridge

Right of way

Ditch with water

Dry ditch

Change in features along the boundary; in this example, change in the position of hedge.

Figure 5-31. Examples of different features erected on or beside a parcel boundary.

6. *Estimated right angle.* It is indicated by a single arc.
7. *Departing survey line.* The abscissa, for a point into which another survey line connects, is underlined once.
8. *Strut.* A strut is a distance measured from a suitable point located on a survey line to a detail point along the hypotenuse of a right triangle having an abscissa difference and a measured ordinate as the two smaller sides. The feet of the numerals are toward the drawn strut. A strut can also be marked as a bracing if the hypotenuse cannot be drawn for lack of space.
9. *Bracing.* A bracing is a distance measured between two monumented boundary points along a line that is not a boundary line. If space permits, the two points can also be connected by a dashed line and the number written as in the case of a strut.

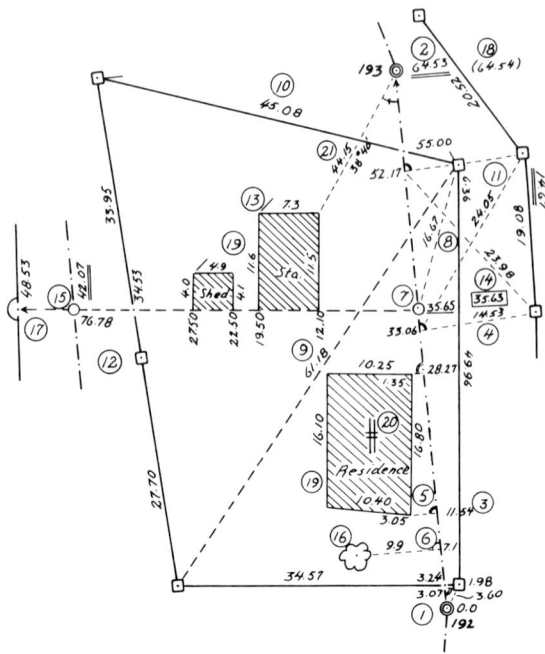

Figure 5-32. Recording of measurements (see text for explanation). Numbers in circles are reference numbers only for the purpose of this text; they are not part of the field sketch.

10. *Boundary length.* A measured distance along a boundary between two monumented boundary points. As a strut or a bracing, a boundary length is always written in the manner shown; the feet of the numerals point toward the boundary.

11. *Ordinate with several points.* An ordinate with more than one point is treated like a survey line that has abscissa values only.

12. *Straight line sign.*

13. *Polar coordinates.* A dashed line indicates the measured distance. The distance is written on one side with the feet of the numerals pointing toward the dashed line; the angle, which is marked by a circular arc with an arrow head, is indicated on the other side, with the heads of the numerals pointing toward the dashed line.

14. *Laid-out readings.* If a certain reading is to be attained at a point, generally an intermittent point of a long survey line, it is laid out and marked by framing the laid-out reading.

15. *Point of intersection.* A point at which two measured distances intersect should be determined and included in the survey as a control.

16. *Natural and topographical features.* They are often determined with lesser accuracy. These features may include trees, bushes, ditches, and temporary structures.

17. *Extension of a survey line.* The measurement is written perpendicularly to the plotted extension.

18. *Computed values.* Computed values are shown as if they had been measured and are placed in round brackets.
19. *Survey of buildings.* This example indicates how buildings can be surveyed. Note that the residential building (marked Residence) has been measured to centimeters and stable (Sta) and shed (Shed) have been measured only to decimeters.
20. *Parallel lines.* The long lines indicate the direction of the lines that are parallel.
21. *Direction indicator.* If polar coordinates are measured for many points, it becomes impractical to write them into the graphical field record. It is more practical to mark the measured values in a numerical field record and indicate only on the field sketch the direction to the station from which the polar coordinates were measured.

Field records in tabular form. Forms for the recording of numerical field data may be designed with two possible objectives: (1) to provide the most effective recording of a certain type of data and (2) to provide a standard form for the recording of all types of data.

The two objectives are mutually exclusive. Depending upon the use of the data, a decision as to the method of recording must be made. In most surveying operations, recording forms are used not only to record the actual measurements but also to carry out the subsequent computations. Angular measurements on a triangulation point, length and angular measurements in traversing operations, distance and angular measurements on a polar station, and leveling are examples requiring field records on forms that also permit computation of intermittent or final results.

The required forms are usually available at the control surveying and mapping agency of the country. There is a definite advantage in countrywide standardization.

Coded field records. If instruments with automatic recording are used, the field records must be even more standardized. As an example, Figure 5-33 shows a single record obtained from a Zeiss Reg Elta 14 punched on a paper tape. Each registration (4) includes the following:

1. horizontal direction r (in grads).
2. zenith angle z (in grads).
3. distance s (in meters).
4. a preset code with 12 positions (ABCDEFGHIJKL).
5. four spaces.

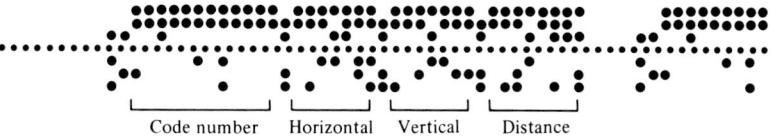

Figure 5-33. Paper tape record of a Zeiss Reg Elta 14 (Carl Zeiss–Oberkochen).

Newer models of electronic tacheometers either employ a cassette magnetic recording (e.g. Wild Tachymat TAC-1) or store data in memory of electronic collectors. The Hewlett-Packard electronic tacheometer HP-3820A uses a hand-held data collector which automatically records survey data. The user can also key in descriptive information. Data is stored as lines of characters, with each line consisting of a 4-digit line number and 16 characters of data. Memory allows storage of up to 4000 lines of data. Once the data is collected, it can be electronically transferred to a calculator or computer for processing.

Densification of Existing Ground Control

It is assumed that a network of permanently monumented and well-referenced control points has been established by triangulation, trilateration (or a combination of both), traversing, or by photogrammetric methods and that plane coordinates for these points are available.

This network of points may not be dense enough for detailed surveying, and in this case, additional points must be established. The required density depends upon the choice of surveying methods for the detailed survey and later updating surveys and upon the required accuracy. These dependencies are correlated.

Because of frequent destruction of survey monuments in urban areas, there is a tendency in various cities not to monument control points permanently but to reference them with great care and precision to markers set on buildings (see Chapter 3).

Control-point requirements for orthogonal and extension methods. In the orthogonal and extension methods, the survey lines (reference lines) should be located as close as possible to the points to be surveyed. To provide optimum conditions for later follow-up surveys, it is important that both end points of each major survey line be monumented or reliably referenced. These points must be properly determined, and this is usually done by traversing in built-up areas.

More specifically, the net of survey lines should (1) be arranged in an unambiguous manner, (2) permit proper error distribution within the surveyed area, (3) hinder traffic as little as possible during actual surveying, (4) be easily reestablished for follow-up surveys, (5) permit the surveying of approximately straight boundaries from a single survey line running along the boundary, and (6) permit, if the orthogonal method is used, the use of short ordinates.

Survey lines should not be too long. If long survey lines must be used, they should be ranged by theodolite, and the intermittent points should be marked so that any one section of a survey line does not exceed a distance greater than 100 m.

If the intersection of two survey lines cannot be avoided, the point of intersection should be determined and possibly marked or monumented.

If it is not possible to survey an area completely from a well-designed net of survey lines tied into the control system, occasional "open" survey lines or open traverses may be used. As a rule, every attempt should be made to avoid the use of these last-resort measures.

Control-point requirements for the polar method. A detailed survey by the polar method is normally a combination of traversing for the determination of the polar stations and detailed survey proper. This arrangement permits the use of forced centering and increases the accuracy of the survey.

The combination of traversing and detailed survey requires extensive reconnaissance because each station must be chosen in the most advantageous position for the detailed survey. In surveying the built-up areas of the city, the polar stations are either traverse points or points determined with the same accuracy as traverse stations.

The spacing of the polar stations should be chosen considering the following points: (1) the accuracy of the instrument and its relationship to the required point accuracy, (2) terrain conditions, (3) control possibilities, i.e., determination of certain points from more than one polar station, and (4) provision of suitable reference points and lines for follow-up surveys.

Precise tacheometers with automatic recording suggest the use of "free-stationing," that is, the polar station not monumented or referenced. In this case, the later processing of the data requires a procedure similar to a block adjustment of aerial triangulation. Obviously, this approach can be applied only to limited areas or in surveys of a temporary nature, since long-term systematic survey work in urban areas always requires a permanent control net of a certain density, which must be either monumented or adequately referenced.

Accuracy Considerations in Detailed Surveys

The accuracy with which the position of details is determined in reference to the nearest control points depends upon the purpose of the survey. A survey for a map, for instance, ought to be conducted such that all well-defined objects can be shown to a specified accuracy within the scale of the map. Assuming this accuracy at ± 0.1 mm as a root-mean-square coordinate error at the map scale, objects to be plotted at a scale 1:1000 should be determined in the field with a root-mean-square coordinate accuracy of ± 0.1 m. For a scale of 1:5000, the corresponding value would be ± 0.5 m. This accuracy can be attained by less precise tacheometric methods.

Since, in an integrated city survey system, original field survey work is performed only once to establish base city maps, considerations of accuracy must be directed to the largest relevant scale (Chapter 9).

A mapping requirement is only one consideration in the establishment of the accuracy of field work. Another important consideration is the definition of physical points or details to be measured. It is obvious that there is little sense in trying to measure points with an accuracy exceeding their natural definition.

The precision with which various types of points can be identified on the ground is given below (5).

Type of point	Identification error in position
Sharp corners of solid buildings	± 1 to ± 2 cm
Boundaries marked with concrete monuments	± 1 cm
Sidewalk curbs	± 1 to ± 2 cm
Permanent fences	± 2 cm
Lamp or other posts	± 5 cm
Wooden fences	± 5 cm
Boundaries marked with wooden posts	± 5 cm
Trees	± 5 to ± 7 cm
Edges of embankments or cuts	± 10 to ± 15 cm
Balks	± 15 to ± 20 cm
Cultivation limits	± 20 to ± 50 cm

It should also be noted that there may be a difference in the importance of positional accuracy depending upon the actual location of points. In some countries, details along the streets are determined with higher accuracy than similar details located in backyards.

The realistic accuracy of current surveying procedures must also be considered. Reference (5) provides the following analysis and numerical values, based on broad operational experience in carefully executed city work:

a. *Rectangular method.* The mean-square error of point determination in position is expressed by

$$m_p{}^2 = m_t{}^2 + m_r{}^2 + m_w{}^2, \qquad (5\text{-}20\text{a})$$

where m_t = root-mean-square error of tape measurements along the control line, ± 1 cm;

m_r = root-mean-square error of ordinate measurements (up to 10 m long), ± 0.5 cm; and

m_w = root-mean-square error due to angular determination using a prism (5′) on a distance of 10 m, including the centering error, ± 2.5 cm.

Consequently, the cumulative error will amount to

$$m_p = \sqrt{1^2 + 0.5^2 + 2.5^2} = \pm 3 \text{ cm.}$$

b. *Polar method.* The mean-square error of point determination in position is expressed by the formula

$$m_p{}^2 = m_d{}^2 + m_\alpha{}^2, \qquad (5\text{-}20\text{b})$$

where m_d = root-mean-square error in distance determination, which for optical precision tacheometers can be expected to be of the order of ± 2 to ± 2.5 cm for average distances of 40 m, and

m_α = position error of about 0.6 cm caused by an error in the determination of direction.

The resulting error m_p is

$$m_p = \sqrt{2.5^2 + 0.6^2} = \pm 3 \text{ cm}.$$

The authors conclude, therefore, that the accuracy of both methods is the same. This result may apply in cities of relatively flat terrain. Hills or rolling ground will substantially reduce the accuracy of the orthogonal method; the polar method is little affected by moderate elevation differences.

The identification error m_i must also be taken into account; the final accuracy of point determination m_f can be found from the formula

$$m_f = \sqrt{m_p^2 + m_i^2}.$$

Assuming, for example, that the boundary point is a permanent fence post ($m_i = \pm 2$ cm), the final accuracy would amount to

$$m_f = \sqrt{3^2 + 2^2} = \pm 3.6 \text{ cm}.$$

In detailed surveys, the relative positioning accuracy of nearby points (neighboring accuracy) is of greater importance than the relative location of points separated by long distances. The reason is obvious: Widths of streets, dimensions of buildings, and sizes of land parcels must be known with better accuracy than the relative location of objects or points separated by 2, 5, or more, km. However, modern control nets permit a positional point accuracy of about ± 2 cm (Chapter 3), and there are urban communities that claim this accuracy in their existing control nets. In any event, it is safe to assume that, in modern control nets, this accuracy could be applied to neighboring control points; consequently, the error caused by tying in a detailed survey to the control net can be neglected if the projection scale of the latter is taken into account, since the effect of the control-net error amounts to only a few millimeters (in our example, $\sqrt{3.6^2 + 2^2} = \pm 4.1$).

Of practical importance is the fact that the error of computed distance D between two points depends only on the relative positional accuracy of those points. In other words, error of a distance derived from coordinates of two end points is independent of the distance between the points. On the other hand, error of a distance measured directly in the field, particularly when a simple means such as tape is used, depends primarily upon the measured distance.

Most existing specifications for cadastral and detailed surveys provide a simplified method for checking the maximum allowable discrepancies (tolerances) between independent length determinations, such as computed distance, and the same distance measured directly in the field or between two or more direct length measurements, by using an empirical formula:

$$\Delta = (c + aD + b\sqrt{D})\text{cm}, \tag{5-21}$$

where c represents the identification uncertainty of the end points, aD represents the effect of systematic errors on length measurement, and $b\sqrt{D}$ represents the effect of accidental errors on length measurement (6). The distance D is in meters, and the parameters c, a, and b depend upon the required accuracy, which is different in densely built-up areas, sparsely built-up areas, and rural areas. It is common, therefore, particularly in cadastral surveys, to specify different accuracy classes. For instance, in some West German states, the following classes are recognized:

$$\Delta_1 = (2.5 + 0.015D + 0.4\sqrt{D})\text{cm}$$
$$\Delta_2 = (5.0 + 0.03D + 0.8\sqrt{D})\text{cm}$$
$$\Delta_3 = (5.0 + 0.05D + 1.2\sqrt{D})\text{cm}.$$

These classes provide the following tolerances, depending upon D:

	$D = 20$ m	$D = 50$ m	$D = 100$ m	$D = 200$ m
Δ_1 (cm)	5	6	8	11
Δ_2 (cm)	9	12	16	22
Δ_3 (cm)	11	16	22	32

More recently, in a German computational cadastre, stricter tolerances were specified,

$$\Delta = (2 + 0.02D + 0.2\sqrt{D})\text{cm},$$

resulting in the following numerical values:

D (m)	20	50	100	200
Δ (cm)	3	4	6	9

The effect of the components on the final Δ is shown in Figure 5-34.

The root-mean-square error of ± 3 to ± 4 cm in position, as quoted above, is, in our opinion, a very good accuracy that more than satisfies most technical and mapping requirements. It is also an accuracy not easy to maintain when simple (but most often encountered) surveying techniques are used. In computational cadastre in urban areas, higher accuracy is suggested. This can be achieved at an operational scale only if suitable electromagnetic surveying instruments and proper computational procedures are used.

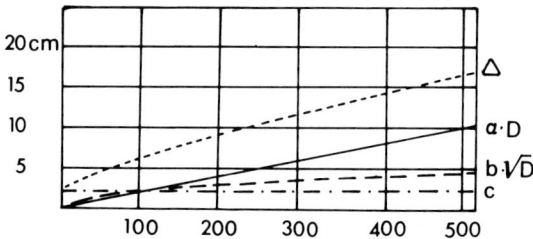

Figure 5-34. Effect of various factors in length measurement.

Surveying by the Extension Method

The term extension method is used to indicate the determination of the location of objects by the extension of their straight elements or boundaries until they intersect survey lines. These extensions are also referred to as range ties.

For this method, a network of survey lines, preferably arranged as triangles, is required. Triangles provide the most favorable arrangement because their sides define unequivocally their form, and the extension can always be tied conveniently into the survey lines. Usually, the field party follows a survey line, measuring the successive intersection points and, at the same time, the dimensions of the surveyed details.

The extension method becomes uneconomical if too many points requiring a large number of survey lines are to be determined. The results are recorded in a field sketch (Fig. 5-35). This record should show all the lines and points surveyed. The features should be drawn approximately to scale in order to avoid errors and misunderstandings during the plotting of the results.

All numbers are written following specific instructions. Uniform symbols, such as those proposed on pp. 195–199, should be used.

Surveying by the Orthogonal Method

For the orthogonal method, the area is divided into rectangular elementary coordinate systems. These systems have a survey line of practical length approximately in the center of the system and many short ordinates perpendicular to the survey line. The ordinates should not exceed a length of from 15 to 20 m to points (such as boundary points and house corners) which are to be read with centimeter accuracy, if the right angles are set up only with the prisms, very convenient in this survey method (p. 186). The ordinates of planimetric details that do not require the same accuracy may be twice this length.

Two different field procedures are used. In the first, the actual field work is divided into three steps. First, the feet of perpendicular offsets are determined and marked. Second, the offsets or ordinates are measured, and also any struts, bracings, or boundary lengths deemed necessary. The survey line is then measured in the running-measurement mode, and all the marked points are

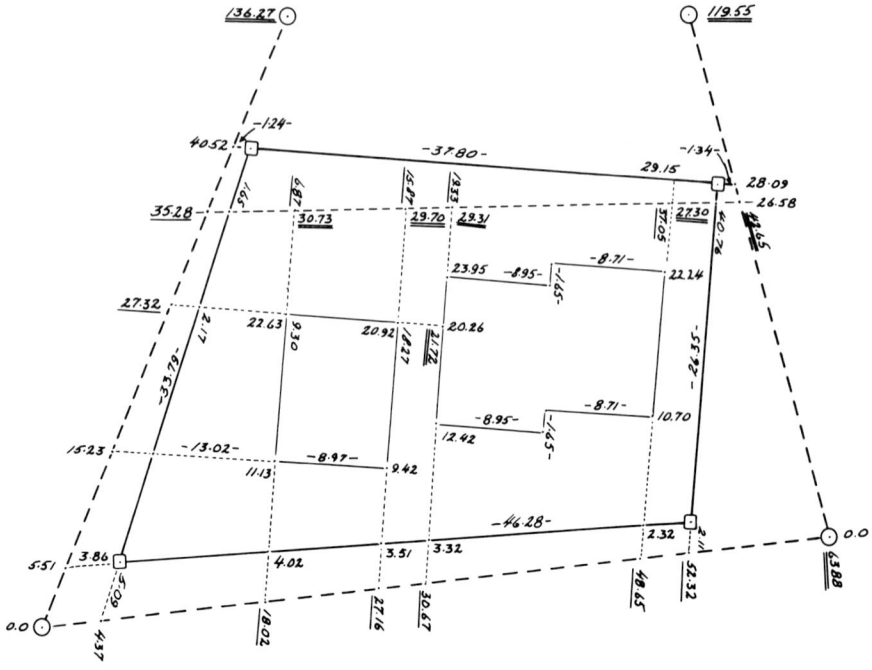

Figure 5-35. Field sketch of a survey by the extension method.

read. Often, the survey line is measured in the opposite direction as a control, without reading all intermittent points.

Another approach frequently used consists of a field party moving along a survey line staked out by range poles and setting off perpendicular ordinates to the planimetric details at both sides of the survey line in the sequence of their occurrence. A prism and a separate, shorter reel tape are used for determinations of ordinates on the survey line and measurements of ordinates, struts, outlines of houses, boundary lengths, etc.

The orthogonal method is particularly efficient if only short ordinates are involved. The results are recorded on a field sketch (Fig. 5-36). The accuracy tolerances given on pp. 203–205 are understood as the maximum positional error for a redetermination of the same point.

Surveying by the Polar Method

The polar method in field survey assures a particularly efficient solution to urban surveys if modern instrumentation is used.

The polar method creates elementary survey systems covering (in a conceptual sense) circular areas. A station and a reference direction are given in each elementary system (Fig. 5-27, p. 192), vectors are measured to each point of interest, and polar coordinates are recorded.

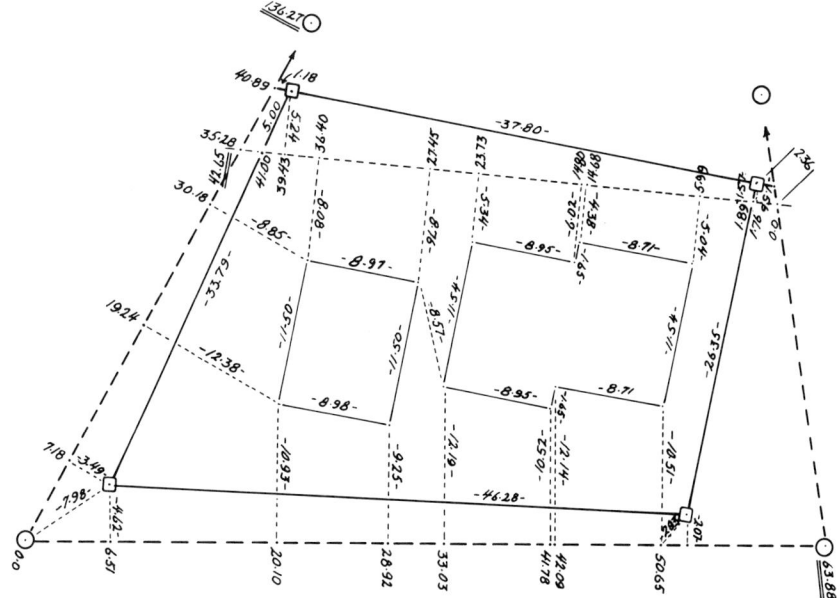

Figure 5-36. Field sketch of a survey by the orthogonal method.

The polar method has the advantage that the reference direction is derived from a visible target at a considerable distance, e.g., from a coordinated church spire. This is important if the survey is not combined with a traverse survey, since the effect of instrument eccentricity decreases as the distance to the reference target increases.

A survey by the polar method can progress rather rapidly since all measurement readings are done on the polar station only. When using optical or electro-optical distance measurements, the method is relatively independent of the terrain. Modern instrumentation permits vector lengths of up to 1 km within the accuracy requirements for legal and most engineering purposes. Even though this facility cannot be taken advantage of in urban areas, it provides a welcome flexibility in field operations.

Distance measurement is of critical importance. All tacheometers used for the polar method require a special target on the point to be measured. This target can be a special staff which is often awkward to handle. For tacheometers utilizing electrooptical distance measurement, a reflector, or a set of reflectors, is usually required. These reflectors are set in a housing. Staffs and reflectors cannot be placed centrally, e.g., over a vertical edge of a house. This difficulty can be overcome by placing the target in a well-defined eccentric position, as indicated in Figure 5-37. However, the eccentric positions used must be recorded.

The efficiency of a polar survey depends largely upon the engineer responsible for the selection of the points to be surveyed. The engineer must produce a

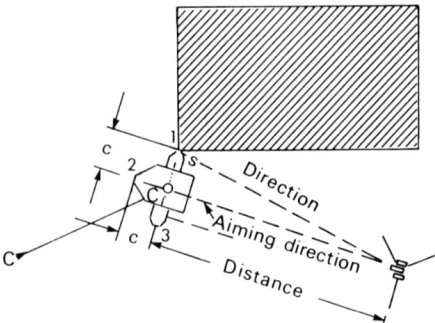

Figure 5-37. Measurement by the polar method. The aiming center C of the tacheometer can be placed in a predetermined eccentric position away from the house edge in the correct direction by reference point 2 or it can become the foot of a perpendicular offset to the measured direction by reference points 1 or 3. All three reference points are points on the reflector housing to be placed directly on a vertical edge. According to Ref. (7).

field sketch showing all survey points with their numbers and ensure that these numbers are also entered in the numerical field record containing all measured quantities or polar coordinates. It is advisable to prepare beforehand a sketch of the major objects in the area to be surveyed, approximately to scale. The preparation may also involve the transfer of existing extension or photo-grammetric surveys into polar coordinates to check the agreement between various operations.

The polar method is well suited to subsequent processing because each point can be independently computed from the polar station.

A relatively dense net of stations is required because of the complexity of urban areas. These stations can be existing control points or stations determined prior to or in conjunction with the detailed survey.

Vertical stadia surveying. Vertical stadia surveying, i.e., surveying by the polar method of lower accuracy, is worth mentioning because of its simplicity and applicability to surveys of topography of the terrain. Since distances are read off a simple vertical rod by using the stadia hairs, the inconvenience of the central placing of the staff or reflector on a vertical edge is eliminated or alleviated.

Vertical stadia surveying in urban areas is particularly convenient if a map with contour lines for small areas is required. The maximum object distance is dependent upon the map scale. The following values are orientation figures only.

Map scale	Maximum distance (m)
1:500	100
1:1000	150
1:2000	200
1:5000	300

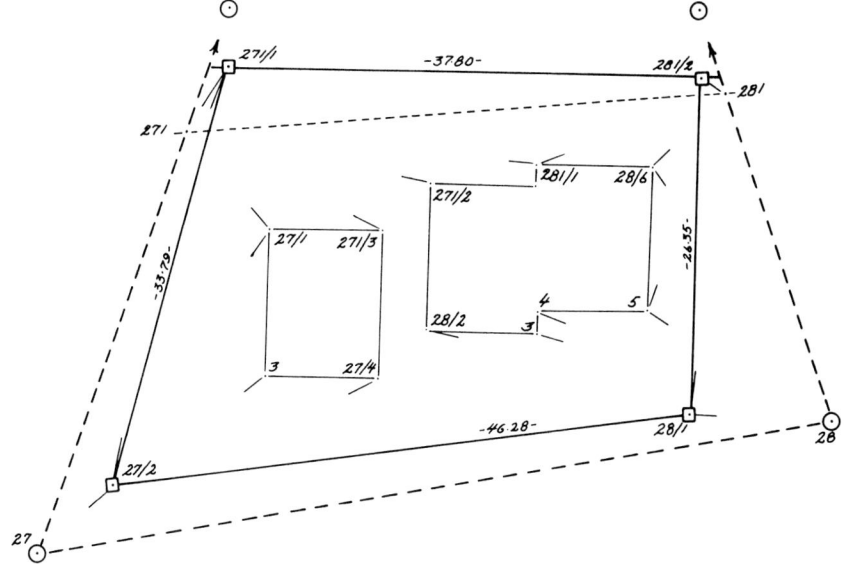

Figure 5-38. Field sketch of a survey by the polar method.

The points to be surveyed are selected by the survey crew chief, who prepares a field sketch indicating point locations, point numbers, and if necessary, topographical features, and the direction of fall lines (Fig. 5-38). Measured directions and distances are recorded in a suitable form for further processing. The results can be plotted directly in the field for each station by means of a plotting table attachment.

Horizontal stadia surveying. When using precision self-reducing optical tacheometers with horizontal stadia, an accuracy of the order of 2 to 3 cm can be achieved if the measured distances are not extended over 100 m. This is sufficient for most technical and legal surveys. However, use of horizontal stadia is cumbersome and is critically affected by atmospheric conditions. Strong winds and insolation (with the associated image vibrations) are particularly adverse conditions.

The field procedure is much the same as in surveys with vertical stadia. In the surveying of vertical features with an eccentric location of stadia, an approach similar to that presented in Figure 5-37 must be taken.

Electronic tacheometers in polar surveys. Electrooptical distance measurements supersede the somewhat awkward and inconvenient horizontal staff. Similarly, the automatic recording of distances and angles improves the efficiency of optical tacheometers. Several instruments that combine distance-measuring capability with the features of the classical theodolite have been developed recently (Chapter 3). These instruments contain such familiar features as telescopes that rotate through the zenith points, clamps and tangent screws for horizontal and vertical movements, forced centering, and a compensator for the vertical circle index.

Automatic recording of readings and automatic processing of survey data results in an instrument for field surveying that is almost universal in application. Automatic recording eliminates some of the potential gross errors, e.g., interchange of distances or interchange of digits within a value. It also speeds up measurements significantly since manual recording of the numerical data is no longer required.

Electronic tacheometers have the same error characteristics as ordinary theodolites. The effects of these errors are usually eliminated through repeated measurements. However, since human recording errors have been removed and single measurements suffice to obtain desirable accuracy, the surveyor should carry out a few test measurements at each station before commencing the actual survey. These measurements permit the correction for instrument errors of a reading taken in only one telescope position.

If the traverse is to be measured together with all detail points, repeated direction measurements for the determination of traverse points are required. On the other hand, the instruments can be used advantageously in the free-stationing mode, where each set of data measured from one station is combined with overlapping sets from other stations and subjected to a block adjustment. The polar station is only one of the many points, and special accuracy precautions need no longer be taken. It has been shown (8) that the free-stationing mode yields roughly the same accuracy as point determination from predetermined polar stations.

An automatic recording of angles and distances eliminates some human errors. It does not eliminate discrepancies between numerical and graphical field records, however, and these may be quite confusing. A graphical plot produced from the automatic records is usually sufficient to localize blunders and eliminate discrepancies.

Further progress in this important new technique is expected.

Controls in surveys by the polar method. Points that are to be determined reliably and with a high accuracy should always be controlled by additional measurements. Several possibilities can be considered. An obvious checking procedure is to carry out the survey from two different stations. This control is thorough and completely independent. However, conditions in urban areas often prevent this approach.

Another, and possibly more practical, approach consists of repeating the whole polar measurement operation. When measuring the second time, a different reference direction should be used for angular measurements, the zero point of the vertical staff should be changed and another vernier used if possible on the horizontal staff.

Another checking procedure is the measurement of distances between the points determined by the polar method. This approach, however, is too laborious and cumbersome to be practical.

It should be noted that the first and third control procedures cannot be used for direct control in the field unless a pocket calculator is used to compute coordinates for each surveyed point. Since the polar method does not provide a simple procedure for checking results directly during field operations, gross errors are likely to occur more frequently than in a survey by the orthogonal method. In addition, the dual field records, which cannot be avoided in the polar method, constitute a further source of gross errors. Two major types of gross errors are the interchange of point numbers and the interchange of distances. Particular attention should be paid to these errors.

Point Numbering

The numbering of detail points is indispensable for the following three reasons:

1. the point number identifies the same point in graphical and numerical field records,
2. the point number is required for electronic processing of field measurements and for coordinate computation, and
3. a point number is used for coding different types of field data.

Although there are several different ways of numbering points, only variants of one numbering procedure will be discussed (9). This procedure consists of assigning two-part point numbers—a base number and a unit number.

The base number refers to a numbering district (such as a polar station with all points measured from it), a jurisdictional area (such as an area within a municipal boundary), or an area outlined on a map or a field sketch. The unit number is a sequential number within the numbering district.

Station number as base number. Station numbers are assumed to be given. A station is considered a monumented point that is part of the control net. Each point measured from the station is assigned a general sequential number that usually does not exceed two digits. The relationship between station and point must be unambiguously defined. Figure 5-39 shows a section of a graphical field record of a polar survey in which these relationships are indicated by small arrows. For the purpose of averaging, identical points measured from several stations should only be given one number. This number should be assigned when the point is measured the first time, for example, the points 2 and 8 in Figure 5-39 are labeled 123/2 and 123/8 since they were measured first from station 123 and then from station 124.

Another solution for recording the relationship between the base number and the unit number is shown in Figure 5-40. Most points have been measured from station 81, but some have been measured from the other three stations. It should be noted that the first and last point of a sequence of points is given a

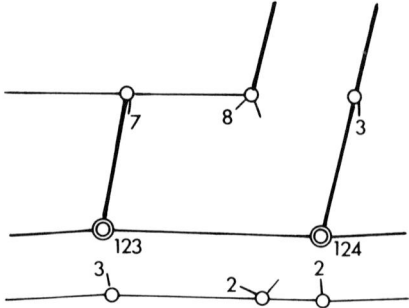

Figure 5-39. Section of a polar field sketch showing two stations 123 and 124 and points measured from these stations. According to (9).

base number and a unit number (e.g., points 13/1 and 13/5); the points in between are given only the unit number (points 2, 3, and 4).

This numbering procedure can also be applied to the orthogonal method by using either the starting point of the survey line or the combined numbers of the starting and end points of the survey line as base number. The procedure can also be applied to photogrammetry, with the photograph number as the base number.

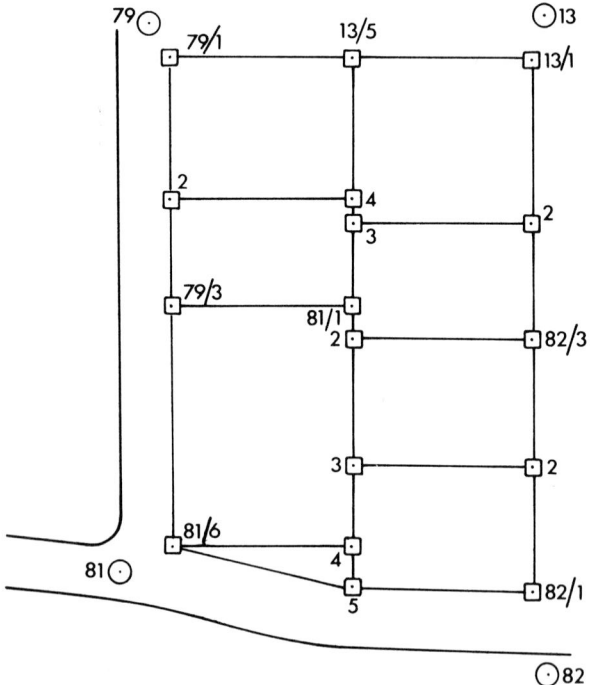

Figure 5-40. Field sketch of a polar survey of boundary points.

Jurisdictional-area number as base number. Jurisdictional areas may be municipalities or parts of municipalities. Depending upon their size and the density of points to be measured, the unit number can become rather large. Points located on the jurisdictional boundary should also be numbered only once. The numbering of control points causes some problems, and this is discussed on pp. 213–214.

Field-sketch number as base number. In this case, an area to be surveyed is broken down into smaller areas that can be shown in a single sketch.

These records, which will be filed as reference for later surveys, should be assigned a number. This number can also become the base number. If this solution is chosen, the numbering of control points that already have a number causes certain problems (see the following section). Also, points will appear on overlapping field records; these points must be given one number only (Fig. 5-41).

The number of points located in each record will often exceed 100 so that three digits are required for the unit number.

Comparison of the different numbering systems. All three numbering systems have advantages and disadvantages.

Preparation. The use of the station number as base number requires the least preparation; only the station number needs to be known. The use of jurisdictional districts also requires that all points located on jurisdictional boundaries must be identified, and these points must be checked lest numbers have already been assigned to them. If field-sketch numbers are used as base numbers, the boundaries of each sketch must be determined before the survey, and all points appearing in more than one record must be identified and checked for prior numbering.

Follow-up surveys. Numbering in follow-up surveys depends on whether the original station was used as a reference. If a station disappears, the numbers given to the existing points remain unique, but the numbering system becomes

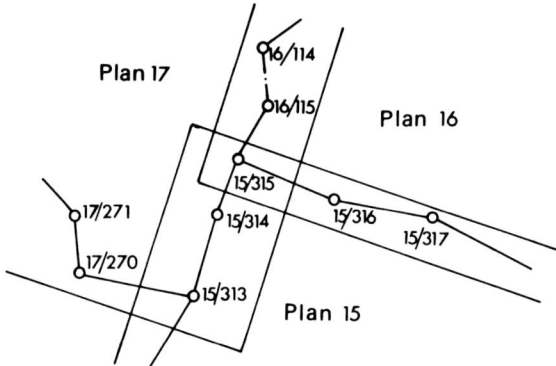

Figure 5-41. Point numbering in an area with overlapping field records. According to (9).

meaningless unless all points referenced to this station are rereferenced and renumbered. This does not apply to referencing to either a jurisdictional district or a graphical field record area.

List of point coordinates. The lists of point coordinates change every time new points are added. Referencing points to stations results in short lists that can be easily extended. If the area shown on one field sketch is used as a reference area, additional points can also be added easily but, because of the larger unit numbers, the point numbering becomes less clear. This is even more true if a jurisdictional district that covers an area considerably larger than that covered by a single field sketch is used as a reference area.

Control measurements. Struts, bracings, and boundary lengths, in addition to the measurements taken to determine a point location, are often measured in order to verify the correctness of the determination of points significant to legal boundaries. These measurements are often entered on special forms in which the use of the station as reference requires frequent changes in the base number, a situation which the other two solutions avoid.

Survey method. The use of the station as reference is obviously well suited to the polar method and photogrammetry but is not equally suitable for other field surveying methods. This differentiation does not apply to the use of either a jurisdictional district or the graphical field record as a reference area for assignment of the base number.

Filing of survey results. The base number can also be considered as a number for the filing of survey results, particularly if efforts are made to assign sequential base numbers. Depending upon the office organization, the station or the graphical field record can serve as a reference for filing. Use of the station is preferable if the emphasis is placed on filing the numerical data and having a set of reference graphical records. The station may disappear however, thus making the file more or less inoperative.

Filing according to the area covered on a graphical field record eliminates the requirement for additional reference records. Also, the reference area is less likely to be redefined.

The use of a jurisdictional district as a reference area does not offer filing advantages. The reader may have other criteria for determining the usefulness of the various reference systems. The selection of any one system will eventually depend upon the organizational structure of the survey as a whole.

The numbering of surveyed points. Point numbers play an important role in filing and further use of the survey data. They should (a) hinder the survey process as little as possible, (b) be as short as possible so that they can be assigned quickly and with a high certainty, (c) be applicable to all surveying techniques, (d) not hinder later electronic data processing, (e) be easy to locate in numerical and graphical field records, and (f) be well adapted to the requirements of follow-up surveys. In addition, numbers should be easily derivable from existing control

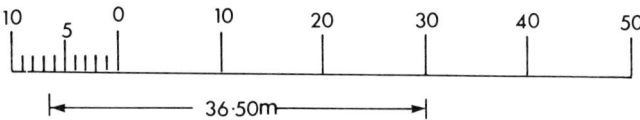

Figure 5-42. Linear scale, 1:500.

point numbers, allow for changes in control point numbers, and permit renumbering of small areas if the existing numbering becomes obsolete.

Since point numbering occurs in the field, the party chief must take all necessary steps to make the numbering system efficient and easy to use without mistakes.

Further Processing of Terrestrial Measurements

Field measurements are practically always further processed after they are recorded. We shall consider the most frequently encountered processing operations and distinguish between plotting, simple numerical processing, and electronic data processing.

Graphical plotting. Graphical processing of the data contained in the field records requires that the control points on which the survey is based have been plotted on suitable material (map sheet).

Survey results obtained with the extension method can be plotted even if only a ruler is available. The distances can be determined from a linear scale (Fig. 5-42) or, more accurately, from the transversal scale (Fig. 5-43). In the first case, only a needle or pencil is required to mark a point after the ruler has been placed at one end point of the distance on a round value (30 m in Fig. 5-42). In the second case, a compass with two pointed branches is required for taking the distance off the transversal scale and transferring it to the plotted plan.

Field data from the orthogonal method can be plotted by first transferring all the abscissa values to the plan and then erecting perpendicular lines at their end points and setting up the ordinates. This procedure separates the abscissas from the ordinates and is therefore somewhat error prone. A preferable method is to plot with an orthogonal plotting device.

Accurate plotting of measured directions requires a precise protractor. If polar coordinates are to be plotted, arrangements must also be made for the setting up of distances. This can be done with a simple protractor having a scale along a diameter or with a somewhat more sophisticated device. It is

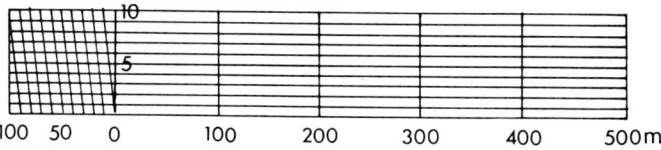

Figure 5-43. Transversal scale, 1:5000.

noted that, in the first case, the protractor must be recentered for each change in direction or held in a centered position by a needle placed in the circle center; in the second case, the protractor and the attached scale rest on a base plate.

When all the points from one elementary survey system, e.g., a survey line or a polar station, have been plotted, struts, bracings, and boundary lengths measured as controls should be checked. In some instances, points may have been determined from more than one survey line or polar station. In these cases, their location will be checked by setting up the data from the second elementary survey system.

When all points have been plotted and checked, they should be connected by lines if so required and the resulting features should be identified by proper signatures. Examples can be found in Chapter 9, which deals with city maps.

Simple numerical processing. Throughout this chapter, reference has been made to various computations required in the course of surveying. These computations include intersection with distance measurements, check of orthogonal elements with struts, intersection with angular measurements, resection, determination of obstructed distances, slope correction, averaging of measured directions, comparison between two orthogonal surveys taken from the same survey line, and comparison between two polar surveys taken from the same station. These are all well-known problems that can be solved by relatively simple computations.

In many survey offices, coordinates of surveyed points must be computed for use in the plotting of map manuscripts and for storage as final surveying results. These are simple but very massive computations involving large numbers of points; therefore, the computational procedure is outlined here. We shall consider only the three most frequently encountered cases: points located on a survey line, points located at a given perpendicular distance from a survey line, and transformation from polar to orthogonal coordinates.

Computation of points located on a survey line. Figure 5-44 shows a survey line AD with two intermittent points B and C located in a conformal plane coordinate system X, Y. The coordinates of points A and D and the measured distances a, b, and c are given. First, $s = \sqrt{(X_D - X_A)^2 + (Y_D - Y_A)^2}$ is computed and compared to $S = a + b + c$, and the difference is checked. The coordinates of points B and C are computed as

$$
\begin{aligned}
Y_B &= Y_A + \Delta Y_A & X_B &= X_A + \Delta X_A \\
Y_C &= Y_B + \Delta Y_B & X_C &= X_B + \Delta X_B
\end{aligned}
\tag{5-22}
$$

with

$$
\begin{aligned}
\Delta Y_A &= \frac{Y_D - Y_A}{S} \cdot a & \Delta X_A &= \frac{X_D - X_A}{S} \cdot a \\[2mm]
\Delta Y_B &= \frac{Y_D - Y_A}{b + c} \cdot b & \Delta X_B &= \frac{X_D - X_A}{b + c} \cdot b.
\end{aligned}
\tag{5-23}
$$

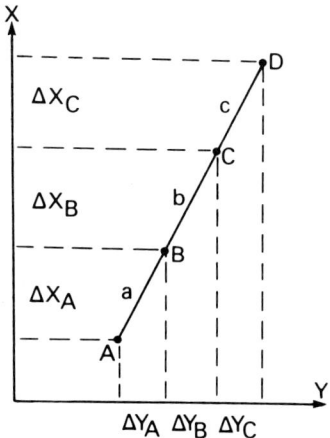

Figure 5-44. Computation of points located on a survey line.

The computation can be checked with the following equations;

$$Y_D = Y_C + \Delta Y_C \qquad X_D = X_C + \Delta X_C. \qquad (5\text{-}24)$$

It is noted that $Y_D - Y_A$, $X_D - X_A$, and ΔY, ΔX are in the scale of the ground plane coordinates and that a, b, and c are actual measurements. Hence the formulas include a scale adjustment.

Computation of points located on a perpendicular distance to a survey line. Figure 5-45 shows a survey line AB and a point C located in a conformal plane coordinate system X, Y. The coordinates of A and B and the measured values a, b, and c are given. First, point D is derived in the manner discussed in the

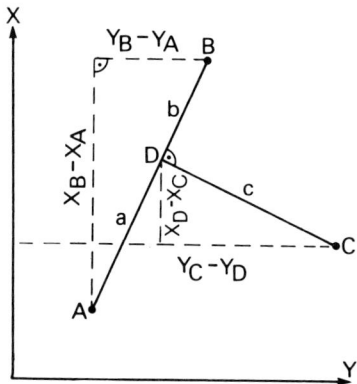

Figure 5-45. Computation of points located on a perpendicular distance to a survey line.

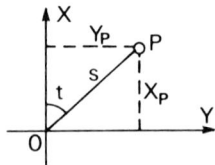

Figure 5-46. Rectangular and polar coordinates.

preceding section. Because direction $D \rightarrow C$ is perpendicular to $A \rightarrow B$, we have, from similar triangles,

$$Y_C - Y_D = \frac{X_B - X_A}{a + b} \cdot c \qquad X_D - X_C = \frac{Y_B - Y_A}{a + b} \cdot c. \qquad (5\text{-}25)$$

The differences $Y_C - Y_D$, $X_D - X_C$, $Y_B - Y_A$ and $X_B - X_A$ are given in the scale of the coordinates; $a, b,$ and c are actual measurements. Hence the formulas include a scale adjustment.

Transformation of polar coordinates into rectangular coordinates. Figure 5-46 shows the relationship between geodetic grid coordinates northing (X) and easting (Y), direction (t), and distance (s) from the origin.
With given direction t and distance s, the grid coordinates within an elementary system centered on station 0 can be derived as

$$Y = s \sin t \qquad X = s \cos t \qquad (5\text{-}26)$$

The reverse transformation can be carried out with

$$s = \sqrt{X^2 + Y^2} \qquad \tan t = \frac{Y}{X}, \qquad (5\text{-}27)$$

the quadrant of t being determined according to the algebraic signs of Y and X.

Electronic data processing. Electronic data processing goes beyond simple computation of required values. Owing to the very high speed and reliability of electronic computations, a much greater volume and a higher complexity of computations can be considered. For instance, overall adjustment of surveys can be carried out that could permit the use of free polar stations or free survey lines and thus save expensive field work.
Ease of electronic computations permits the inclusion of the computation of check measurements (bracings, boundary lengths) into overall adjustment, whereby higher accuracy of final results can be expected.
Computers are also used for the storing of survey and associated data and automatic plotting, editing, and updating of maps. Owing to the large volume and complexity of surveys in urban areas, stringent accuracy requirements, and frequent alterations, electronic data processing is becoming increasingly attractive on a worldwide scale.

References

1. Szymański, J. *Instrumentoznawstwo Geodezyjne* (Geodetic Instrumentation), Warsaw, 1956.
2. Apel, H. Die Ausführung von Fortführungsvermessungen mit Hilfe polarer Messungselemente, *Allgemeine Vermessungs-Nachrichten*, Heft 10, 1968.
3. *Zeichenvorschrift für Katasterkarten und Vermessungsrisse*. Hessisches Landesvermessungsamt, Wiesbaden, 1969.
4. Dekker, H., and Quee, H. A System of Complete Automated Calculation and Mapping, *Proceedings of the XIVth Congress of FIG, Commission 5*, Washington, D.C., 1974.
5. Bramorski, K., Gomoliszewski, J., and Lipiński, M. *Geodezja Miejska* (Urban Geodesy), Warsaw, 1973.
6. Wahl, B. J. Technical Features Essential to a Modern, Multi-Purpose Land Data System, *Proceedings of the North American Conference on Modernization of Land Data Systems*, Washington, D.C., 1975.
7. Aeschlimann, H. Kern DM 500, ein neues elektronisches Tachymeter, *Vermessung, Photogrammetrie, Kulturtechnik* No. 3, 1974.
8. Ackermann, F. Ergebnisse einer Programmentwicklung zur Blockausgleichung grossräumiger Polaraufnahmen. *Technik der elektrooptischen Tachymetrie*, Sammlung Wichmann, Karlsruhe, 1973.
9. Anon. Anleitung zur Durchführung von Grundbuchvermessungen mit automatischer Datenverarbeitung, *Schweizerische Zeitschrift für Vermessung, Photogrammetrie und Kulturtechnik*, No. 7, 1970.

Additional Readings

Bartels, et al. *Das Vermessungswesen in der Praxis*. VEB Verlag Technik, Berlin, 1957.
Birchal, H. F. *Modern Surveying for Civil Engineers*. Chapman & Hall Ltd., London, 1955.
Breed, C. B. *Surveying*. John Wiley & Sons, New York, 1971.
Brinker, R. C., and Taylor, W. C. *Elementary Surveying*. International Textbook Company, Scranton, Pa., 1956.
Davis, R. E. *Elementary Plane Surveying*. McGraw-Hill Book Co., New York, 1955.
Davis, R. E., and Foote, F. S. *Surveying Theory and Practice*. McGraw-Hill Book Co., New York, 1953.
Deumlich, F., and Seyfert, M. *Instrumentenkunde der Vermessungstechnik*. VEB Verlag Technik, Berlin, 1957.
Elektronische Tachymetrie, Herbert Wichmann Verlag, Karlsruhe, 1971.
Grossmann, W. *Vermessungskunde, I*. Walter de Gruyter, Berlin, 1972.
Grossmann, W. *Vermessungskunde, II*. Walter de Gruyter & Co., Berlin, 1971.
Grossmann, W. *Vermessungskunde, III*. Walter de Gruyter & Co., Berlin, 1969.
Gruber, O. von. *Optische Streckenmessung und Polygonierung*. Herbert Wichmann Verlag, Berlin, 1955.
Jordan, W. et al. *Handbuch der Vermessungskunde, Band II*. J. B. Metzlersche Verlagsbuchhandlung, Stuttgart, 1963.
Kissam, P. *Surveying for Civil Engineers*. McGraw-Hill Book Co., New York, 1956.
Legault, A. R., McMaster, H. M., Marlette, R. R. *Surveying, An Introduction to Engineering Measurements*. Prentice-Hall, Inc., Englewood Cliffs, N.J., 1956.
Middleton and Chadwick. *A Treatise on Surveying, I*. E. & F. N. Spon Ltd., London, 1955.
Middleton and Chadwick. *A Treatise on Surveying, II*. E. & F. N. Spon Ltd., London, 1955.

Rayner, W. H., and Schmidt, M. O. *Surveying, Elementary and Advanced.* D. Van Nostrand Co., Princeton, N.J., 1957.

Skelton, R. H. *The Legal Elements of Boundaries and Adjacent Properties.* The Bobbs-Merrill Company Publishers, Indianapolis, Ind., 1930.

Smith, J. R. *Optical Distance Measurement*, Crosby Lockwood & Son Ltd., London, 1970.

VEB Carl Zeiss Jena Nachrichten, Heft 5–8, April, 1957.

Technik der elektro-optischen Tachymetrie, Herbert Wichmann Verlag, Karlsruhe, 1973.

Whyte, W. S. *Basic Metric Surveying.* Newnes–Butterworths, London, 1969.

Zill, W. *Vermessungskunde für Bauingenieure, I.* B. G. Teubner Verlagsgesellschaft, Leipzig, 1956.

Zill, W. *Vermessungskunde für Bauingenieure, II.* B. G. Teubner Verlagsgesellschaft, Leipzig, 1958.

Chapter 6

Utility Surveys

INTRODUCTION

The term "utility" encompasses the city water and sewer system and all installations connected with gas, heating, electricity, and telecommunication. The utility survey deals with the mapping and descriptions of the overhead and underground servicing lines such as ducts, cables, pipes, and associated elements such as manholes, poles, catch basins, transformers, and hydrants.

The term "infrastructure" is sometimes used instead of the word utility. Generally, however, infrastructure has a broader meaning. It identifies, under one heading, the whole physical plant necessary to support community life.

As the utilities of modern cities become not only more complex and costly but also more essential, the need for a complete, accurate, and up-to-date integrated system for supplying information on existing underground and overhead utilities is becoming apparent. There is a particular need for positional information on all pertinent elements in a form such that an operations manager, a planner, or an administrator can be presented with a clear picture of the current situation and any proposed changes. Surveyors play a key role in the establishment of a relevant information system since the *location* of various utilities is the basis of the information system. This does not imply that there are not many other important features that must be recorded. It is recognized that type of material, load capacity, date of installation, maintenance record, and the like are all important items to be included in the surveys. *Nevertheless, only the precise location of various installations permits proper correlation of the pertinent data in an unequivocal manner and formation of an efficient information system.*

The location (positioning) of utility elements is not a difficult task from the point of view of surveying techniques; however, it is time consuming, and problems may be encountered in the location of old underground lines. The main problem is how to organize the collection, storage, and retrieval of the large amount of data on positions and important features of various elements so that the information can be continuously updated and is complete, accurate, and easily accessible. Moreover, the construction of underground and overhead utilities in urban areas is doubling approximately every 10 years.

The initial cost of establishing an information system is high, but the experience of cities that already have some kind of working system shows that the cost is quickly recovered. The integrated information system provides benefits to the municipality by shortening the time necessary for the planning and designing of new utilities, allowing construction in areas where existing utilities have not been used to their full capacity to be avoided, minimizing damage to existing lines by reconstruction and current engineering projects, and allowing better and more rapid planning of new general urban projects and improvements, easier maintenance and faster repair of the services, and a significant increase in the general safety factor for construction workers.

The need for a systematic documentation of underground utilities was recognized in some European cities more than 60 years ago. The first systematic cadastre of underground utilities was established in Switzerland in the small city of Olten in 1915, followed by the initiation of mapping by the city of Basel in 1917. Almost all large cities in the world have some kind of documentation of the underground conduits, but only a few have a reasonably efficient *integrated* information system that is well maintained, up to date, and satisfies the need of agencies, planners, engineering companies, and other users.

Much research and experimental work is being done on the subject in various countries. An international study group on mapping of underground utilities was established several years ago within Commission VI (Engineering Surveys) of Féderation Internationale des Géométres to bring together those interested in these problems.

The large number of cities that are involved in the development of "integrated utility information systems" and the large number of conferences, symposia, and discussions on the topic are the best indicator of the importance of this problem to modern cities.

SURVEYING OF UTILITIES

Surveying Methods and Accuracy

Determination of the position of utilities is usually accomplished by locating the axes of the lines and central points, such as manholes, catch basins, transformers, hydrants, exchange boxes, and poles. If the width of the service line or diameter of other details exceeds 0.5 m, the edges or outside contours should be mapped.

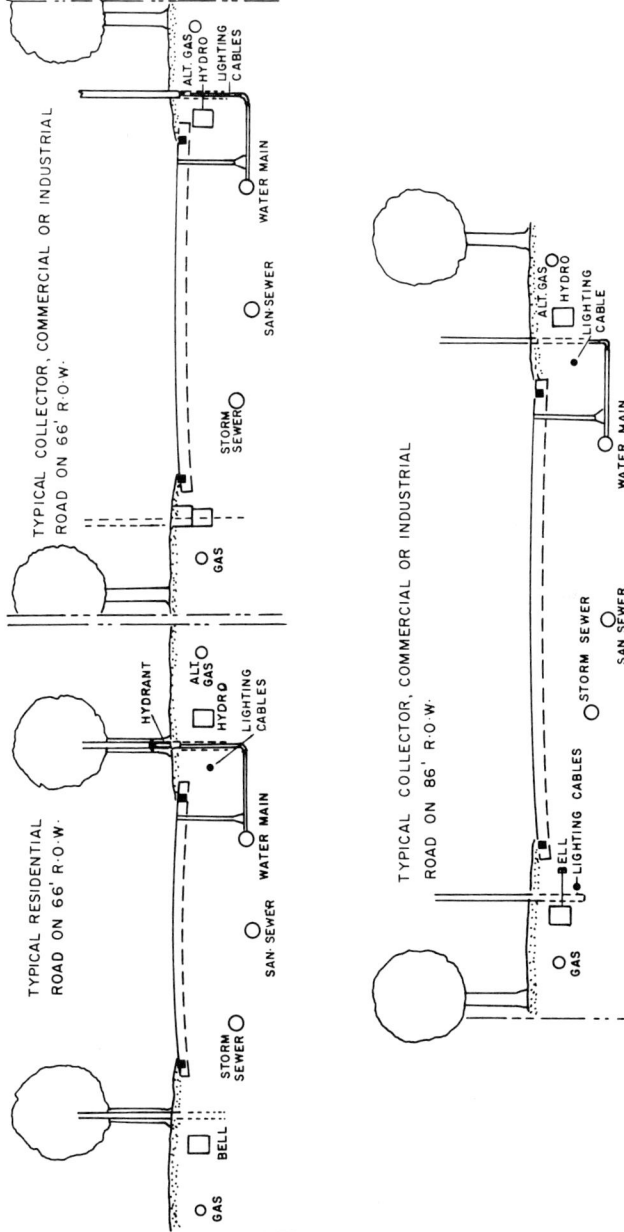

Figure 6-1. Typical distribution of service lines in the city of Ottawa.

Location of the axes should include all axial points at which the lines change their horizontal or vertical direction. If several cables are placed together in one line, the location of the center line should be found, supplemented by a description of the number and type of cables included in the line. If the line is very irregular, as it may be with electric or telecommunication cables, the location of the axis may be generalized by an averaged straight line unless the deviations of the line are larger than ± 20 cm.

Most cities have regulations on the minimum and maximum depth of utility lines as well as the allowable minimum distances between individual lines and distances to building foundations. Figure 6-1 gives a typical distribution of service lines in the city of Ottawa for typical cross sections of streets and roads. These regulations cannot, however, be taken as the basis for an eventual relocation of utility lines once they have been constructed and covered. The experience of many cities shows that the actual location, so-called "as-built," usually differs considerably from the design, particularly in the horizontal plane.

Some larger cities find it economical to build utility tunnels in which several different lines are placed together. Positioning becomes very simple because only the tunnel is being mapped, supplemented by a cross section of a typical distribution of individual conduits in it (Fig. 6-2).

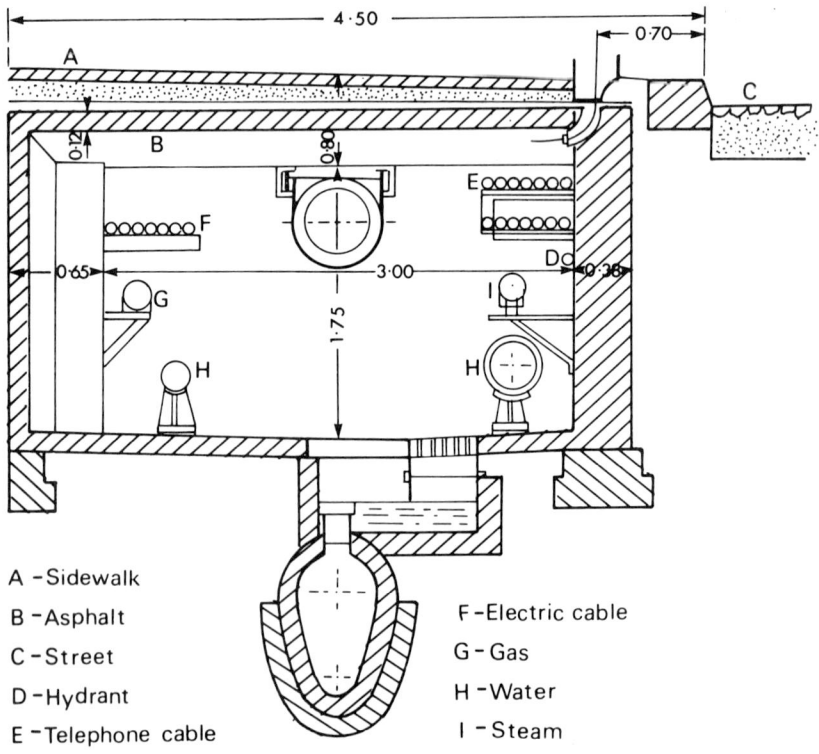

A – Sidewalk

B – Asphalt

C – Street

D – Hydrant

E – Telephone cable

F – Electric cable

G – Gas

H – Water

I – Steam

Figure 6-2. Utility tunnel in the city of Hamburg.

The accuracy specifications for the location of utilities differ from one city to another, ranging from a few centimeters to 30 cm. Most users wish to know the location of the axial and detail points with a standard deviation of the order of 10 cm in both the vertical and horizontal directions with respect to surrounding identifiable details on the surface, such as corners of buildings, curbs, or surveying control points. The only exceptions are drains with a gravitational flow, such as sewer lines for which the elevation in flat areas must be known with a standard deviation of the order of a few millimeters.

Position information on utilities must refer to their location "as-built." Therefore, the surveyor in charge of supplying the information must obtain it by a direct location survey. This does not create any technical problems in the case of overhead lines or details of underground installations that are visible on the surface. Conventional ground survey techniques such as the polar method, with EDM or optical distance measurements, or the orthogonal method, are used. Photogrammetry can supply from 20 to 40 % of the information.

Underground utilities should be surveyed immediately after construction, before they are covered. Many cities have very strict regulations in this regard. However, this creates some organizational problems (see the following section). Moreover, cities that are developing an integrated surveying system (cadastre) must first collect the positional information on all existing underground features. This information can be compiled from the following sources:

> *branch materials* from individual agencies based on existing sketches of the location, design plans, maps (if any), and, finally, the memory of the persons involved in the construction of the utility;

> *direct field surveys* that may require very tedious measurements of utility lines under very cramped and sometimes dangerous conditions, which use as an access manholes, catch basins, or specially excavated ditches.

The branch material is usually unreliable, and it should never be accepted as the sole source of positional information. Even if old branch maps of utilities exist, they must be verified by field surveys. These maps are not oriented to the city coordinate system. Branch materials, however, are indispensable as supplementary information for field surveys.

Field surveys of utilities are most time consuming. According to the city of Bern, Switzerland (1), it takes over 20 man-hours of field work to position utilities over the area of 1 ha. This does not include the time spent in searching for supplementary information from branch materials. The utilities involved are sewage, telephone, cablevision, electricity, gas, and water.

Electronic locators are of great help in positioning old utilities. A detailed discussion of different types and accuracy of the locators is given in Ref. (2). Katona (3) gives a good concise review of the practical applications of electronic locators. Many models are available on the market, for instance, LR-1 and STH-2, produced in Poland, Telmes TT-2150, produced in Hungary, or ESP (electromagnetic subsurface profiling) system, developed by Geological

Survey Systems Inc. in the United States. Locators allow the determination of the horizontal components of the axial points of the lines with a standard deviation of the order 10 to 20 cm and determination of the depth with a standard deviation of 20 to 40 cm. Their use is usually limited to a certain type of utility, depending on the model of the locator and conditions under which it must be located. These details should be discussed with the manufacturers before purchase.

Organization of Data Collection

The positional data must be collected by competent survey technicians who, beyond having a good knowledge of surveying techniques, understand the functioning of the utilities and their construction elements. In some cities, the collection of positional data is done separately under the supervision of specialized surveyors employed by respective agencies.

On the other hand, the whole idea of an integrated information system is that uniform positional data on the utilities of individual branches, in a common reference system, should be available in one central office and kept continuously up to date. Experience shows that optimal results are obtained when the collection of positional data, processing, and updating of the integrated information systems is done by one central surveying office.

An extreme example of the centralization of the utility information system can be found in the city of Warsaw (4), where the city survey office is responsible not only for mapping and updating but also for staking out utility projects.

In many cities, such as Montreal, Canada, and Bern, Switzerland, a separate branch of the municipal survey office is responsible for mapping all newly constructed utilities as built and for keeping the integrated information system up to date. Each agency is obliged to inform the surveying office that on a specified day in a specified place new elements will be laid down. The survey office is obliged to send its survey crew to carry out the as-built positional surveys before the utilities are covered. Expenses of the surveys and of updating the information system are shared between the individual agencies and the city. This type of organizational survey seems to work well.

CADASTRE OF UTILITIES

The cadastre contains the integrated positional information and descriptions of the utilities, which are correlated with the property boundaries. There are no generally accepted rules or specifications on the content of the description portion of the cadastre. This depends on the local needs of the users and of the utility agencies. Therefore, the content should be discussed and agreed upon before the establishment of the cadastre.

The utility cadastre may also have either graphical or numerical (computerized) characteristics, or a combination of the two. Most old working systems were based on large-scale "strip maps" along the streets, in scales of 1:500, 1:250, or the like, that were not always correlated with topographic and property maps. In the 1960s, most cities with advanced cadastres started changing from the "street map" system to grid maps of 1:500, 1:250, 1:200, based on enlargements of a common base city map, which is usually at scales of 1:1000 or 1:500.

Recent years have brought intensified research in the computerization of information systems and a trend toward numerical storing and processing of the data, but most working systems are still based on conventional graphical procedures and manual operations.

In an idealized system (5), every relevant piece of information on all utilities in a given area, such as a metropolitan region, is stored in a computer memory. From this stored information, computer-drawn maps can be created or any selected data can be called up on a visual display unit (VDU) and a hard copy (a paper print) made. The selection may encompass all details in a specific area, such as a street intersection, or contain only details of one specific utility. In the computerized approach, the utility data bank can be developed into an information system capable of providing answers to questions such as, "if the population density in a block were to increase by 50%, what would be the overload on the water system, the sewer system, and the power supply?"

In spite of a strong trend toward general computerization of utility data in advanced countries, development in this field is still in its initial stage. There is general agreement, however, that the cities that are just starting to develop information systems have to base their work on conventional maps of underground and overhead information, employing as the foundation a city base map of the surface situation. Nevertheless, whether manual or computer-supported mapping is considered, graphical maps of various utilities at uniform scales are absolutely indispensable as a basic component of the information system. The primary role of a computer is to store and process the large amount of utility data and to simplify the production and updating of utility maps.

The eventual digitization of utility information may be obtained either by digitization of existing maps or by direct computer processing of the field surveys. In both cases, the derived coordinates must be connected to the city control network so that the information can be correlated with other information on the city fabric.

A study (5) of different numerical approaches shows that sufficient systems (hardware and software) are in existence to demonstrate that it is possible to do virtually anything with computerization and display of spatial data—for a price. Unfortunately, hardware and software are not sufficient to build a successful system. Other essential elements are needed. Two of the most important are (a) the procedures and the channels to collect data in a complete and correct form and (b) the availability of personnel to continuously operate the system and keep it up to date.

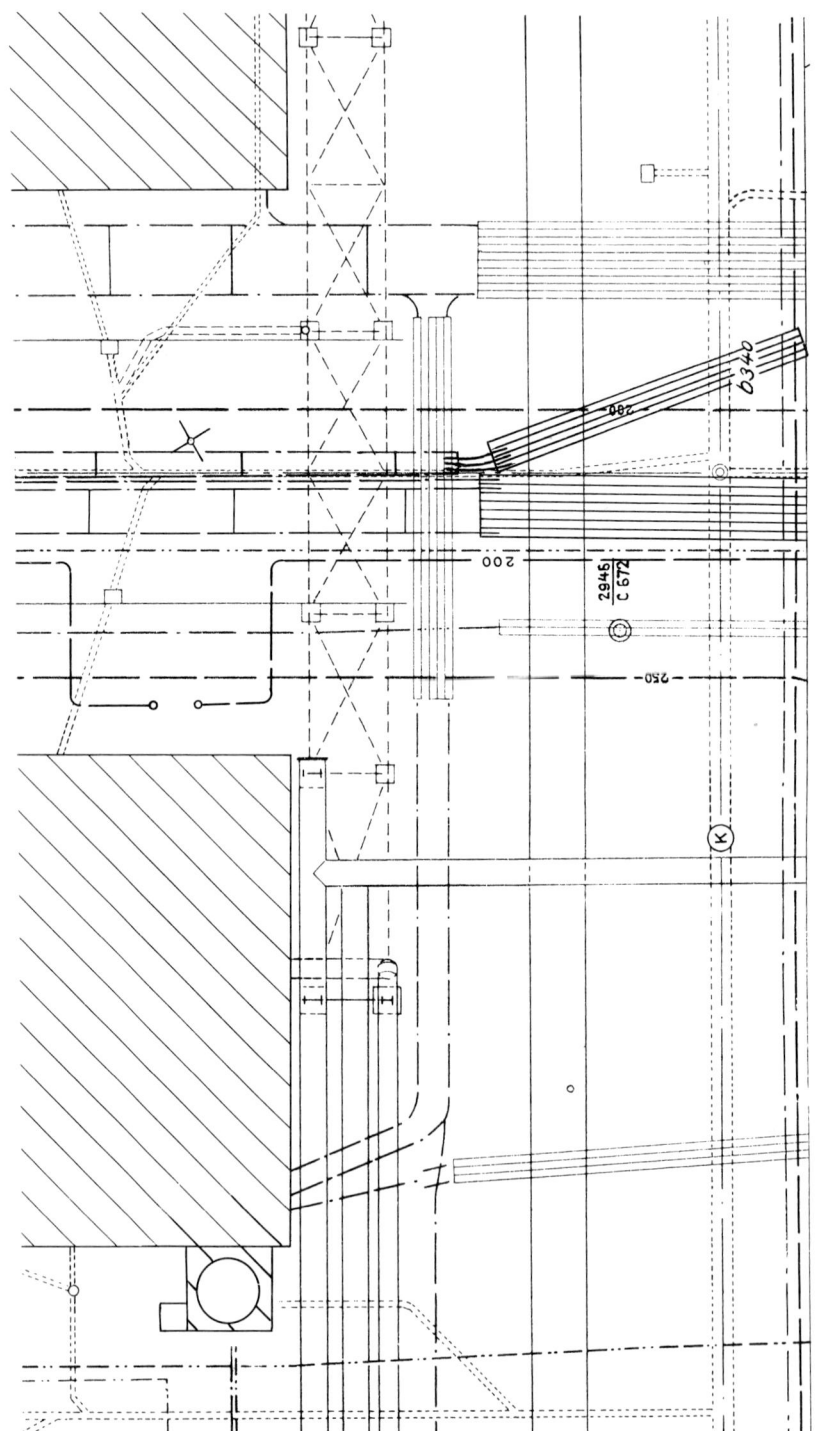

Figure 6-3. A portion of the composite utility map of the city of Basel (original scale 1:200) (courtesy of the City Survey Office of Basel).

In a computer-supported system, the primary operational phases are input, storage and processing, and output. Each phase has, at its disposal, the following basic equipment:

Input

    a. Stereoplotters with digital read-out and recording systems.
    b. Digitizers.
    c. Field surveying equipment.

Storage and processing

    a. Computer processing unit.

Output

    a. Visual display unit.
    b. Computer-driven plotting tables.

Input and output are the main factors determining the cost of the system. Computer specialists are now talking of "zero-cost hardware," i.e., hardware whose cost will be negligible compared to that of input, output, and software. The latter feature, especially, is of significant complexity and cost.

The city of Basel in Switzerland has a well-advanced, computer-supported utility information system. Its numerical cadastre, however, has been based on existing, high-quality (Fig. 6-3) conventional maps that have been digitized.

Another example of a utility information system is that used by the city of Warsaw (6). The system was designed as part of a general survey data bank and is based on data available from the city survey department and utility agencies. The diagram in Figure 6-4 shows the flow of information in the utility subsystem. The only information that is computerized is that needed for decisions concerning a large (of the order of 50 ha) urban district and for answering questions on the possibility of connecting a new customer to the utility network.

A set of maps at 1:500 and 1:250 is the basis of positional information. The maps, which are made by the branch of the city survey department responsible for proper positioning of details, consist of original plotting manuscripts, field survey files, and thematic overlays showing horizontal and vertical control, topography of the surface, underground and overhead utilities (existing and planned), urban projects, and property maps. These transparencies form the basis for making composite multicolor maps for different purposes. Utilities are numbered according to the districts and urbanistic regions of the city and the type and function of the lines. This method of numbering has sometimes proven inconvenient when searching for a detail of the utility on a map. In future, the coordinates of characteristic points of installations will be used as indicators. All utility surveys in Warsaw are connected to the city control network.

An example of a more "manual" approach in setting up a cadastre of underground and overhead utilities is a project in Hungary (3). Initially, some thought was given to establishing a numerical cadastre, but when it was realized that this would require a sizable investment, conventional mapping has been adopted

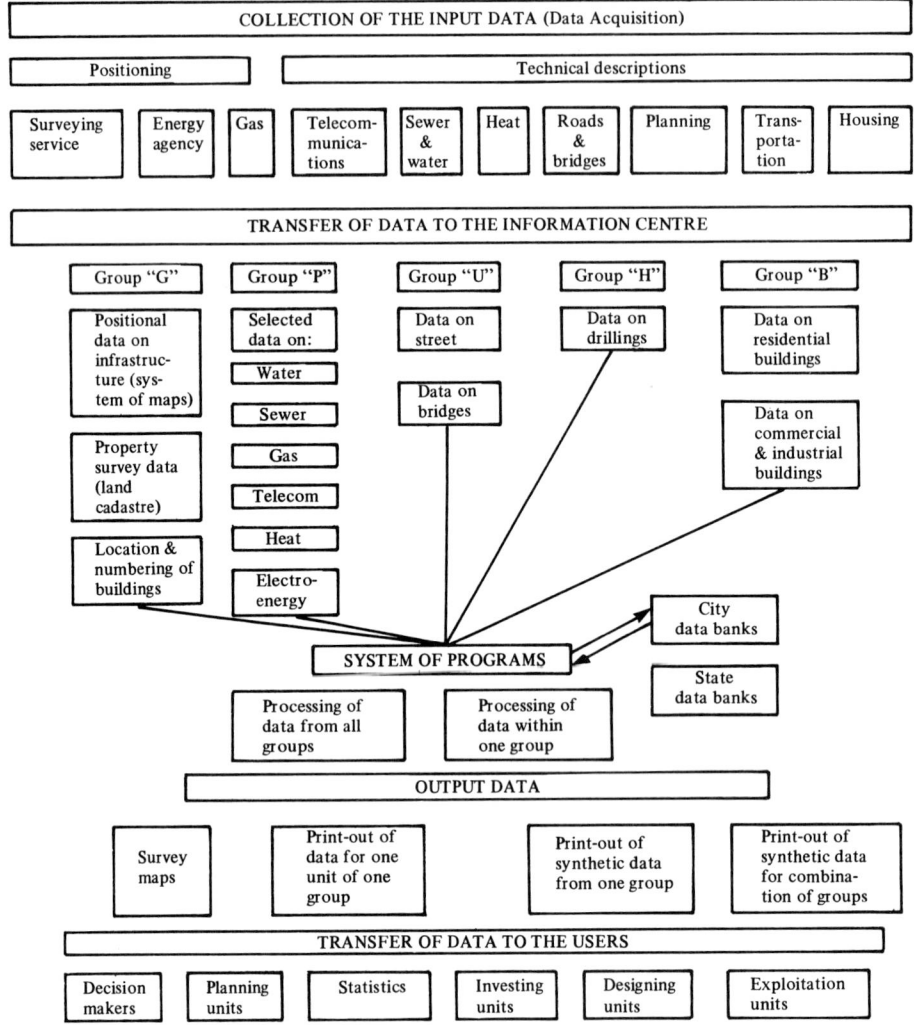

Figure 6-4. Scheme of the Infrastructure Information System in the city of Warsaw (6).

as an intermediate solution. This "graphical" cadastre is arranged to permit conversion to a numerical system.

The contents of the cadastre are as follows: transparencies of individual installations at the scale 1:4000, branch maps of individual utilities at 1:500, composite map transparencies of utilities at 1:500, drawings of various details, and descriptive data sheets (description files).

The updating and maintenance of the utility cadastre is done in the central registry office (responsible for the composite maps) and in the branch registry offices. All agencies concerned must by law supply information on proposed

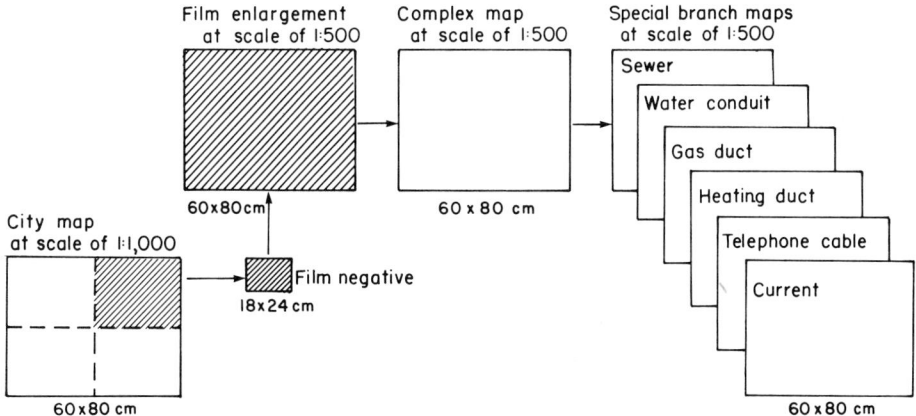

Figure 6-5. The sequence of map production (3).

new utilities to the central registry office, which coordinates work between individual agencies. All field surveys and mapping must be carried out in a uniform manner according to national instructions and specifications. New surveys are done by individual agencies, which have to forward their pertinent records and branch maps to the central registry office once a month to update the composite maps.

Utility mapping is done in the city coordinate system using, if needed, enlarged maps from existing base city maps at 1:1000 and 1:2000. Figure 6-5 shows the sequence in which the maps at 1:500 are derived. The composite map shows utilities in different colors.

The Hungarian system requires good coordination between individual agencies and their cooperation in updating the composite map.

The city of Bern (population 160 000 in 1972) has a simpler and (what seems to be) smoothly running cadastre of utilities (1). Mapping started in 1954 in the form of street maps at the scale of 1:200. In 1968, the city had 300 maps on long sheets with a width of 30 or 60 cm. Difficulties in the use of street maps led to a change in the system. In 1969, it was decided to remap the entire city in a "raster" (grid) system. The maps at 1:200 are enlarged photocopies of the property cadastral maps at 1:500 taken from the so-called "eternal books" of the land registration system. The enlarged photocopies are produced in pale gray, and only such details as survey points, buildings, and curbs, i.e., details useful in the field location of the utilities, are drawn in black ink. Horizontal positioning only is shown because of rigid rules about the depth of individual lines. These surveys and the updating of maps are done centrally by a section of the city survey office that has three survey crews available to individual agencies whenever a new utility is to be laid down.

Mapping is done either before the installations are covered (in new developments) or on the basis of visible surface details, complemented by branch maps and sketches. Different colors are used for different utilities, and those that are mapped from branch maps only are drawn in black. Various symbols allow the use of black-and-white copies of multicolor maps.

Figure 6-6 (see insert opposite page 246) is an example of the multicolor map of the city of Bern. The colors are blue for water, violet for sewer, yellow for gas, red for electricity, and green for telephone.

From the study of the various utility information systems and the trends in this area, the following recommendations are made for cities that have no type of integrated information system on utilities.

1. Lack of resources to implement a computer-supported information system should not be used as an excuse for postponing the development of a less sophisticated, conventional utility information system. Owing to the continuous and rapid expansion of utilities, delays cause serious losses to the urban community and make it more difficult and expensive to establish the system.
2. Uniform utility maps should constitute the basis for the development of utility information systems in urban areas.
3. Scales at 1:500 or larger should be required for urban utility mapping. Grid section maps prove to be more practical than strip maps along streets for handling an information system for the entire city. Enlargements of the 1:1000 base city maps are adequate for mapping. However, the selection of appropriate line thickness and the size of various symbols on base maps are important.
4. In many cities, it has been found that transparent sheets of a stable plastic material are best for the direct plotting of a variety of information (as in updating) and for direct reproduction (copying). The enlargements of base city maps are reproduced directly on transparencies. The underground and overhead utilities are usually plotted directly on the transparancies, which are used as masters to produce the comparatively small number of copies required. Updating is carried out on the original transparent manuscript.
5. The information used for the manual production of utility maps should be arranged in a form suitable for future computer-supported approach.
6. The development of the information system should be based on the policy that the three basic components of the system are of equal importance. These basic components are: collection of information on existing utilities, display and accessibility of information, and continuous maintenance and updating of the system. All three components must function equally well or the system will not work. A less detailed system, that is exact and up-to-date is preferable to more detailed system that has incorrect and outdated information.
7. The experience in some cities proves that there are many possible ways of organizing and maintaining an integrated utility information system. It seems to be clear, however, that the best operating systems are those for which the city survey office takes the responsibility for establishing and maintaining the information system.

8. The following steps should be considered for the organization of a utility information system.
   a. The setting up of a mapping committee consisting of representatives of the utility and planning agencies and the city survey office. The committee should provide general guidelines on the operational scheme of the system and the specifications to be followed.
   b. Detailed elaboration of standards and specifications to be used in the actual surveying and mapping.
   c. A utility mapping committee that continues to exist after the establishment of the information system in order to suggest modifications and improvements as the situation develops.
   d. The special attention that should be paid to the mapping of new work carried out as the various installations are put in place.

References

1. Friedli, J., and König, A. Twenty years of experience in the inventory of underground utilities, *Proceedings of the International Symposium on Underground Utilities*, Lodz, Poland, 1974.
2. Bramorski, K., Gomoliszewski, J, and Lipinski, M. *Geodezja Miejska*, PPWK, Warsaw, 1973.
3. Katona, S. New uniform registration system of public services in Hungary, *Proceedings of the XIV FIG Congress*, Washington, D.C., 1974.
4. Klopocinski, W., Leitungskataster und Projektierung unterirdischer Leitungen in Warschau, *Proceedings of the XIII FIG Congress*, Wiesbaden, 1971.
5. Hamilton, A., Chrzanowski, A., and MacNaughton, N. *Infrastructure Information Requirements in the Maritime Provinces: An Analysis*, Technical Report No. 37, Department of Surveying Engineering, University of New Brunswick, Fredericton, Canada, 1976.
6. Wroblewski, A., and Rybicki, T. Kataster der Geländeerschliessung als Ausschnitt der Datenbank einer Stadt, *Proceedings of the XIV FIG Congress*, Washington, D.C., 1974.

Additional Readings

Andrecheck, B. T. The city of Ottawa underground public utilities central registry, *The Canadian Surveyor*, 5/1972.
Calek, F. Dokumentation unterirdischer Leitungen in einigen FIG-Ländern, XIII FIG Congress, Wiesbaden, 1971.
Dragonetti, A., and Baj Agnoletto, E. Livellazione di precisione per il controllo di pavimentazioni soggette a forti carachi, *Boll. Soc. Ital. Fotogram. Topogr.* 4/1972.
Friedli, J. Aufbau und Nachführung eines Leitungskatasters für grössere Städte, *Geodesia*, 1/1972.
Hage, S. Erfahrungen bei der Vorbereitung des Aufbaus des Leitungskatasters in Dresden, *Vermessungstechnik*, 11/1973.
Jope, R. Der Bestandriss—Dokumentationsbestandteil des Leitungskatasters, *Vermessungstechnik*, 11/1973.

Katona, S., and Papp, G. Über die Lage der Kartenevidenz der unterirdischen Leitungen in Ungarn. XIII FIG Congress, Wiesbaden, 1971.

Kluge, W., and Steizig, W. Aufbau des Nachweises technischer Versorgungsleitungen in Städten (Leitungskataster), *Vermessungstechnik*, 2/1973.

König, A. 15 Jahre Leitungskataster der Stadt Bern 1972, Bern.

Krumphanzl, V., and Michalcak, O. *Inzenerska Geodezie* (Engineering Surveys), Vol. II, Kartografia, n.p., Prague, 1975.

Lienhard, K. Leitungsortung Vermess.—*Mensuration, Photogrammétrie, Génie rural—Vermessung, Photogrammetrie, Kulturtechnik*, 7/1973.

Luckert, K. Der Einsatz von Leitungssuchgeräten bei der Ortung vorhandener Wasserversorgungsleitungen für die Herstellung von Bestandplänen, *Vermessungstechnik*, 1/1973.

Schelle, P., and Lorenz, W. Die komplexe Leitungskarte—Standardisierung ihres Inhalts und dessen Darstellung, *Vermessungstechnik*, 6/1973.

Schmidlin, W. Die Bedeutung des Leitungskatasters für grössere Städte. XIII FIG Congress, Wiesbaden, 1971.

Schmidlin, W. Der Leitungskataster des Kantons Basel—Stadt, Vermess, *Mensuration, Photogrammétrie, Génie rural—Vermessung, Photogrammetrie, Kulturtechnik*, 7/1973.

Simek, J. Die Dokumentation unterirdischer Leitungen in der CSSR. XIII FIG Congress, Wiesbaden, 1971.

van Weelden, J. F. Leidingenregistratie in de huidige praktijk (Leitungsregistration und heutige Praxis), *Geodesia*, 10/1972.

Chapter 7

Remarks on Cadastral Surveys

INTRODUCTION

The purpose of the cadastre is to ascertain the location, size, type, and use of real property and to record data pertaining to value and ownership rights. Real property may consist of a land parcel either with or without a building on it or an apartment in a building. In the latter case, the property is easy to define by specifying the building, the floor, and the location of the apartment on that floor. The size of the apartment can be obtained from existing drawings of buildings or it can be measured with sufficient accuracy by a measuring tape.

Definition of a land parcel is a much more intricate matter. Boundaries are not necessarily marked by permanent features, and as a result, defining a land parcel by its verbal description may be not only difficult but impossible. Even if a description of boundaries is sufficient to define the land parcel (usually because the boundaries follow some distinct physical features of a permanent character), a description alone is not a practical or acceptable answer to the problem of defining individual land parcels; even seemingly permanent features, such as buildings and walls, can be destroyed, whereas the land remains. Therefore, the description of property boundaries must not depend solely upon accidental physical objects located along or close to the boundaries. The only reliable and safe definition of property boundaries is achieved by suitable measurements tied into a permanent control net. Whereas *ownership rights* and their character and extension are matters of specific customs or laws and regulations in a country, *the geometric definition of a land parcel is strictly a surveying problem*, concerned with the location of points in three-dimensional space and governed by the universal rules of surveying techniques, which are identical in all countries. Obviously, a number of techniques of varying characteristics and accuracy can be considered. But once the accuracy requirements are agreed upon, a competent surveyor must take the responsibility for the totality of the cadastral work, since reliable determination of land parcels is the most essential component of any cadastral system.

BASIC FUNCTIONS AND CHARACTERISTICS OF AN URBAN CADASTRE

A cadastre system has three basic functions:

1. Identification of real properties and their owners and provision and maintenance of basic data for taxation purposes. The cadastre was instituted in early times as an instrument of tax collection, and it has maintained this function throughout its history. An equitable distribution of real-property taxes depends upon the value of the properties; therefore, a cadastre must contain at least the basic data essential for the levying of real-property taxes. This function is known as the *fiscal* function of the cadastre.
2. Location of boundaries and the recording and maintenance of information that defines ownership rights and their limitations. Depending upon the legal characteristic of the cadastre in operation, the survey data, together with other documents available in the cadastral office, may provide the sole guarantee of the ownership of a specific land parcel and the only information concerning its precise boundaries. This function is known as the *juridical* function of cadastre.
3. The surveying and mapping products resulting from cadastral operations, aimed at satisfying the two previously mentioned functions, are of great practical use in the planning and execution of various projects and form the basis of a more general information system. This is particularly true in urban areas of rapid development. This function is rapidly moving into the focus of cadastral operations, and as a result, the cadastre is acquiring a *multipurpose character.*

To fulfill these functions, a cadastre must have a certain content and must meet the following operational and quality requirements (1):

1. The core of a cadastral system consists of technically correct surveying (field or photogrammetric) of real properties. In urban areas, only surveying that is based on a permanently monumented control net is acceptable. A survey not based on a reliable control net will not meet the basic cadastral requirements and so will not provide the basis for integration of all surveys within the urban community. Consequently, such a system would be technically and economically unsatisfactory and wasteful.
2. The results of surveying (field or photogrammetric) must be converted into maps or stored in a form suitable for instantaneous conversion into maps by modern computer-based storing, display, and plotting techniques. The maps must be relatively complete to be technically useful

and economically justified. In addition to property boundaries, they must include *buildings and all structures*, as well as the main natural features of the terrain. For areas outside cities, photogrammetric procedures including orthophoto and stereoorthophoto techniques should be considered. The map is the key factor that determines whether the cadastre has a multipurpose character and broad technical and economic significance.

3. The control net provides uniform accuracy; it is the controlling mechanism that permits the building up, with time, of the surveys of individual properties into an entity covering a whole urban community. The control points, therefore, must be permanently monumented or adequately referenced and maintained and protected by law. New control points of proven quality established during surveys should be tied into the existing net.

4. With the map there must be descriptive records, usually consisting of a register of property owners and a land-parcel register. Information on individual parcels, such as their size, value, mortgages, and easements, should be included in these records. Common identifiers and reference keys must permit instantaneous identification and correlation of data on the maps and in the registers. Files outside of the cadastral system containing related information must also be identified.

5. Cadastral records (survey data, maps, and descriptive registers) must be "final," i.e., they must depict true ownership conditions at a given moment and should have legal validity. Consequently, a law must be passed and enforced that no transfer of ownership or change in ownership rights or conditions can occur before it is duly recorded and certified by the cadastre.

6. The finality of the cadastre requires that all the information maintained in the cadastral system be continuously and "automatically" updated. Without this paramount feature, the cadastre loses its value.

7. A modern cadastre constitutes the logical and most suitable basis for a much wider information system (2, 3), with many separate data files under the jurisdiction of authorities outside the cadastral office. A multitude of operational principles and techniques must be taken into account to incorporate various data into a smoothly operating, integrated information system.

It should be obvious that the cadastre is essentially a surveying operation governed by the principles and technical rules of surveying. Since individual rights, the social order, and public confidence are at stake, special care must be taken in the determination of boundaries. On the other hand, similar care and accuracy are needed in general urban surveying because of various technical requirements. Consequently, cadastral and technical surveying in urban areas should be integrated into one operational system under a single supervisory authority. However, should the responsibility for cadastral surveying be separated from the city survey office for local reasons, coordination between the two activities should be assured, including automatic exchange of pertinent data in order to prevent wasteful and expensive duplication and confusion. Modern

computer-based technology facilitates this approach, but this kind of coordination is needed even more in surveys of cities in developing countries, in which
only modest technical and financial means are available.

FIELD WORK

A surveyor preparing for field work must equip him- or herself with all relevant
and previous survey documents on boundary location. The actual surveying
work can proceed according to accepted technical rules and specific cadastral
instructions. In cadastral work, strict adherence to certain prescribed procedures, such as the form of recording measurements, is imperative not only to
ensure uniform quality of the work but also to facilitate the control, use, and
exchange of the survey results. In all properly organized cadastral systems,
detailed specifications are issued that prescribe the main procedures, the
accuracy that must be met, and the form of recording and presenting the data.

One basic rule, common to all cadastral systems, must be strictly observed:
The survey procedures of important points, boundary points in particular, must
contain a self-checking feature. In other words, in any determination of boundary
points, a control measurement should always be provided to eliminate blunders,
such as the misreading of the tape by 0.1 m or 1.0 m. Figure 7-1 illustrates a
right and wrong approach. Moreover, control measurements allow establishment of the accuracy of field determinations and an immediate rectification of
errors where required. Availability of electronic pocket calculators makes
practical the direct checking of the results in the field. The simple computational
formula is

$$\sqrt{a^2 + b^2} = c.$$

When using the orthogonal method, the inclusion of supplementary control
measurements is both simple and efficient. Should a discrepancy be discovered,
the cause can easily be established and eliminated.

The situation is more complex when the polar method is used. In this case, the
control is provided by repeated determinations or by measurement of distances
between surveyed points. In boundary surveys, the sides of land parcels are
always measured directly if physically possible; consequently, the control
measurements are usually available. Whatever the nature of control measurements, they can be used in the polar method only after the results of surveys are
completed in the office. If discrepancies are found to be in excess of prescribed
tolerances, repetition of the field survey is required.

This should not be taken as an argument against polar methods, particularly
when used in conjunction with electromagnetic distance-measuring equipment
having automatic recording of angles and distances, which significantly reduces
the number of blunders in the recording of field measurements. The latter
equipment is quite new and expensive, so that the experience available so far is

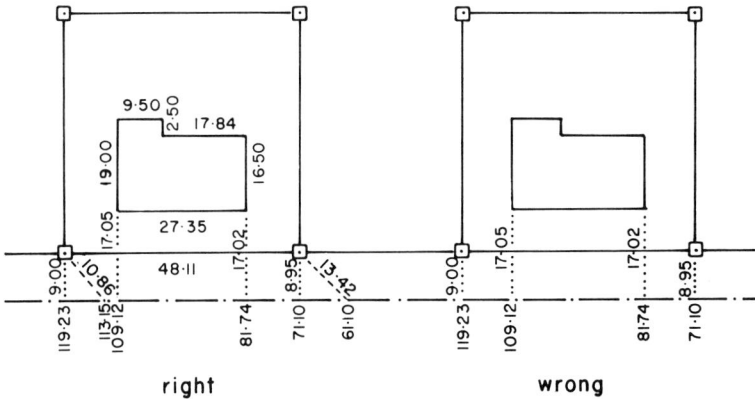

right   wrong

Figure 7-1. Self-checking survey procedures.

rather limited. However, the very precise electromagnetic distance-measuring devices with automatic recording are expected to become an important tool in city surveys, including cadastral surveys.

GRAPHICAL, NUMERICAL, AND COMPUTATIONAL CADASTRE

Depending upon the surveying method and its accuracy, three types of cadastre are generally recognized.

1. Graphical cadastre. Initially, graphical cadastre based on the plane table technique was in operation. With the introduction of the orthogonal and polar methods, and more recently photogrammetry, the plane table technique has been all but abandoned, and there is no need to consider it further. In graphical cadastre, the primary product is a large-scale map. Obviously, numerical values such as coordinates, distances, and angles can always be read from the map, but only a limited *graphical* accuracy can be extracted. Photogrammetric graphical methods, including orthophoto techniques, may be of interest to the reader, and more information on the subject can be found in the chapter on photogram-metry (Chapter 8). In city areas, a graphical cadastre does not provide the required accuracy and should not be considered. However, if property bound-aries can be determined with a satisfactory accuracy by other procedures, photogrammetric graphical methods could be used to complete the cadastral boundary map with terrain details that should be shown on the map.

2. Numerical cadastre. This term was introduced to define a cadastral system based on field surveys, providing numerical geometric elements (lengths and angles) which, when referred to the control net, locate boundary points on the terrain. The accuracy of such a system is superior, and if it is desired, both the location and size of each individual parcel can be determined to any reasonable degree of accuracy, as opposed to the graphical cadastre, in which accuracy is limited beforehand by the primary graphical product. The boundary points in a numerical cadastre are usually carefully monumented, and the monuments generally constitute the primary evidence on the actual location of the property boundaries.

Reliable monumentation of property boundaries can have an effect on the general approach to the surveying accuracy problem. It can be argued that once the land parcels are *physically* and permanently marked, the excessively accurate and costly surveying of boundaries is not essential and that the accuracy specifications, at least outside of densely built-up areas, could be relaxed.

Numerical cadastre does not eliminate the possibility of using photogrammetry as a surveying method, since photogrammetry is capable of providing precise numerical data. The visibility and definition of boundary points in aerial photographs is the limiting factor.

3. Computational cadastre. Assuming that the surveying accuracy is high (compatible with available advanced surveying techniques), still another solution can be proposed. Instead of setting and maintaining boundary monuments, which in urban areas is not an easy task, measurements referred to the control net could be accepted as sufficient to determine the precise position of boundary points at any time. Otherwise, there is little difference between a numerical and a computational cadastre as far as surveying procedure is concerned, and the somewhat confusing term "computational" merely indicates that, because of the high surveying accuracy, the computed coordinates sufficiently define the location of boundary points and that monumentation is no longer required.

Acceptable accuracy standards are an arbitrary matter in this approach, but since the measurements are the only elements in boundary determination in a computational cadastre, the natural tendency is to specify a rather high accuracy, such as ± 2 cm (relative accuracy), in point coordinates. The adjective "relative" is used in the sense of the accuracy with which a point can be referred to other points within a limited area, more specifically, to the nearest control points.

ACCURACY CONSIDERATIONS

The question of accuracy occupies a very central place in discussions on cadastral surveys. Accuracy hinges on many factors, some of a psychological nature, and must be evaluated in the context of specific local conditions and requirements. Only the main factors having a bearing on the approach to accuracy specifications will be discussed here. It is suggested that the reader acquaint himself with the brief discussion on field surveying accuracy on pp. 201–205 and with the remarks on the use of photogrammetry in urban cadastre on pp. 306–307.

In technical discussions on the subject, frequent reference is made to "neighboring" and "absolute" accuracy, terms that require some explanation.

The owner of a property is primarily concerned with the precise and reliable location of property boundaries as referred to surrounding terrain features. The notion of "absolute" location of the property within a local or national coordinate system is mainly foreign to the owner and of little interest to him or her. An obvious and relatively simple way of defining land property and securing its boundaries, particularly in built-up areas, is to tie in boundary points to such relatively permanent features as buildings, street curbs, and monumented boundary points. The tying measurements (and the sides of city lots, for that matter) are usually short, and even when simple measuring tools such as a tape are used, measurements can be determined with an accuracy of from 1 to 2 cm. The accuracy with which the relative position of nearby points (usually the distances between them) is established is called in cadastre "neighboring" accuracy; it is of particular importance because it refers to surveying data that are understood and often used by property owners.

Assuming that suitable reference points are available and that the referencing system is updated as the need arises, the above procedure is capable of locating boundary points, but it is not necessarily capable of determining the size and form of the land parcel, particularly if some of the essential geometrical elements, such as sides or diagonals of the parcel, cannot be measured because of physical obstacles. However, this latter fact is of little concern to the owner, who thinks primarily of his or her property in terms of a physical entity defined by boundary corner points.

This approach to securing property boundaries, with all its shortcomings and faults, is still in common use in countries that have not yet embarked on a technically sound, and economically more rewarding, integrated survey system based on a geodetic control net. Particularly in urban areas, the community is also concerned with the precise location of property boundaries, but for reasons different from those of the individual owners. The community is interested in maintaining social order by setting up a cadastre that in a simple and clear way determines who owns what in the realm of real estate. From the viewpoint of accuracy requirements, various technical requirements and enforcement of regulations—essentially all building regulations—are of primary importance. With the growing scarcity of space in urban areas and the complexity of the city fabric, the precise location of buildings and utility plants is essential. Since building regulation lines are either identical with property boundaries or the two are mutually dependent, the technical accuracy requirements concerning buildings apply also to property boundaries. There is, however, a difference in the kind of accuracy of cadastral surveys expected by landowners and urban authorities. The latter must insist on a uniform accuracy that extends over all the community, or a substantial part of it, because street curbs, building lines, tunnels, viaducts, water mains, and cables must follow certain courses closely related to property boundaries but usually exceeding the dimensions of individual parcels. Such uniform accuracy over larger areas, which we may call "absolute accuracy," can be achieved only if the cadastral surveying is based on an adequate control net, as explained in Chapter 3.

Acceptance of an integrated survey and mapping system in urban areas, of which the cadastre constitutes the basic component, and technological progress in the surveying field paved the way for computational cadastre, which combines the neighboring accuracy requirements with those of a more general character, referred to as absolute accuracy requirements. Starting from a dense net of control points with a satisfactory positional accuracy (of the order of ± 2 cm, for instance), in computational cadastre it is assumed that the accuracy of boundary point determination is such that their monumentation is not necessary, since their positions can be reestablished on the ground at any time. Hamburg in West Germany is frequently used as an example of such a cadastre in operation (4). The Maritime Provinces of eastern Canada are another example of the use of a computational cadastre and the development of a so-called guaranteed boundary system based on coordinates of property corners and a full knowledge of the variances and covariances of the coordinated points (5).

Conceptually, this type of mathematical definition of boundary points is straightforward, elegant, and appealing. No doubt it represents the ultimate solution to cadastral problems, particularly in urban areas. However, it requires very high accuracy in surveying and must be based on a dense and even more accurate control net. Any change, even if minor, in coordinates of control points affects the position of boundary points, a most undesirable effect. The accuracy which is required is not only expensive but is only possible if very competent technical cadres are available. It would appear, therefore, that a mixed system, as represented by a numerical cadastre, is a practical solution under prevailing circumstances.

Further considerations should not be overlooked. First, with the exception of the obvious nicety and operational superiority of a very accurate system, it is difficult to defend an exaggerated accuracy in the cadastre, particularly the faulty argument, frequently encountered, that the high accuracy in cadastral work in cities is imposed by the high price of land. Naturally, high land values in urban areas make owners particularly sensitive to questions concerning property boundaries and property rights. However, the actual value of a property is only very loosely connected with its precise size. The price of land is a negotiable quantity depending upon many factors, primarily upon the market situation at the moment.

It should also be remembered that the accuracy of boundary determination is only one of several attributes that decide whether the cadastral system in operation fulfills its function. Completeness, continuous updating, and finality of cadastral documents are additional important features. High accuracy requirements in cadastre may be justified only after other equally important requirements are met.

In conclusion, the designated accuracy of cadastral work in urban areas should not be an abstract quantity derived from the performance of modern surveying equipment and procedures but should reflect realistic local needs and conditions. As a general guideline, the accuracy of high-quality technical surveys of important planimetric details and engineering surveys can be suggested. This would mean a positional accuracy of boundary points, as referred to a neighboring control net, of the order $m_{x,y} = \pm3$ cm $- \pm4$ cm (pp. 201–205).

CADASTRAL MAPS

Cadastral surveying must be continuously kept up to date to be meaningful. Cadastral surveys—and in particular the cadastral maps that are the graphical presentation of the surveys—are therefore most suitable as the base of the general integrated survey in urban areas. The original field sketches, with all recorded measurements, are kept as original evidence from which the precise large-scale map is produced. An example of such a map at 1:1000 is given in Figure 7-2.

Usually in cities, for strictly cadastral purposes, a map sheet (overlay) is produced that contains only the information needed in the cadastre system, whereas all other planimetric and topographic details (contour lines, etc.) are included in separate overlays according to the accepted map system. Modern computer and display techniques permit the concept of a map as computer-stored information system that can be presented in graphical form on request. In such an approach, changes in cadastral information are stored in the computer and any map produced on request truthfully represents the actual situation of the day.

In the initial graphical cadastre, pertinent geometric information on land parcels was obtained by graphical field surveying procedures—plane table techniques. Even if the resulting cadastral map was supported by direct field measurements, such as the length of boundary lines, the location accuracy of boundary points depended entirely upon the graphical accuracy of the map. Therefore, full attention was paid to the geometric accuracy of the cadastral map from the very beginning, and this attribute, with time, became a distinct feature of a cadastral map. Specific knowledge and techniques have been developed concerning suitable materials, compilation and updating of cadastral maps, and the technique of derivation of correct map coordinates, always taking into account any map deformation (shrinkage) that has occurred.

The introduction of numerical cadastre has resulted in an increased accuracy of cadastral maps. In this approach boundary and other points can be defined by field measurements, with points referenced to the control net and relative position established through coordinates. One might think, therefore, that a precise cadastral map is no longer justified. However, a precise large-scale map has undeniable merit and convenience. Requests such as the establishment of intersection points of a given line with streets, property boundaries or other planimetric features require lengthy computation, the results of which, to be meaningful, must be plotted. The availability of a relatively precise map permits the carrying out of this kind of operation graphically with the utmost simplicity. Graphical operations are irreplaceable, particularly in planning work, and the growing availability of efficient electronic computers does not alter this situation.

Figure 7-2. Example of a cadastral map at the scale 1:1000 (courtesy of the City Survey Office of Munich).

The computers, however, do help to store and process data, including rapid automatic drafting of maps with the required accuracy of about ± 0.1 mm. Consequently, computers should be regarded as a welcome aid in establishing cadastral maps, but not as their substitute.

References

1. Blachut, T. J. What constitutes a land records system—a cadastre? *Proceedings of the North American Conference on Modernization of Land Data Systems*, Washington, D.C., 1975.
2. Blachut, T. J. Cadastre as a basis of a general land inventory of the country, *Cadastre*, National Research Council of Canada, 1974.
3. Ziemann, H. Technical and legal aspects of a cadastre-based land information system, *The Canadian Surveyor*, January 1975.
4. Wahl, B. J. Technical features essential to a modern, multipurpose land data system, *Proceedings of the North American Conference on Modernization of Land Data Systems*, Washington, D.C., 1975.
5. McLaughlin, J., Chrzanowski, A., Thomson, D., and MacNaughton, N. *Maritime Cadastral Accuracy Study*, Land Registration and Information Service, Fredericton, N.B., Canada, March, 1977.

Additional Readings

Blachut, T. J., and Villasana, A. *Cadastre*, PAIGH publication, National Research Council of Canada, Ottawa, 1974.

Dale, P. F. *Cadastral Surveys within the Commonwealth*, Her Majesty's Stationery Office, London, 1976.

Dowson and Sheppard. *Land Registration*. Great Britain Colonial, Her Majesty's Stationery Office, London, 1952.

Hearle, E. F. R., and Mason, R. J. *A Data Processing System for State and Local Governments*. Prentice-Hall, Englewood Cliffs, N. J., 1963.

Herbin, R., and Pebereau, A. *La Cadastre Français*, Les éditions Francis Lefebvre, Paris, 1953.

Norman, P. E. *Photogrammetry and Cadastral Survey*, I.T.C. Publications, A-33, Delft, 1965.

Proceedings of the Conference "Concepts of Modern Cadastre," Ottawa, *The Canadian Surveyor*, March 1975.

Proceedings of the North American Conference on Modernization of Land Data Systems. North American Institute of Land Data Systems, Washington, D.C., 1975.

Proceedings of Reunion Panamericana "Los Levantamientos Integrados y el Desarrollo en Los Paises," Revista Cartografica, No. 28, 1975.

Ziemann, H. *Land Unit Identification. An Analysis*. National Research Council of Canada publication P-PR46, 1976.

Chapter 8

Use of Photogrammetry in Urban Areas

INTRODUCTION

There are two basic attitudes towards the use of photogrammetry in urban surveying and mapping. In cities with an organized, integrated surveying and mapping program, the use of photogrammetry is relegated to secondary applications. Indeed, if the basic city map consists of a very precise, detailed cadastral map, kept continuously up to date by the operational structure of cadastre, there is no need to use photogrammetry except for special applications and projects or to provide supplementary pictorial information.

In most cases, however, cities do not have comprehensive and continuing survey programs for providing maps. As the demand for maps becomes more acute, photogrammetry is called upon as a rapid and financially acceptable, if only partial, remedy.

Recently, some countries have put into practice a combination of terrestrial and photogrammetric techniques (1, 2). The increase in photogrammetric accuracy (particularly when using numerical techniques), the facility of storage and processing of numerical data, the pictorial and graphical display of information, and the improvement of automatic drafting capabilities make such a combination of techniques an attractive and comprehensive solution. In this approach, it is conceivable, for instance, that the most important details, such as boundary corners, curb lines, and building outlines along streets, are measured in the field, and all other details photogrammetrically.

Photogrammetry yields important advantages, even though the field survey accuracy, particularly the "neighboring" accuracy of points close to each other, is better than photogrammetric accuracy, mainly because of difficulty in the

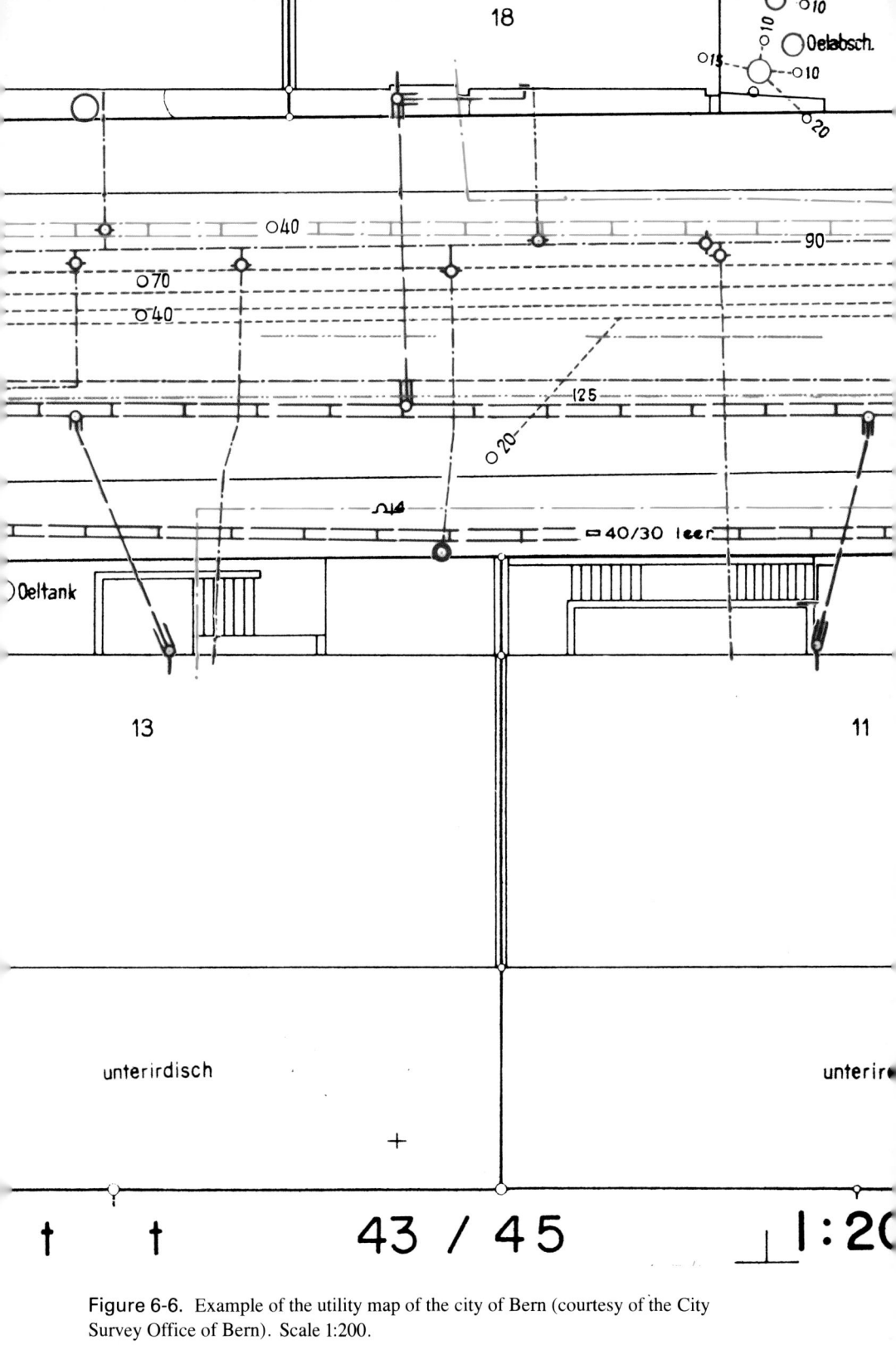

Figure 6-6. Example of the utility map of the city of Bern (courtesy of the City Survey Office of Bern). Scale 1:200.

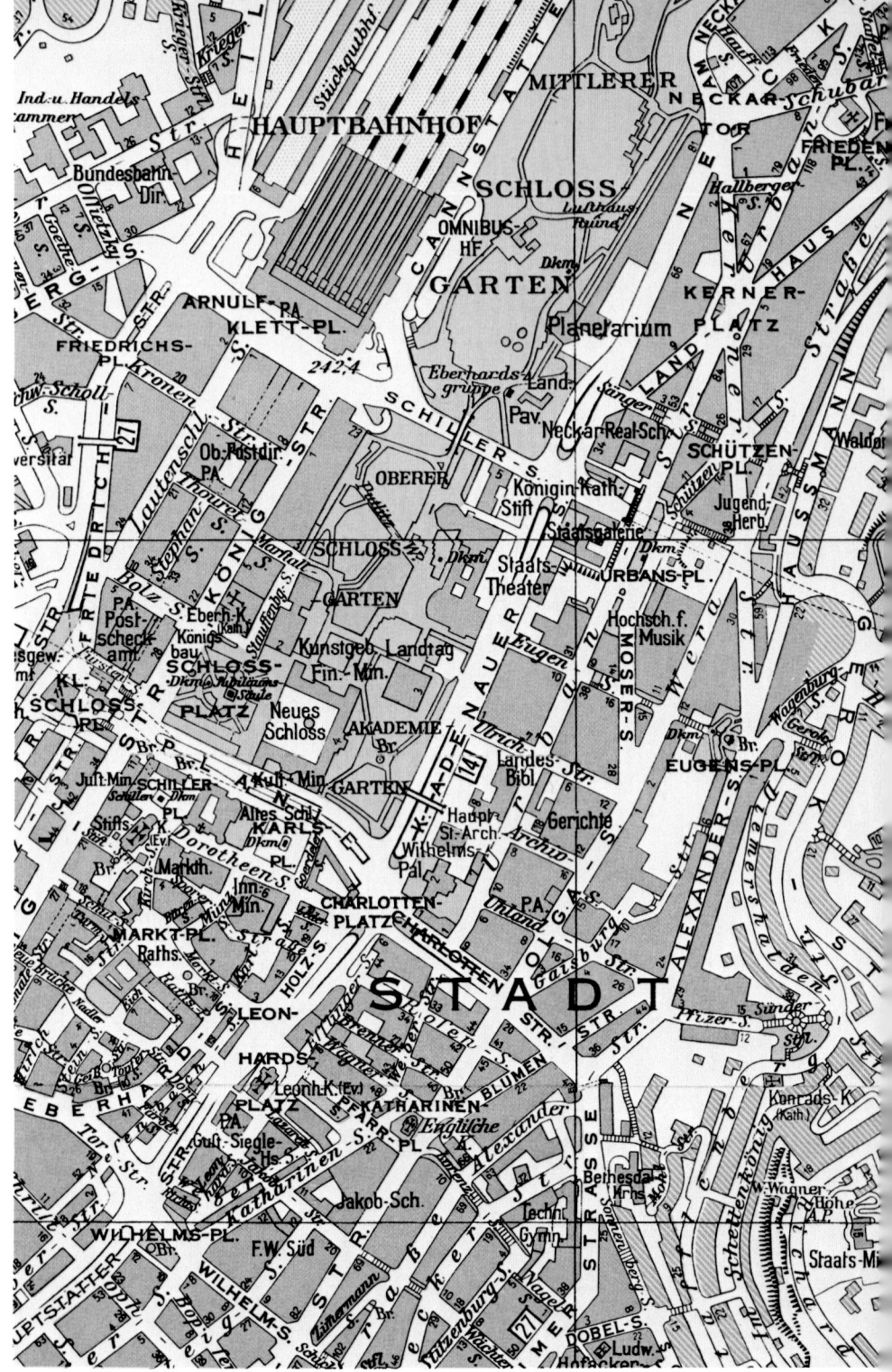

Figure 9-6. The 1:10 000 city map (courtesy of the City Survey Office of Stuttgart).

precise definition in aerial photographs of points to be measured. The photo-grammetric measuring (surveying) and mapping process is fast and does not interfere with city traffic. Moreover, photogrammetric measuring operations are affected by physical conditions or obstacles only when these obscure the points to be measured on aerial photographs. Otherwise, the points on the ground can be separated by rivers, walls, or buildings and can differ vastly in elevation—points can be located at ground level or on top of buildings or bridges. The only concern is the visibility and definition of points on the photographs.

In contrast, in field procedures, these factors lower not only the speed of the work but also frequently its accuracy. In addition, the photogrammetric measuring accuracy of well-defined points is essentially uniform and is primarily dependent upon the scale of aerial photographs. Consequently, by using an appropriate flying height, an accuracy of a few centimeters, which is satisfactory for many applications, can be achieved. Finally, the pictorial information provided by photogrammetry and the rapidity with which photogrammetric data can be supplied are additional factors making the use of photogrammetric techniques in urban areas a most attractive proposition.

NOTIONS CONCERNING AERIAL PHOTOGRAPHY

General Considerations

The complexity of urban areas and the high value of land impose high quality requirements on the photographic image and its geometry. The fundamental rule that "only what is visible on the photograph can be measured" is particularly valid since any town or urban area contains a multitude of minute details, mainly man-made features, whose position must be precisely determined and recorded. Objects such as poles and posts, manholes, catch basins, hydrants, transmission lines, pillars, sidewalk curbs, and fences fall into this category. Metrically correct photographs, taken by high-resolution cameras, are important in any photogrammetric project. This is particularly true in urban areas because of stringent accuracy requirements. The first basic consideration, therefore, must be the good photographic and geometric quality of aerial photographs.

The aerial cameras and the photographic processing equipment used in photogrammetric work are largely automatized. It must not be assumed, however, that modern equipment alone can ensure the required quality of aerial photographs. The question is rather complex owing to a large number of intervening factors, some of which are beyond easy control under operational conditions. Since the scope of this book does not permit a thorough discussion of the subject, a reader interested in more details should consult specialized publications (3–6).

Proper attention to the planning and execution of flight missions must also be kept in mind. Low-altitude flights and the limited visibility of ground points, e.g., ground outlines of buildings along streets, require careful planning of the flight pattern and meticulous execution of the flight. Any deficiency in this respect affects the cost and overall quality of the final product.

The more detailed questions reviewed in the following sections are closely related to these general considerations.

Black-and-White and Color Photographs

At present, black-and-white photographs prevail in mapping operations, including the mapping of urban areas. When compared with color emulsions, black-and-white photographs offer these advantages: Higher sensitivity ("fast" emulsions). Wider range of acceptable exposure. Simplicity in processing. Simple production of glass diapositives. Higher resolving power. Lower cost.

The sensitivity of the photographic emulsion is important if the image motion is to be reduced to a minimum. Large-scale photography over cities requires flights at low altitude. The forward motion of the aircraft may produce blurring at the scales to be considered. If the ground velocity of the aircraft is v (in kilometers per second), the exposure time is t (in seconds) and the scale of photograph is $1/m$, the elongation e of the image in the photographic plane can be computed as

$$e = vt/m. \tag{1}$$

If for instance, the ground velocity of the aircraft is 200 km/hr, the exposure time is 0.01 s, and the scale of photographs is 1 : 5000, then

$$e = \frac{(200)}{(60)(60)} \cdot \frac{(0.01)}{(5000)} \text{ km} = 0.11 \text{ mm}.$$

Under the above conditions, the images of all objects would be elongated in the flight direction by 0.11 mm, which would cause a noticeable blurring of the photographic image.

To avoid this blurring effect, in planning photographic flights, the resolving power of the camera/film combination, which defines the size of the object that can be resolved in the image plane under stationary conditions, must be taken into account. The resolving power numbers vary for different camera/film combinations, but for a mapping camera, the somewhat optimistic figure of 50 lines/mm at the image center can be assumed; in the corners of the usual 23 × 23-cm wide-angle aerial photographs, 10 to 15 lines/mm resolution is a typical number. Brock (3) has suggested that the effect of image movement should not exceed 0.6 of the resolution in the stationary image. Consequently, the acceptable exposure time in our example should be

$$t < \frac{(0.02)(0.6)(5000)(3600)}{(2 \times 10^8)} \text{ s} \cong 0.001 \text{ s}.$$

In a practical case, a somewhat different figure may be derived, primarily because the resolving power number, which depends upon many factors, may

Table 8-1. Exposure Index and Resolving Power of Various Kodak Aerial Films

Kodak aerial film	Color negative 2445	Black-and-white		
		Plus-X 2402	Double-X 2405	Tri-X 2403
Aerial exposure index	32	80	125	250
Resolving power for test-object contrast of 1.6:1 (lines/mm)	40	50	50	25

differ from the 50 lines/mm used in the example. Nevertheless, the example illustrates how stringent the requirements are, particularly since the forward motion of the camera is usually accompanied by an angular motion caused by aircraft unsteadiness and vibration.

The angular movements of the cameras are complex phenomena depending upon many parameters, such as the aircraft type, aerodynamic characteristics, and specific vibrations, the type of camera mount, atmospheric conditions during the flight, and the pilot (automatic or human). It has been estimated that the combined angular velocities amount to about 0.015 rad/s. In our example, this would represent about 2 mm/s of the image motion. In a more recent study, Carman (4) determined image velocity resulting from aircraft vibration for five different aircraft. It varied from 4 to 10 mm/s—significant values. To reduce image velocity to an acceptable size, a relatively short exposure time must be used. This implies sensitive emulsions; black-and-white emulsions offer characteristics superior to color material. Since an increase in sensitivity generally lowers the resolving power of the emulsion, the choice of emulsion is limited. Table 8-1 illustrates the situation, assuming that Kodak photogrammetric material can be regarded as characteristic of the market material available.

Range of acceptable exposure. For a specific emulsion, the required exposure time depends upon the amount of light that is reflected from the photographed object through the lens to the photographic emulsion. Usually the photographed scene consists of many elementary areas of different brightness, each of which requires a different exposure time for its optimum photographic rendition. Since this is impossible, there is a danger that the exposure time selected may be too long for bright portions of the scene and too short for dark portions. The relationship between exposure time and darkening of exposed film presented by the characteristic curve, is basic for a thorough understanding of the photographic process (Fig. 8-1).

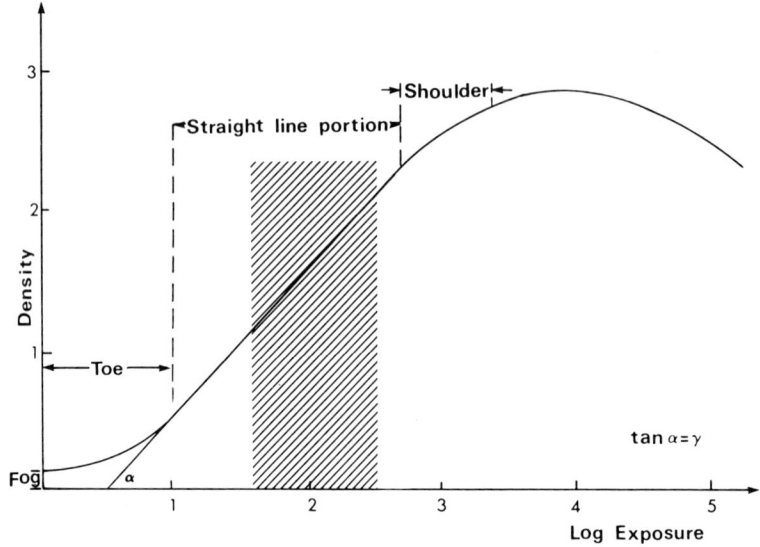

Figure 8-1. Characteristic curve.

The straight-line portion of the characteristic curve is of particular interest because if the exposure time is within this interval, the photographically recorded image will have the same relative brightness differences as the photographed scene. However, the scene has a certain *range* of brightness, which is indicated by the hatched area in Figure 8-1. If the straight-line portion is relatively long, the photographic reproduction will ensure the correct contrast rendition, and in addition, certain errors in the selection of exposure time will not affect this relationship. This feature, called "exposure latitude," is characteristic of photographic emulsions and allows intended or unintended departures from the optimum exposure time without the risk that the results of what are often difficult and expensive photographic missions will have to be rejected.

As a rule, the exposure latitude is greater for black-and-white than color emulsions. Other factors, such as simplicity in processing, also make the determination of the optimum exposure in black-and-white photography less critical than in color photography and consequently make the use of black-and-white photographs a simpler matter.

Simplicity in processing. Black-and-white film is easy to process; it is also easy to obtain other products, such as glass plates or film diapositives, paper prints or enlargements. The manual development of aerial black-and-white film can be carried out in less sophisticated laboratories. In the latter case, however, uniform results, even within one film roll, are not guaranteed.

Glass-plate diapositives. A most important feature of black-and-white photographs is the simplicity in producing glass-plate diapositives. In plotting and measuring operations, the dimensional stability of the photographic image is of great importance; the simplest way to ensure this is to produce contact glass-plate diapositives from the original film negatives. There is a wide range of material to choose from that includes glass plates of varying thickness, flatness, and photographic characteristics.

It must be stressed that for photogrammetric operations in which the best possible accuracy is sought, as in most operations over urban areas, *a parallel-light printer* should be used for the production of contact glass diapositives. It has been reported in the technical literature that this type of printer can significantly reduce the magnitude of printing errors.

In contrast, production of color glass diapositives is more complex and expensive; Kodak, for instance, used to produce color glass diapositives by *glueing color film positives* to the surface of glass plates. This approach has been discontinued, probably because of the prohibitive cost of such diapositives; only the usual film diapositives are available at present. When producing diapositives on film, the film base is thicker than that of original aerial photographs. This provides copies with improved dimensional stability, but which are still inferior to a photographic glass plate (7).

It should also be remembered that some photogrammetric plotters, particularly those based on optical projection, require photographs in the form of photographic glass plates.

Dimensional stability and geometric accuracy. The particularly critical component as far as the dimensional stability of aerial photographs is concerned is the photographic emulsion, which exerts considerable force on the film base as ambient conditions vary. Since black-and-white emulsions are approximately three times thinner than color emulsions (7 μm versus 20 μm), better dimensional stability should be expected from black-and-white photographs.

Additional factors of a physical and physiological nature, such as the definition of the photographic image (resolving power), geometry of the photographic image (a one-layer emulsion versus a three-layer emulsion in color photographs), interpretability and physiological perception phenomena, must be considered in evaluating the final accuracy of measurements obtained from aerial photographs. Some researchers (8) claim equal accuracies for both types of photographs. A report on quite extensive tests carried out under the auspices of the American Society of Photogrammetry, Committee on Color Photography (9), states: "the tests performed show that the aerial color film tested can be used to obtain precision and accuracy approaching or surpassing black-and-white aerial film." These results refer to Kodak black-and-white and color photographs taken by a Wild RC8 camera. More specific comparative data are listed in Table 8-2.

Resolving power. The resolving power is another most important characteristic of the photographic film. A high resolving power, combined with a relatively high speed, is particularly valuable. Black-and-white emulsions show better performance, but the resolving power of color emulsions has been vastly

Table 8-2. Results from Black-and-White (B/W) and Color Photographs

Author, year, and instrument	Scale of photographs		Film type		Number of stereopairs		Check points		Root-mean-square errors μm			
									B/W		Color	
	B/W	Color	B/W	Color	B/W	Color	B/W	Color	m_p	m_z	m_p	m_z
M. C. van Wijk, 1964 A7	1:12 000	1:12 000	Kodak Plus X	Ultraspeed	6	6	80	80	10.5	9.5	12.5	11.5
	1:10 000	1:10 000		Anscochrome	6	6	78	78	11.5	10.5	14.0	13.5
	1:8000	1:8000			6	6	74	74	12.0	14.5	16.5	13.0
	1:6000	1:6000			6	6	60	60	12.5	13.5	12.5	8.5
									11.6	12.0	13.9	11.6
T. J. Blachut, 1977 Anaplot	1:10 000	1:10 000	Kodak Double X	Kodak Ektachrome 2448	1	1	71	46	7.3	9.1	6.7	9.3
T. J. Blachut, 1977 Anaplot	1:10 000		Black-and-white diapositives from above color negatives		1		46		9.2	11.9		
H. Ziemann, 1976 Zeiss stereocomparator PSK	1:10 000	1:10 000	Kodak Double X	Kodak Ektachrome 2448	1	1	63	50	8.0	8.0	8.6	10.0
	1:10 000		Black-and-white diapositives from above color negatives		1		49		9.0	12.0		

improved in recent years and is now approaching that of black-and-white film. The comparative figures for Kodak materials are given in Table 8-1.

Despite inconveniences, color photographs, particularly if observed stereoscopically, give a superior presentation of the terrain and greater interpretability of its various details. Urbanists and other experts outside the surveying and mapping field prefer color photographs. For this reason, color photographs are often taken in addition to black-and-white photographs. With this approach, color film transparencies or color paper prints usually suffice; thus the need for producing expensive color glass-plate diapositives is eliminated. In plotting operations, another approach is to use black-and-white glass-plate diapositives produced from original color photographs.

Cost consideration. Differences in the price of photographic materials may have only a negligible effect on the total cost of the photographic operation, which depends chiefly upon the cost of the actual flight and the preparations for it. If, for example, the town to be photographed is located in a remote area with difficult access and only short periods of suitable weather for photographic missions, the overall cost may be very high, and any additional expense for a given type of photographic material and its processing may be of no importance. Conversely, if a local company can carry out the flight and the prevailing weather conditions are favorable, the added cost of the photographic material and of the laboratory photographic work can be of some significance.

The effect of the somewhat higher cost of color material is even less important if the total cost of the mapping operation, including the field work, plotting, and cartography involved, is considered.

This situation may soon change drastically as new types of films, such as K-C films based on different principles and technology (10), reach the market and satisfy requirements in the photogrammetric field.

Effect of Atmospheric Conditions on Aerial Photographs

Photographic missions must be carried out under favorable conditions. This means uniform and adequate illumination, and in the geographical zones where it is applicable, missions should be planned for seasons when the trees are bare.

The most important fact is that adequate sun illumination permits a short exposure, which is essential for good-quality photographic images. Photographs are usually taken on sunny days because sun illumination increases contrast and consequently improves the resolution of photographic images. Details such as lampposts and hydrants can be recognized more easily if they cast shadows. In addition, a scene with more contrast provides the photographic image with a certain brilliancy pleasing to the observer.

Shadows that are helpful in recognizing "vertical" objects on the terrain are undesirable if they are produced by objects such as massive buildings and trees. Occasionally, if the exposure time is too short or the film is not adequately

Figure 8-2. Samples of aerial photographs of an identical section of a city.

processed, the shadows may be very dark and may not permit terrain details in shadow to be visible in photographs. To reduce the number of details "covered" by shadows, it is recommended that photographs be taken as close to midday as possible, when shadows are shortest.

A more effective remedy is to take photographs under an overcast sky, when the sunlight is diffused by clouds. The photographic image is less "brilliant," but it is free of disturbing shadows.

To provide quantitative information of the optimum type of photographs and atmospheric conditions, the National Research Council (NRC) of Canada, together with the Ministère des Terres et Forêts of the Province of Québec, carried out a study over the city of Hull in 1968 and 1969. Using the same camera and airplane, three photographic missions were flown according to the following specifications: (a) full sunshine, black-and-white film, Kodak Double X; (b) overcast sky, black-and-white film, Kadak Double X; and (c) full sunshine, color film, Kodak.

From the original color photographs, black-and-white glass-plate diapositives were also produced. In Figure 8-2, identical sections from different types of photographs are shown. Pertinent data from the results, published by J. Sima (11), follow.

All photographs were at a scale of 1:4000, taken with a Wild RC8 camera, $f = 152$ mm, 23×23 cm. The observations and measurements were made on a Wild A-7 Autograph using $10 \times$ magnification in the observation system.

First, using 292 objects typical of urban areas, of which 8 types belonged to objects on the ground surface and 9 types were above the ground surface, Sima tried to determine an "identification index" for various aerial photographs. He used the following classification system:

1. Very good definition, sharp outline of the object presenting precise geometrical form.

    Score 10

2. Good definition, sufficient to permit positive geometric outline of the object.

    Score 6

3. Acceptable definition, geometric outline not fully guaranteed, but no danger of gross errors of identification.

    Score 2

4. Identification not possible.

    Score 0

The results of this analysis are contained in Table 8-3. The best readability is offered by black-and-white photographs taken under an overcast sky. Poorer results were obtained from black-and-white and color photographs taken on sunny days. Black-and-white diapositives produced from color negatives seemed to offer the poorest readability.

Since the definition of points measured photogrammetrically is the most important factor affecting their planimetric accuracy, some correlation between

Table 8-3. Quality of Stereoscopic Identification

Date of flight	1/11/68,	23/3/69,	15/4/69,	15/4/69,	Number of
Sky conditions	clear sky	overcast sky	clear sky	clear sky	objects
Diapositives			Color negative		
1:4000	B/W	B/W	B/W diapositives	Color	
			Average value of visual classification		
On the ground					
Targeted ground control points	9	9	9	7	11
Nontargeted ground control points	8	8	5	8	13
Catchbasins	6	6	5	7	14
Manholes	5	7	5	6	10
Crossovers	3	7	3	2	6
Rails	6	10	6	7	18
House corners	6	9	4	5	18
Wall and fence corners	7	7	6	7	14
Average	6	8	5	6	104[a]
Above the ground					
Service poles	7	6	6	7	31
Suspended signal lights	9	9	2	8	5
Other poles and posts	6	5	4	5	20
Individual trees	3	4	4	5	20
Fire hydrants	5	6	4	6	11
Chimneys	8	9	5	8	21
Flat-roof corners	8	8	6	7	38
Sloped-roof corners	6	7	3	6	29
Wall and fence tops	6	4	5	7	13
Average	6	6	4	6	188[a]
General evaluation of ability to precisely identify	Good	Best	Poor	Good	292[a]

[a] = Total.

Table 8-4. Planimetric Accuracy as Determined from Photographs Taken Under Different Atmospheric Conditions

Date of flight Sky conditions	1/11/68, clear sky	23/3/69, overcast sky	15/4/69, clear sky	15/4/69, clear sky	
Diapositives (1:4000)	B/W	B/W	Color negative B/W prints	Color	Number of
Groups of objects		Root-mean-square error m_{xy} derived from the distance errors			distances measured
Targeted ground control points	±3.1 cm	±1.6 cm	±2.3 cm	±2.6 cm	5
Nontargeted points Catchbasins and					
manholes	8.8	7.1	13.2	12.4	4
Rails	5.8	4.6	4.9	5.7	9
House corners	8.7	6.4	9.0	7.3	9
Wall and fence corners	19.2	8.4	7.4	11.5	7
Service poles	8.9	11.0	8.6	10.7	16
Other poles and posts	10.4	8.8	11.6	11.0	6
Flat-roof corners	6.3	6.5	8.5	7.1	17
Average accuracy of nontargeted points (66 distances)	±9.7 cm	±7.5 cm	±9.0 cm	±9.3 cm	

the quality of readability, as established in Table 8-4, and the planimetric accuracy of the respective points should be expected. In the study conducted at the NRC, Sima compared a number of distances determined first directly in the field and then photogrammetrically; from these differences, he derived root-mean-square errors in horizontal positions of photogrammetrically determined end points, using the following formulas

$$\Delta_d = d_{\text{field}} - d_{\text{photogr}}$$

$$m_d = \pm \sqrt{\frac{[\Delta_d \Delta_d]}{n}}$$

and

$$m_{x,y} = \frac{m_d}{\sqrt{2}}.$$

The points at both ends of the measured distance should be of the same type so that the same photogrammetric pointing accuracy can be assumed. The advantage of this procedure is that distances can be measured in the field with great accuracy, thus permitting reliable assessment of photogrammetric results.

Moreover, since the points are relatively close to each other, neighboring accuracy is involved, which is often of primary importance.

The results which were given by Sima in Table 8-4 support those of previous investigations in part. The following conclusions seem to be justified:

The accuracy of targeted points was much higher than natural points.

Photographs taken on overcast days provided the best accuracy.

Color photographs, including black-and-white diapositives from original color photographs, offered slightly better results than original black-and-white photographs taken on sunny days.

The relatively good results from black-and-white diapositives printed from color negatives seem to contradict the rather poor readability characteristic of these diapositives. The somewhat subjective nature of the part of the experiment, concerned with readability classification and the resulting uncertainty of the final results, must be appreciated, however. Also, the differences between some of the results are too small (9.7, 9.0, and 9.3 cm) to be of conclusive significance.

PLANNING OF PHOTOGRAPHIC MISSIONS

Choice of Camera

A number of aerial cameras are used in reconnaissance and general surveying and mapping missions, but only high-performance photogrammetric cameras should be considered for urban work. General economy, the complex content of city areas, and the high accuracy requirements called for in urban surveys dictate this approach. Cameras of sturdy construction, high resolution, and consistent and uniform geometric quality of images should be chosen. High-quality photographs are of paramount importance to the whole photogrammetric operation, and the slightly higher expense of a good camera is a wise investment in light of the resulting benefits, which include ultimate financial savings due to the better quality of photographs. Obviously, it is not suggested that every city concerned should purchase a suitable camera. The city should insist, however, that the cameras used by the contracted operator be of the highest quality and that the photographs meet the standards laid down in recognized specifications such as those issued in Canada, which can be received on request.[1]

The next important consideration concerns the type of lens and, more specifically, its angular field. At present most photogrammetric cameras have a standard picture size of 23 × 23 cm. By varying the camera constant (focal distance), the angle of the field of view can be modified (Fig. 8-3). Some standardization of camera constants in photogrammetric aerial cameras are presented in Table 8-5. As can be seen in Figure 8-3, photographs taken from the same

[1] Specifications for Air Survey Photography, 1967 (amended January 1970), by Interdepartmental Committee on Air Surveys, Dept. of Energy, Mines and Resources, Canada, Ottawa.

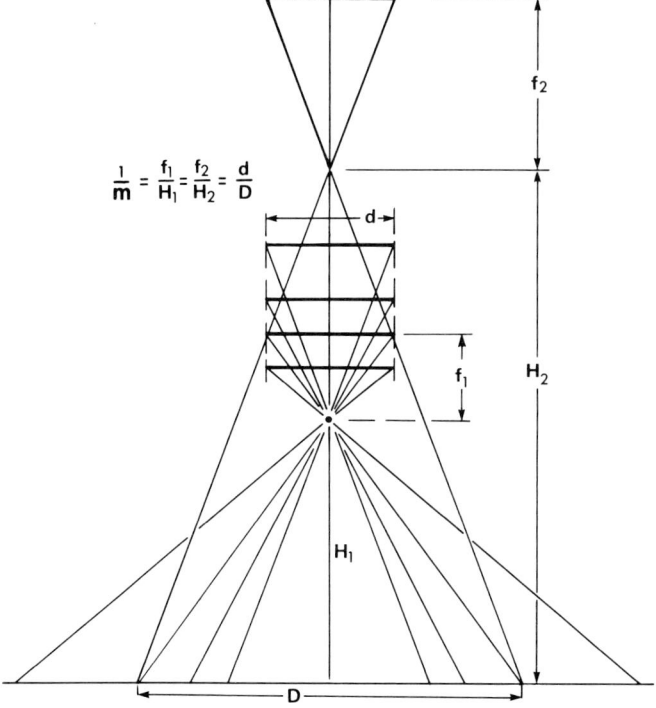

Figure 8-3. Angle of camera's field of view and the resulting ground coverage.

altitude by cameras of different focal lengths have different scales and cover ground areas of variable size, following the relationship

$$D^2 = (d \cdot m)^2 = \left(d\frac{H}{f}\right)^2,$$

where d is the side of a square photographic frame and m is the scale coefficient. It also follows that photographs of the same format and identical scale will cover the same size of ground area irrespective of focal length. Since the flying

Table 8-5. Typical Photogrammetric Lenses

Type of lens	Format	Camera constant (mm)	Angle of field of view	Base–height ratio for 60% overlap
Narrow angle	23 × 23 cm	300	47°	0.3
Normal	23 × 23 cm	210	64°	0.4
Wide angle	23 × 23 cm	150	83°	0.6
Super wide angle	23 × 23 cm	90	115°	1.0

heights for urban work are within a comfortable flying range (with the possible exception of very low flights), the choice of the most suitable type of camera is flexible. Two contradictory factors must be considered.

The first is the accuracy of height determination. It has been proven that height determination depends upon, in addition to the scale of photographs, the so-called base–height ratio, or the ratio between the average distance between the successive camera stations in the air and the flying height over the ground. A larger ratio permits more precise photogrammetric determination of heights. Consequently, in the projects where terrain heights are of particular importance, preference should be given to wide-angle cameras.

Secondly, much better insight into built-up areas is provided by normal or narrow-angle cameras equipped with lenses of $f = 210$ mm, 300 mm, or longer. The tendency toward longer focal distances will probably be further strengthened by the growing interest in orthophoto maps and orthophoto products in general. The factor that usually overrules the decision is the fact that most conventional plotting equipment is not capable of handling photographs taken with cameras of 300 mm or longer focal length. As a result, the popular wide-angle photographs, $f = 150$ mm, restricted to their central portion if necessary, are generally used. This requires large longitudinal (80%) and side (60%) overlap. Modern analytical plotters remove previous restrictions on size of focal length and can handle photographs of arbitrary focal lengths with ease.

Scales and Respective Flight Arrangements

Because of accuracy requirements and the loss in visibility caused by trees, buildings, and other structures, in urban areas, photographic missions must be planned with particular care, and special measures that are uncommon in the usual topographic operations must frequently be taken.

The scale of the photographs, which is dependent upon the purpose of the project, must be decided at the outset of the project.

Medium- and small-scale photographs over urban areas. Certain concepts and definitions are being introduced to systematize the subject. The boundary line between medium- and large-scale photographs is of little importance. As a guideline, photographs at scales not sufficient for the recognition and plotting of discrete points and features characteristic of city *base maps* are classified as medium-scale photographs. For example, photographs at $1:15\,000$ or $1:20\,000$ are considered to be medium-scale photographs. They are flown along *parallel flight lines*, either following generally accepted flight directions or according to other local considerations such as shape of the area to be covered.

Photographs in this category are used for aerial triangulation, production of medium- or small-scale maps (conventional or photomaps), and production of photomosaics or other photographic material for all studies except cartographic work, the planning of future cartographic work, and preliminary studies of various engineering projects.

Except for considerations mentioned on pp. 247–258, photographic missions in this category are carried out following the same rules as those of a general topographic mapping project.

Scales smaller than 1 : 20 000 are considered to be in the small-scale category. Photographs at these scales are used primarily in general studies, particularly those of a city and its surrounding region, and for mapping at small scales. The missions are carried out according to rules accepted in topographic mapping.

Large-scale photographs over urban areas. Large-scale photography, usually 1 : 10 000 and larger, is used for actual basic city surveying and mapping. Three distinct approaches can be singled out: (1) parallel flight-lines arrangement, (2) flight lines parallel to streets, and (3) single, "aimed" photographs. Depending on conditions, any combined approach can be taken.

1. Parallel flight-line arrangement. This is the most common approach for obtaining complete photographic coverage. To ensure the best possible coverage of the mapped area, it is good practice to use 80 or even 90 % longitudinal overlap and, when planimetry is particularly complex, 60% side overlap (Fig. 8-4). A 60% *side overlap* increases the number of flight lines and makes the flight mission more expensive. An increase from 60 to 80% in longitudinal overlap has a negligible effect on the overall complexity and cost of the mission but offers the possibility of selecting the most suitable stereopairs in plotting particularly complex areas. It should be understood that, to form stereopairs with 60% overlap for plotting purposes, every second photograph is selected from the strip with an 80% longitudinal overlap. The usual procedure is to produce paper prints from all photographs and decide from which negatives

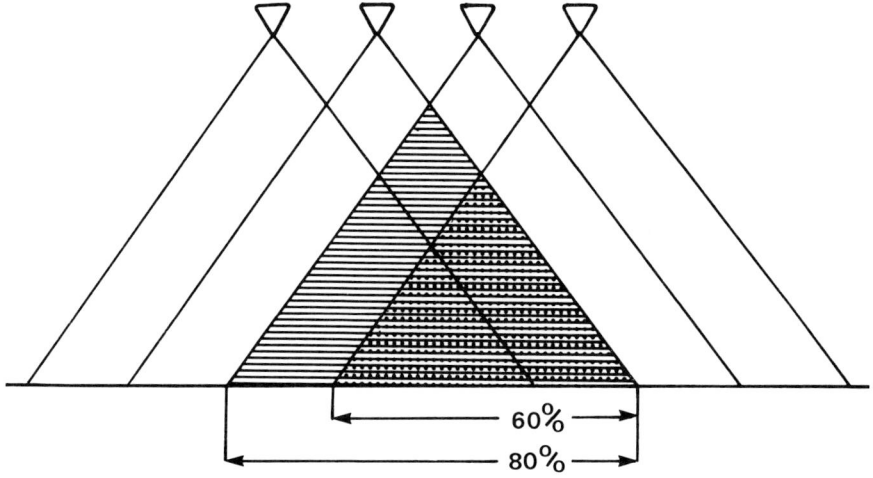

Figure 8-4. Varying longitudinal overlap.

the plotting diapositives should be made. Should diapositives of additional photographs be required during the plotting phase, they can be ordered accordingly, which avoids the expense of producing large numbers of redundant diapositives.

When 80% longitudinal and 60% side overlap are used, there is a good chance that most ground details will be able to be plotted, even in cities with high buildings. Therefore, for the production of a city's basic large-scale maps, such as 1:1000, a longitudinal overlap of 80% and a side overlap of at least 30% are recommended.

General plotting is carried out using photography with a 60% longitudinal overlap because of accuracy considerations, but where necessary, stereopairs having an 80% overlap can be used to fill in details not visible stereoscopically in 60% photographs. In Figure 8-5, for example, ground line I of the building can only be plotted from stereopair 1–2; similarly, ground line II can only be plotted from stereopair 4–5. Any other stereocombination does not permit a full stereoscopic plotting of these details.

Another important requirement in the planning of flight-line patterns for large-scale city mapping is that the map sheet be plotted from a minimum number of photographs. To accomplish this, the following conditions should be met:

> Flight lines should be parallel to the reference grid.
>
> The map sheets should be located within a single strip of photographs, if possible.

The second condition, if applicable, requires precise planning and execution of the flight program. Any effort in this direction is essential for the smooth execution of the mapping project and overall cost reduction, particularly when orthophoto methods are used.

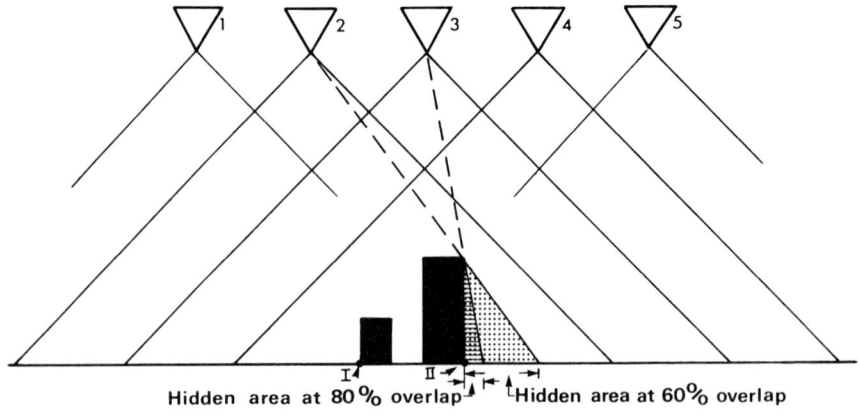

Figure 8-5. Use of 80% longitudinal-overlap photographs.

2. Flight lines parallel to streets. This approach is often taken with specific projects or as a complementary measure along particularly important, primary streets of the city. If the street is wide enough that minor departures of the airplane from the flight line over the street axis do not jeopardize the unobstructed view of the ground line of the buildings at both sides of the street, the advantage of such an arrangement is particularly obvious.

3. Single, "aimed" photographs. These are made to obtain photographs over specific urban areas such as squares, bridges, railroad stations, buildings, and parks. "Aimed" means that an attempt is made to obtain photographs over preselected points, as marked on maps or photomosaics. The camera operator either may choose to commence a sequence of photographs along short flight lines with preselected longitudinal overlap, so that individual photographs will be exposed over the determined nadir points, or may prefer to expose each single photograph individually. This decision will depend upon various circumstances, such as the purpose of the photographic mission, the characteristics of the object to be photographed, and the characteristics of the aerial camera (the facility with which single and consecutive photographs can be made).

In some countries (notably Poland), aimed photographs over city areas are made from helicopters, since these craft can move very slowly or even hover over selected points. Initially, helicopters had very strong vibrations, which made them unsuitable as vehicles for aerial cameras. Maintenance of the orientation (swing) of photographs is another difficulty. Nevertheless, the improvement in characteristics and flight performance should make helicopters acceptable as carriers of aerial cameras.

Targeting of Points

Good definition of the points to be determined photogrammetrically is the basic condition for achieving high and uniform accuracy in photogrammetric work. Points that are considered in the mathematical sense permit the location of various physical features in the accepted reference system, or are part of the reference system itself. Examples are building corners, centers of lampposts or manholes, property corners, and traverse points. To define a point in a satisfactory manner, ground objects or features must have the following characteristics:

Symmetric form (round, square, crosslike).
Suitable size (the photographic image of the object must be slightly larger than the measuring mark of the photogrammetric plotters).
Sufficient contrast with the surrounding area.

Very few natural or man-made ground features meet the above requirements. Moreover, since control points to be established photogrammetrically, as well as boundary and similar points, are frequently indistinguishable on the terrain, they must be specifically marked on the ground to make them visible on aerial photographs. Therefore, the question of targeting arises in most photogrammetric operations.

The main difficulty in targeting is operational in nature. Targeting of a large number of points requires considerable time. On the other hand, precise scheduling of the flight mission is most difficult in many regions because of unpredictable weather conditions and other operational factors. As a result, a lengthy time lapse may occur between the completion of targeting and the actual execution of the photographic flight, with the risk that many targets may be removed or destroyed in the meantime. This danger always exists in densely populated urban areas, even if the time lapse between targeting and flight is very short.

Missing targets that mark property boundaries or control points to be determined photogrammetrically cause additional field work, but the absence of targets on control points determined in the field may jeopardize the usefulness of the flight mission. Particular attention, therefore, must be paid to targeting points that are intended to be used as primary control. First, a number of points greater than the required minimum should be established and targeted. A good practice is to establish and target "satellite" points in close proximity to the actual points, which permits the location of satellite points in suitable spots, where the probability of their destruction is reduced. Use of "point clusters" instead of singles also increases the overall accuracy of photogrammetric work. This approach is strongly recommended when targeting a limited number of particularly important points.

Since destruction of targets is particularly easy if they are removable, painted signs of suitable shape and which are placed on a permanent surface should be used whenever possible. The abundance of paved surfaces in city areas and the availability of fast-drying and permanent paints makes this approach logical. Painting should be done during the night. It should be remembered, however, that targets painted on empty streets or squares during the night may be obscured by pedestrians and cruising or parked cars in daytime.

If possible, targets should be placed at ground level. The main reason for this is the difficulty of stereoscopic pointing at a high-contrast spot (white target image), which can result in an erroneous setting of the measuring mark in height. Since any z-error also affects x- and y-coordinates of the point, it is recommended that, during measuring and plotting operations, the measuring mark be set at the height of the surrounding terrain first and then centered over the target.

In urban areas, most control points and ground details, such as catch basins and manholes, are at ground level and are easy for paint targeting.

If points are located within a soft soil area and paint cannot be used for marking, other techniques must be considered. One is to place a target of a convenient size and form centrally over the points (Fig. 8-6). Targets are usually made of stiff material, such as plywood, pressboard, or plastic or metal sheets painted matte white. They should be fixed to the ground by suitable means; this is often done by nailing the targets to wooden pegs driven into the ground. Stiff targets

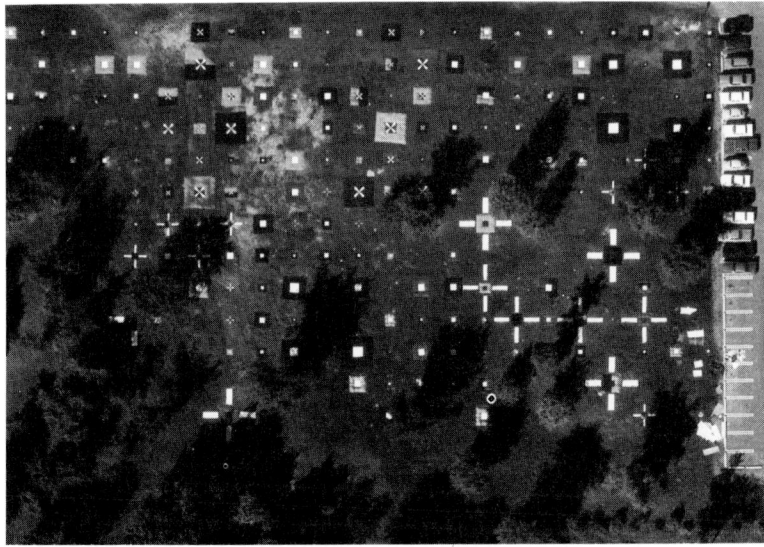

Figure 8-6. Examples of targeted points.

are occasionally used as "elevated" targets if targeting at ground level is not possible, and this is often the case when boundary points are located along walls, fences, or hedges.

It is important that targets remain flat until aerial photographs are taken. Otherwise, the image of the target could be seriously distorted, and results would be impaired.

The greater the contrast between the surrounding terrain and the target, the better the definition of the point in aerial photographs. The darkness of the surface as perceived on the ground might be most deceiving. For example, concrete and asphalt usually reflect sufficient light to be recorded in photographs as bright surfaces. Consequently, white signs on a concrete or asphalt surface do not photograph as well as is often assumed by the field crew; therefore, the targets should be somewhat larger than the theoretically computed sizes.

The size of targets is another important consideration. It is essential that the images of targets in aerial photographs be larger than the measuring mark seen in the photogrammetric plotter or measuring instrument. Otherwise, the measuring mark would cover the image of the target, and accurate pointing would not be possible. For comfortable pointing, the image of the target should have a diameter 40 to 50 % larger than the measuring mark. Starting from this requirement, the minimum size of targets to be used could be computed if it were not for the photographic and optical factors involved in the process. The limited resolving power of the camera–film combination, which is particularly affected by a low contrast between the targets and the terrain background, requires that the targets are from 2 to 3 times larger than the size determined by a strictly geometric consideration.

By painting the area surrounding the targets with a black, *nonreflecting* paint, the size of the targets can be reduced. It is simpler, however, to paint larger targets white in the first place than to use two colors for smaller targets; this is the general approach followed, unless the contrast between the background and the target is not sufficient to identify the point on the photographs.

PHOTOGRAMMETRIC DETERMINATION OF SUPPLEMENTARY CONTROL POINTS: AERIAL TRIANGULATION

General Considerations

The basic control net in urban areas must provide points of a density and accuracy sufficient to be compatible with the stringent surveying requirements in cities, including cadastral survey. The control points, at least the most important of them, must be permanently monumented or referenced, so that they can be easily and accurately located in the field. Only then can various surveying operations within the area be integrated in one uniform system, the required accuracy be assured, and a meaningful accumulation and completion of the existing surveying and mapping data for a continued updating process be achieved. Since an accuracy of the order of ± 2 cm is expected from the basic control net, precise field techniques for its establishment must be used.

There are, however, situations in which there is a need for rapid determination of *supplementary* control or reference points and, if properly handled, the accuracy provided by photogrammetric means is acceptable. Moreover, a large number of urban communities in various countries either lack even rudimentary maps or have primitive maps of limited accuracy, or perhaps only photomosaics. In such cases, photogrammetrically determined control points are not only acceptable but often offer the only immediate solution.

Another example where photogrammetrically determined control points constitute an important part of carefully designed city surveying and mapping systems is the situation in which higher accuracy for surveying details along streets and lower accuracy for off-street details is specified. Of course, the prescribed tolerances in both cases are relatively strict. Whereas the accuracy of the first group of points implies the use of the most precise field surveying methods, the relaxed tolerance for second group permits an efficient use of photogrammetry for the determination of suitably located instrument stations for the intended field operations.

It should also be noticed that by selecting very large-scale aerial photography, such as 1 : 2000, very high accuracy in photogrammetric determination of control points can be achieved. This approach, however, may not be practical, on the grounds of general efficiency and economy.

Depending upon the purpose and accuracy requirements, photogrammetry may be used to densify a relatively extensive control net covering either the whole or large part of an urban community. The procedure used in this case is called *block triangulation*, since a block of photographs consisting of more than one strip is involved.

If a single strip of photographs is sufficient to cover the area in which supplementary control points should be established, it is referred to as *strip triangulation*.

The simplest case in city survey is *densification of control points in a single stereopair*. Some authors also define this work as aerial triangulation. Since any photogrammetric plotting or numerical determination is based on a spatial triangulation process, the extent to which the term "aerial triangulation" can be used is an arbitrary matter.

Aerial triangulation of any extent can be carried out by using either analytical or analog methods, depending on the instrument available, and the process is consequently referred to as either *analytical* or *analog* aerial triangulation.

Since aerial triangulation only permits densification of the existing control net, the initial step in any photogrammetric triangulation project is the establishment and targeting of the ground control net. The spacing of ground control points depends upon accuracy requirements, type and quality of measuring instruments, quality of photographic material, and the triangulation procedures used.

If densification of the ground control net is restricted to single models, each must be completely controlled. In strip triangulations, adequate ground control must be available at both ends and in the center of the strip. In block triangulation, peripheral control (horizontal and vertical) and some control in the interior of the block (particularly vertical control) is required.

If more than one pair of photographs (single stereomodel) is used in the aerial triangulation process, a certain accumulation of errors takes place. This may best be demonstrated in the example of a triangulation of a strip of photographs. Depending upon the procedure used, each consecutive photograph or model is connected to the previous one until a "model of a strip" is formed. The triangulated points are recorded in a continuous coordinate system common to the entire strip. As a result of the unavoidable accumulation of various errors, however, the strip is deformed, and the deformation grows with the number of photographs in the strip, following the general pattern indicated in Figure 8-7.

A large part of this deformation is systematic in nature. The object of the adjustment of the triangulation is to remove deformation errors from the triangulation results. Initially, simple graphical, numerical, or even mechanical procedures were used in the adjustment process. At present, computational programs are available for this purpose. Densification of a ground control net by using properly controlled single models is a relatively straightforward and safe operation; however, a strip or block triangulation is a most complex procedure requiring superior theoretical knowledge and experience. Only a carefully planned triangulation project covering all phases starting with the photographic mission and including supporting field surveys, targeting of existing and to-be-determined control points, and selection and adjustment

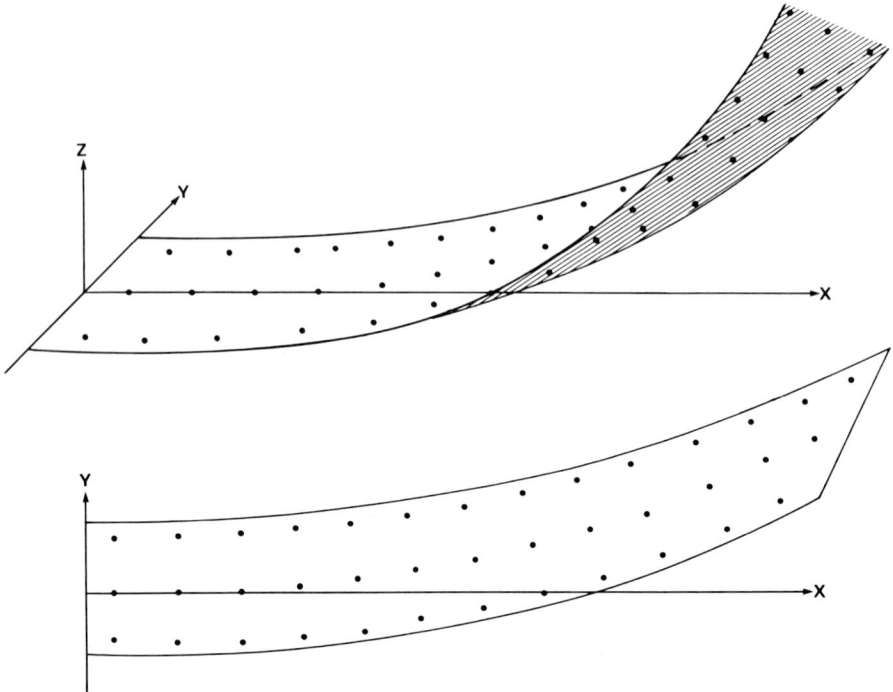

Figure 8-7. Vertical and horizontal deformation of a triangulated strip.

of the triangulation technique can provide satisfactory results. Therefore, municipalities would be well advised to entrust this work only to the most competent operators or agencies of proven reputation. All contracts must insist on an adequate field check of triangulation results, before final payment for the work.

The outline of photogrammetric triangulation procedures in the paragraphs that follow is intended only as a general introduction to the subject.

Use of Targeted, Natural, or Marked Points in Aerial Triangulation Process

Various types of points involved in the aerial triangulation process are used. The first group constitutes ground control points on which the actual aerial triangulation is based. Careful targeting of these points is absolutely necessary to ensure acceptable results. Also, a recommended practice in precise triangulation is to target the tie points between strips.

If the photogrammetrically determined points are to be used as ground reference points, they must also be permanently monumented and targeted for photographic flight. If permanent monumentation is not practical or is not intended for secondary control points, the selected points must be carefully targeted and referenced to the surrounding permanent details, so that their position can be precisely reestablished on the ground when necessary.

If aerial triangulation points are not to be used as future control points in field surveys, all kinds of "natural" points that are well defined in aerial photographs can be considered. Usually, points are selected within areas of common overlap between successive photographs and adjoining strips, so that they can double as pass points.

Control points for purely photogrammetric operations need only be marked on diapositives by using one of the commercially available point marking devices. "Pricking" of points on the transparencies (negative or diapositive) is a practical measure. In addition to clearly defining points to be used, pricking of points eliminates time-consuming, and sometimes ambiguous, point description by sketches or marking on paper prints.

Instruments for Analytical Aerial Triangulation

Triangulation instruments in which the image coordinates of points to be triangulated are measured and their ground positions determined by computation belong to this category. The analytical approach has many advantages over analog methods:

> There are no restrictions insofar as type of photographs is concerned (focal length, distortion characteristic, format, field of view) as long as the photographs satisfy general accuracy considerations.
>
> Coordinate measurement is a precise, simple, and relatively rapid operation.
>
> Any known corrections can be applied to improve the accuracy and thus increase the general efficiency (including economy) of photogrammetric work.
>
> Mathematical computations provide more precise solutions than those obtained by analog means; they can be carried out using sophisticated programs developed elsewhere.
>
> The introduction of the analytical plotter, a particularly precise and efficient instrument, has provided a powerful tool that meets the most demanding photogrammetric surveying and mapping requirements, particularly those encountered in city work, including aerial triangulation.

The actual triangulation process must be adapted to the type of measuring equipment used. Careful planning and preparation of the work on the measuring instrument is essential for the general efficiency and quality of the work.

Use of stereocomparators. Use of stereoobservation extends the photogrammetric technique to any arbitrary terrain point. Obviously, triangulation points intended for use as reference points in future field work must be identifiable on the terrain and should be permanently monumented and targeted. If, however, the points are to be used as a control for photogrammetric work exclusively, they need only be unambiguously defined in the photographs. In such case, suitable image points can be used as triangulation points, even points that are well defined in only one of the pictures of the stereopair. *Réseau points* or points pricked on every second photograph are occasionally used for this purpose. The choice of the latter approach simplifies the marking of triangulation points considerably.

The several stereocomparators on the market include the most precise, with instrument errors not exceeding 2 μm, and the less sophisticated, with instrument errors of from 5 to 7 μm. Since the final results depend upon the quality of photographs and the performance of the operator, the difference in accuracy of the measuring instrument within the range mentioned has a limited effect unless all elements of the whole operation are rigidly controlled. Ease and speed of operation, reliability, and good optical characteristics (which of course are indispensable if high accuracy is expected) are equally important factors that must be carefully considered before the definitive choice is made.

Use of monocomparators. Only one photograph is viewed and measured on a monocomparator at a time (12–14). Since photogrammetric determination of ground position requires measurement of identical points on the two photographs forming the stereopair, the corresponding points must be pricked on consecutive photographs by point-transfer devices. If points are targeted, the use of monocomparators is simple and efficient. However, pricking of a certain number of points to be used as tie points between overlaps or strips cannot be avoided. The greatest attention must be given to this operation since the operator of the measuring instrument has no means of checking whether the points are pricked properly, and any inaccuracies in pricking will have a direct effect on the final quality of the work. Pricking must be done in stereomode with identical points marked on three corresponding photographs (or more, if the point is to be transferred to the adjacent strip).

Monocomparators lack flexibility when compared to stereocomparators. However, the rather modest cost of these instruments, and the very high efficiency of some, should be considered.

Use of analytical-type plotters. In more recent years, another powerful tool has been added to the storehouse of photogrammetric instruments. This is the analytical plotter, invented at the National Research Council of Canada, and subsequently manufactured under licence by OMI in Italy. Production of the third generation of this instrument, known under the name of Anaplot, has been started in Canada (Fig. 8-8). Also, a number of companies manufacturing photogrammetric equipment exhibited their prototypes of analytical plotters at the International Photogrammetric Congress in Helsinki in 1976.

Analytical plotters are very precise and versatile instruments in which the basic geometry of photogrammetric processes and related manipulation of data are provided by a computer, which constitutes the heart of the instrument, and not by rigid physical means. As a result, the basic operations typical of photogrammetric work (determination of internal and external orientation parameters and the reconstruction of models) are carried out with mathematical

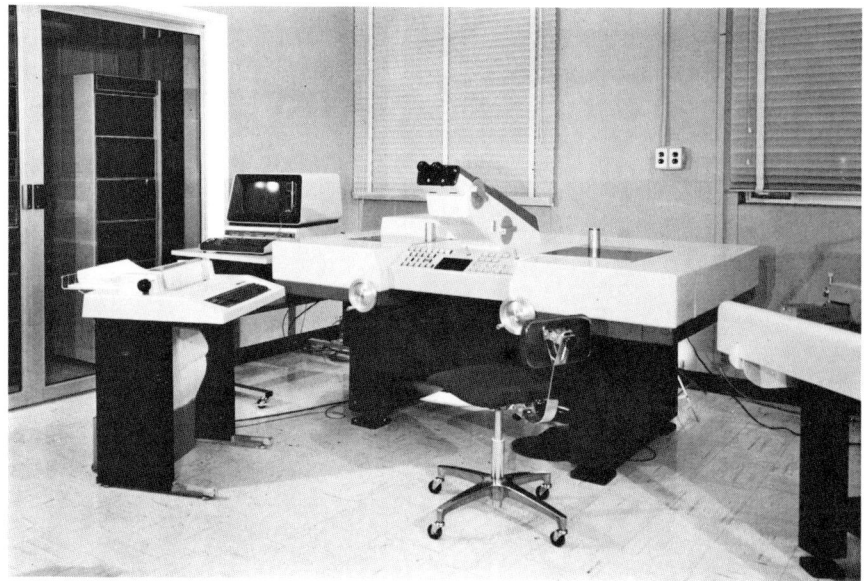

Figure 8-8. Anaplot, analytical-type photogrammetric plotter.

exactness and without any accuracy loss (noiseless processing). Moreover, some operations are automatized, and customary subjective decisions of the operator are replaced by computed results following mathematically optimized solutions. The instrument is also almost "universal," in the sense that it accepts photographs made by cameras of arbitrary geometric characteristics (any size within 23 × 23 cm, arbitrary focal distance and lens distortion) as long as the geometry of the photographic images can be defined.

When referring specifically to the photogrammetric determination of supplementary control points, or aerial triangulation, the analytical type of plotter offers interesting advantages. Owing to the speed and precision of relative (and absolute) orientation, some seconds after coordinates have been measured, each model can be rapidly inspected for the residual vertical parallaxes. Thus from the very outset, gross errors in measurements that usually afflict the analytical triangulation operations are eliminated. Similarly, when using an analytical plotter, rapid intermediate controls of other critical phases of the work can be exercised. The computer of the system can also be used for all kinds of off-line computations encountered in photogrammetric and surveying work.

An important and frequently used triangulation operation for which an analytical plotter provides the best solution is the densification of ground control points in individual stereomodels. In the usual analytical procedures, the coordinates of points to be determined are read from the photographs, and utmost care must be taken to assure that the vertical parallax is properly eliminated in each individual point. This is a tedious and time-consuming operation, and any misjudgements or errors in measurements can be detected

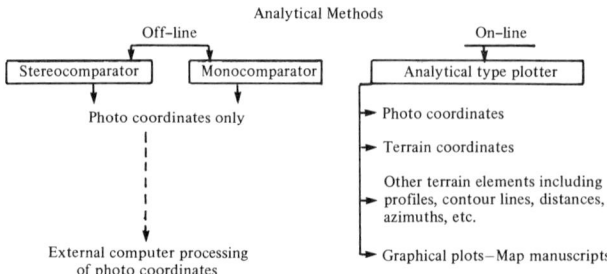

Figure 8-9. Diagrammatic presentation of analytical methods.

only after the completion of the computations. When using an analytical plotter, the perfectly oriented and adjusted stereomodel is presented to the operator instantaneously after the coordinates of a few relative and absolute orientation points are measured. The operator can proceed with the direct measurement of *ground coordinates* of any points in the oriented model, which is a most efficient and safe procedure, free of the most frequently encountered errors in analytical techniques. This mode of operation is known as *real-time*, or *on-line*, operation, as opposed to the *off-line* mode, in which final results are derived from separate computations carried out at different times. The diagram in Figure 8-9 represents the general structure of analytical methods. It should be kept in mind that more extensive triangulation projects always require final off-line processing, even if the computational facilities of the measuring instrument are satisfactory for the purpose.

Analog Methods of Aerial Triangulation

If aerial triangulation is carried out on analog plotters and the results are in the form of model coordinates, it is spoken of as analog aerial triangulation. Various procedures have been used, and the many publications on the subject date from the time prior to the introduction of comparators, and electronic computations and the development of analytical methods (15–17). Even today, it may be necessary to use these methods, particularly if modern measuring and computational facilities are not available. Analog aerial triangulation is capable of delivering acceptable results if it is carried out competently on precise analog instruments.

As in the analytical approach, analog aerial triangulation can be restricted to one single model as well as being extended to a strip or a block of strips.

Analog triangulation of a single model. This is the simplest case of aerial triangulation. To check excessive model deformations, identification and other possible errors in ground control, and the calibration state of the plotter, the number of ground control points used should be larger than the required minimum. It is good practice to use at least five points, of which four should be located in the corners of the stereomodel and one in its center. The procedure is as follows: The model is carefully oriented to the available control points, and the coordinates of the points to be newly established are recorded. If desirable, and if the plotter is equipped with a precision plotting table, the position of the points can be plotted at the same time on a dimensionally stable material, such as transparent polyester sheets or aluminum-mounted paper.

The numerically recorded X-, Y-, and Z-coordinates are model coordinates only; they must be transformed to the ground coordinate system by using the control points. Recording coordinates from an approximately oriented model (absolutely) and deriving ground coordinates of triangulated points through computations in which the model is leveled and precisely scaled is the correct approach and provides the best possible accuracy.

Analog strip triangulation by independent models. This general procedure was first suggested in 1948 (18) for stereoplotters that did not permit continuous strip triangulation owing to restrictions imposed by the design characteristics of the plotter. After electronic computers had been put into service, more refined procedures, also designed for the "universal" plotters, were frequently used. A number of variants have arisen which depend upon the characteristics of the plotters used and the availability of computational facilities. The underlying thought is that the individual models are reconstructed as precisely as possible on the stereoplotter. Connections between the individual models are established (a) by graphical-mechanical means or, if more precise instruments and approach are used, (b) by computation. The first approach is quite cumbersome and lacks accuracy and is not adequate for urban surveying and mapping. The latter approach, which can be considered to be a semianalytical procedure, is capable of providing good numerical results.

Analog continuous strip triangulation. This is a classical aerial triangulation procedure, efficient and precise by the standards of analog methods. It requires universal stereoplotters with the capability of introducing base "in" and "out" to reverse the direction of triangulation on the plotter. Another requirement is a coordinate recording capability. Unlike the previous method, after the first stereopair 1–2 is relatively and (at least approximately) absolutely oriented, the following stereopair 2–3 is relatively oriented by using only the orientation elements of the photo carrier, in which the third photograph is placed without changing the orientation of the second photograph. By doing so, the stereopair 2–3 and the resulting stereomodel automatically acquire an approximate absolute orientation. When each successive photograph is connected in a similar way, a triangulation strip is formed in which the coordinates of all measured points are in a continuous coordinate system.

Analytical Aerial Triangulation

General remarks. The analytical methods of aerial triangulation determine the ground coordinates of terrain points by computational process, using the coordinates of their images as measured in aerial photographs. The orientation of each photograph of a strip or block at its moment of exposure is computed in a rectangular three-dimensional coordinate system. With this orientation, rays from corresponding images of terrain points in different photographs intersect, and the coordinates of all these points of intersection are computed. This computation is performed either separately for pairs or strips of photographs, in an arbitrary rectangular coordinate system, or simultaneously for all photographs, directly in the ground control system. In the first case, the computation is followed by a transformation to the ground control system.

To check on errors and to improve the accuracy of the computed coordinates, more points are measured and more ground control points are provided than the minimum needed to compute the orientation of the photographs. For this reason, the triangulation requires an adjustment in which the method of least squares is used to minimize discrepancies or residuals.

The first step in the computation is the conversion of the measured coordinates to photograph coordinates, with the origin of the coordinate system in the principal point in the photograph. The conversion may include a rotation to make the coordinate axes coincide with the axes defined by the fiducial marks.

Analytical triangulation has the potential of producing considerably more accurate results than triangulation on analog instruments. To realize this potential, the photograph coordinates must be corrected for known errors. Calibration of the measuring instrument provides corrections for systematic screw or other measuring errors, for instance, and for lack of rectangularity of the axes of the measuring device. Calibration of the camera provides corrections for radial lens distortion and possibly for tangential distortion. Corrections for film distortion can be computed from the measurement of the fiducial marks. Here prudence is necessary, especially in the case where four fiducial marks are located in the corners of the photographs, because the distortion at the locations of these marks is generally not representative of the distortion throughout the photograph. Rather than correcting the photograph coordinates by a projective transformation, it may be advisable to apply only correction factors for shrinkage separately in the two coordinate directions. Much more effective corrections for film distortion can be computed if the camera contains a register glass with réseau in or near the image plane, and each measurement of an image point is accompanied by the measurement of one or more neighboring réseau point (19). A less precise but simpler procedure is to derive corrections to image points from a few suitably distributed réseau points.

In addition, the photograph coordinates should be corrected for the effects of refraction and, possibly, earth curvature. It is often sufficient to compute symmetrical radial corrections based on the assumption of exactly vertical photographs.

Analytical triangulation of models and strips. The photograph coordinates, together with a third coordinate along the perpendicular in the principal point, define a rectangular coordinate system x, y, z, with the origin chosen in the projection center. Each vector $\mathbf{x}$ from the projection center to image point has components $x, y,$ and f.

In the coordinate system in which the triangulation is computed, the orientation of a photograph is specified by the coordinates of its projection center and the rotation of the photograph. The latter is defined by an orthogonal matrix $\mathbf{R}$. The nine elements of this matrix are functions of only three independent parameters. Suitable parameters are rotations of the photograph about mutually perpendicular axes or special parameters of which the nine elements are rational functions—which is more advantageous for electronic computation. Assuming, for each photograph, a second coordinate system with origin in the projection center and axes parallel to those of the triangulation system, the coordinates x', y', z' of an image point will be the components of the vector in this system; that is

$$x' = \mathbf{R}x.$$

The earliest and simplest method of analytical triangulation is by strip formation with respect to an arbitrary coordinate system X, Y, Z (20). This method follows closely the procedure used on analog instruments. An arbitrary orientation of the first photograph is assumed, each photograph in succession is relatively oriented, and the resulting model is scaled with respect to the preceding one.

For relative orientation, each pair of corresponding image points in two successive photographs gives rise to one condition equation that states that the above vectors x'_i and x'_{i+1} for photographs i and $i + 1$ and the vector $\mathbf{B}$, which connects their projection centers, must lie in one plane. In vector notation

$$\mathbf{B} \cdot x'_i \times x'_{i+1} = 0.$$

This equation contains the orientation parameters of photograph $i + 1$ as unknowns. Two orientation parameters occur in the vector $\mathbf{B}$, and three occur in the vector x'_{i+1}. The equation is nonlinear with respect to these parameters. To compute the parameters, approximate values are assumed, the equations for all pairs of image points are linearized, and the linear equations are used to compute corrections to the approximate values in an iterative procedure. After this, triangulation is completed by intersecting the vectors x'_i and x'_{i+1} to corresponding image points.

For strip triangulation, an arbitrary orientation of the first photograph and the length of the base are specified only for the first model. For all following models, the orientation of the first photograph is obtained from the preceding model, and each model is scaled from points in the common overlap with the preceding model.

If the models have been measured independently, the x, y coordinate system of a photograph may not be the same in the two models. In this case, strips can be formed from these models in the same way as from models obtained by analog triangulation, i.e., by connecting the models computationally with the help of common points.

Triangulation must be followed by transformation to the geodetic coordinate system and adjustment. The procedures are similar to those used for strips or models obtained by means of analog instruments.

Adjustment of single strips and blocks of strips. The adjustment of strips which have been triangulated either in an analog instrument or analytically consists in the polynomial transformation of the strip coordinates to the geodetic system (21). The transformation must minimize the residuals at the ground control points and, in the case of more than one strip, the discrepancies at the tie points between strips.

The strip coordinates can be affected by random and systematic errors as a result of film deformation or instrumental errors, for instance. Both accumulated random errors and systematic errors produce strip deformations. Typical deformations are torsion, longitudinal height curvature, curvature in the horizontal plane, and scale variations. Suitable second- or higher-degree terms must be included in the polynomial transformations to correct for these deformations. If the photographs show differential distortion, this can be eliminated from the strip coordinates by a suitable scale correction of one of the planimetric coordinates.

In the case of individual strips, the simplest transformation formulas are as follows. The strip is first transformed to the geodetic system by a similarity transformation. In matrix-vector notation, with

$$\mathbf{X} = \lambda \mathbf{R} \mathbf{x} + \mathbf{C}.$$

Here λ is a scale factor, $\mathbf{R}$ is a rotation matrix, and $\mathbf{C}$ is a translation vector. The parameters in this transformation are computed by linearizing the condition equations for the ground control points and computing a least-squares solution in an iterative procedure. An auxiliary coordinate system is then assumed in which the new X-axis lies along the axis of the strip. Corrections ΔX, ΔY, and ΔZ to the new coordinates can now be defined independently as functions of the new X and Y,

$$\Delta X = a_1 + a_2 X + a_3 X^2 + (a_4 + a_5 X + a_6 X^2)Y$$
$$\Delta Y = b_1 + b_2 X + b_3 X^2 + (b_4 + b_5 X + b_6 X^2)Y$$
$$\Delta Z = c_1 + c_2 X + c_3 X^2 + (c_4 + c_5 X + c_6 X^2)Y.$$

The parameters in these functions are computed by inserting in these equations the residuals at ground control points after the similarity transformation.

This procedure is the analytical equivalent of an early graphical procedure in which corrections along the edges and axis of the strip were computed from residuals at ground control points situated in three lines across the strip. The analytical procedure has the advantage that the ground control points need not be located in these lines, but can be more freely distributed through the strip. If sufficient ground control is available and the strip deformation shows irregularities, terms with X^3 and perhaps $X^3 Y$ can be added to the formulas.

The simple adjustment does not take into account the relation that should exist between X-, Y-, and Z-corrections: A local change of tilt in the strip causes changes also in the X- and Y-coordinates, and a local change of scale causes changes in Z if the terrain is not flat. A more sophisticated adjustment will therefore apply corrections that, locally, produce a transformation that differs as little as possible from a similarity transformation.

In practice, this makes the adjustment an iterative procedure in which planimetric coordinates and heights are adjusted in turn and in which in each iteration the effect upon the heights or planimetric coordinates is taken into account. In principle, this adjustment can be set up in two different ways. In each iteration, the computed parameters either are used immediately to transform the strip or are added as corrections to the parameters computed in earlier iterations. In the latter case, the final values of the parameters will be used for a once-and-for-all transformation of the strip. In each of these two procedures, the details of the transformation can be set up in various ways and are to some extent a matter of individual preference.

If the strips are short, it will often be possible to restrict the polynomials to second-degree terms. For longer strips, third-degree terms may be needed, especially for the correction of longitudinal height curvature. Terms higher than third degree are seldom advisable, because they may cause large errors in uncontrolled areas of a strip.

The block adjustment of strips is performed by polynomial transformation of all strips of a block. In this adjustment, ground control points and tie points must be weighted differently, in a way such that a good fit is obtained at the control points as well as between the strips. For each strip, suitable degrees of the polynomials must be selected. For the height adjustment especially, it must be ascertained that the available ground control and the support from adjoining strips are sufficient to make the term in the transformation formulas well defined.

As in the case of individual strips, this adjustment is an iterative procedure, but now each iteration can be performed either on strip after strip or on all strips simultaneously. The first method has the advantage that a program for block adjustment can be easily developed from one strip adjustment and requires little computer storage. To obtain a reasonable speed of convergence of the iterative adjustment, however, the first iteration must already produce a reasonably good positioning of the strips. This requires that at least one strip, which is transformed first, has sufficient control for independent positioning.

In the adjustment of a block of strips, the planimetric transformations should preferably be conformal. Each pair of coordinates can be written as a complex number, $\xi = X + iY$, and the transformation can be written as a polynomial in ξ with complex coefficients. The main advantage of this is that the auxiliary coordinate system with X-axis along the strip axis is not needed for the planimetric transformation, and the transformation of residuals to and from this system is therefore avoided. This approach also saves considerably on storage for the computation of the parameters and on computation time.

After the adjustment, the two sets of coordinates of each tie point obtained from overlapping strips are averaged. This and the averaging of tie points between models, as well as the higher degree polynomial transformations, will cause discrepancies when the individual models are transformed to the adjusted coordinates. If the results of the adjustment are to be used for plotting, the planimetric discrepancies must be within the plotting accuracy. In topographic applications, this requirement can be easily met by the use of perimeter ground control. To meet the required height accuracy, however, bands of height control points are needed across the strips at distances of eight or fewer photographs. In more demanding applications, such as cadastral mapping, it is more difficult to meet the requirements with a polynomial adjustment of strips, and an adjustment of independent models or bundles may be preferable.

Block adjustment of independent models. The adjustment of independent models is performed by subjecting each model of a strip or block to a similarity transformation. The only deformation that is introduced as the result of adjustment is through the averaging of the coordinates of tie points after the adjustment. Because models are much smaller units than strips, the discrepancies after the adjustment are generally small, and the introduced deformation is negligible.

The adjustment of the models of a single strip dates back to the time before the advent of the electronic computer (18). The independent models are first connected into a strip, either during the triangulation or later by analytical means. Sufficient ground control is required in the first and the last model of the strip; these models are transformed using the two control sets, which makes it possible to compute closures in tilts, scale, azimuth, and translation. The intermediate models are transformed by interpolation.

At first, block adjustments of models were applied to planimetric coordinates only, starting from at least approximately leveled independent models, with leveling accomplished by similarity transformations, which again is done most economically by using complex numbers. A simultaneous adjustment is needed, because an iterative adjustment that operates on model after model would require an excessive number of iterations. Even though the transformations are linear, two iterations of the simultaneous adjustment are required to obtain accurate results—one for approximate positioning and one for the final adjustment.

Three-dimensional block adjustments of models also exist. Because the three-dimensional similarity transformation is nonlinear with respect to the parameters, these adjustments require an iterative procedure, which is most economical when it operates on planimetry and heights in turn. If this is done, the correlation that exists between these two factors must be taken properly into account.

Block adjustment of bundles. The simultaneous adjustment of the photographs of a strip or block is known as the adjustment of bundles. The orientation of a photograph and the vectors $\mathbf{x}$ and $\mathbf{X}$ are defined the same way as in the case of strip triangulation. However, the condition of coplanarity is not used. Instead, each measured image point gives rise to one vector equation that states that the image point, projection center, and corresponding terrain point must lie on a straight line. This is the collinearity condition. It is expressed by the equation

$$\mathbf{X}_p - \mathbf{X}_c = \lambda \mathbf{R}x,$$

in which $\mathbf{X}_p$ and $\mathbf{X}_c$ are the position vectors of terrain point and projection center, respectively (i.e., the vectors whose components are the coordinates of these points in the triangulation system), and λ is a scale factor.

The unknowns in this equation are the orientation parameters of the photograph (contained in $\mathbf{X}_c$ and $\mathbf{R}$), the scale factor λ, which is different for each image point, and the coordinates of the terrain points, unless it is a completely known ground control point. The parameters of all photographs and unknown terrain coordinates are computed simultaneously. Again, approximate values of all unknowns must be determined, the equations must be linearized, and the linear equations must be used to compute corrections to the approximate values in an iterative procedure.

With six unknowns for each photograph and three unknowns for each terrain point, the system of normal equations that must be formed and solved in each iteration of the bundle adjustment becomes extremely large, and writing a program for this adjustment is a task of considerable magnitude. Until recently, only a few photogrammetric organizations had produced such a program (22), and in most cases this task required several man-years. Most of the time was consumed by efforts to organize and process the data in order to make the adjustment reasonably efficient and economical.

The following two features are employed to contribute to the economy of the computation and to remain within the available storage space. First, the measurements are sorted according to the terrain points, and the corrections to the approximate coordinates of the terrain points are eliminated from the normal equations during their formation; alternatively, but less conveniently, the setting and the elimination procedure can operate on the photographs and their orientation corrections. Second, the matrix of normal equations is partitioned into 6×6 submatrices corresponding to the photographs. These submatrices are arranged so that nonzero submatrices occur as closely as possible to the diagonal of the matrix; the zero-submatrices are neither stored nor operated upon.

The normal equations can be solved by ordinary Gaussian elimination and back-substitution, operating on the 6×6 submatrices. Some organizations are interested also in the accuracy of the computed orientation elements. Information on this is contained in the inverse of the matrix of normal equations, which has as its elements the variances and covariances of the orientation corrections. Computation of the variances and covariances greatly increases computation time and storage requirements.

It is also possible to add corrections to the photograph coordinates for unknown systematic errors. The collinearity condition must then be linearized with respect to the parameters in these correction terms, and these parameters must be added as unknowns to the normal equations. This does not complicate the formation and solution of the normal equations very much, and it is to be expected that this step will appreciably improve the results of a bundle adjustment. As yet, however, there is no consensus of opinion on the form that these corrections should take.

Accuracy of block aerial triangulation. The accuracy that can be achieved in polynomial block adjustment depends strongly upon the number of ground control points that is used, their distribution, the length of the strips, the degrees of the transformation, and even upon the program that is used.

Illustrative of the latter are, for instance, the results of tests of the Working Group on Block Adjustment of the International Society of Photogrammetry in 1972 (23). In one of these tests, the fictitious data that were used were affected by random errors only, and medium-dense ground control was used along the perimeter of the block, together with a few additional interior height control points. Five participants who used polynomial block adjustment obtained RMS (root-mean-square) values of the errors at check points which ranged at photograph scale from 21 to 70 μm in horizontal position and from 32 to 53 μm in height. Those fictitious data affected also by systematic errors produced RMS values that ranged between 23 and 152 μm in horizontal position and 36 and 131 μm in height.

In practice, the lower limits in these ranges can be reached only with rather dense perimeter control. The height control must be properly located to prevent torsion of the strips, and some internal height control must be present. In less favorable circumstances, RMS values of 50 μm must be expected. If, however, a test gives values larger than 100 μm, there is reason to suspect the results.

Investigations on the accuracy of the block adjustment of independent models show that this adjustment tends to give more accurate results than polynomial block adjustment. This appears to be the case especially for large blocks with extremely little or extremely dense ground control. The one participant in the Working Group who used this adjustment obtained in the above test, with data affected by random errors, RMS values of 17 μm in planimetry and 28 μm in height. The data which included systematic errors gave RMS values of 29 μm in planimetry and 37 μm in height.

Although few results of bundle adjustments have been published as yet, it is already evident that this adjustment is the most accurate method of photogrammetric densification of ground control.

One test with a block of 180 super-wide-angle photographs, dense perimeter control and additional internal height control gave an RMS value of 13 μm at photograph scale for the errors in the position of check points. A test with a block of 130 wide-angle photographs gave RMS values of 26 μm in position and 20 μm in height. In the second test, an additional adjustment was performed in which corrections for unknown systematic errors were included. This reduced the RMS values to 8 and 14 μm, respectively.

In the above test of the Working Group on Block Adjustment, three participants used the bundle adjustment. They obtained RMS values at check points of 10 and 11 μm in planimetry and from 23 to 25 μm in height, with data affected by random errors only. The data affected also by systematic errors produced RMS values of 20 and 30 μm in planimetry and 31 and 34 μm in height.

In the aerial triangulation work carried out on large-scale photographs with optimized overall conditions (high-quality and rigidly controlled photographs, larger than usual longitudinal and side overlap of photographs, sufficient density and correct distribution of ground control points, well-planned and precise instrument and computational work), much higher accuracy than that quoted has been reported by some authors. It must be noted, however, that despite all precautions and careful work, occasional gross errors creep into photogrammetric determinations. They are mainly restricted to individual points and are usually caused by the operator's misjudgement or by occasional errors of an accidental character, such as local image deformation. To eliminate the possibility of these gross errors, fully independent photogrammetric determinations must be considered, which implies an independent set of photographs. This approach is used in some countries (pp. 307–308).

The continued improvement of photogrammetric techniques encouraged researchers to develop further and experiment with aerial triangulation in view of the growing need for large-scale work, including urban surveying and mapping. Use of photogrammetry for densification of the horizontal control net in urban areas is particularly attractive, since precise targeting of points to be determined is not objectionable and does not present the same difficulty as is the case, in cadastral surveying, for instance. This is due to the relatively limited number of points involved and their convenient location for targeting. A number of recent publications (24–27) prove that coordinate accuracy (m_x, m_y), based on photographs at 1:10 000 scale and larger, is of the order of from 5 to 7 μm in the photographic plane. This accuracy can be further improved by using multiple determinations from independent photographs, as demonstrated by D. C. Brown (27), F. H. Ackermann (24, 25), C. C. Slama (50), and others.

The accuracy of height determination is usually lower than that reported above, but is still satisfactory for most topographical work. However, from vertical reference points in urban areas, much higher accuracy is expected; this can be achieved best by field leveling.

The desired accuracy of a photogrammetric product and the accuracy specifications for supporting control points should be properly balanced. It has been shown by Kratky (28) that the accuracy of control points used to support a photogrammetric stereomodel does not necessarily have to be higher than the basic accuracy of the operation used to derive the model. If the accuracy of horizontal control is approximately the same as that of the plotting, the resulting final standard error in derived x and y model coordinates increases, on the average, by only 14% compared to the ideal conditions of extremely accurate control. Likewise, the error in elevation would increase by only 12%. This interesting finding should be taken into account when planning triangulation work, so as to make it as economic as possible.

The Procedure Offering Maximum Accuracy in Photogrammetric Determinations

In photogrammetric coordinate determination of control points (aerial tri-angulation) or other terrain points in urban areas, the highest possible accuracy is often sought. To achieve it, three basic requirements must be met:

1. High-quality photographs taken with a réseau camera.
2. The use of analytical methods.
3. Correction of image geometry for image deformation.
4. Multiple determinations from independent (additional) photographs.

The high-quality photograph is a requirement that is much more frequently quoted than met. High-quality photographs combine the best possible definition of the photographic image with minimum image deformation. Well-planned flight missions carried out with modern cameras suspended in a camera mount capable of damping particular aircraft vibration and properly selected film and exposure time and subsequent correct processing of film and diapositive copies can provide the desirable, high-quality photographs that constitute the basic condition for the highest possible photogrammetric accuracy. Since, as will be noted in discussing point 3, unpredictable image deformations cannot be avoided, even in the best controlled and executed photographs, a camera equipped with *réseau plate* should be used. To maintain the initial quality of aerial photographs, the glass-plate diapositive copies should be produced on a *parallel-light printer*.

To extract all available geometric quality from aerial photographs, analytical methods must be used. They offer the best possible accuracy in measurements and correct mathematical processing of measured quantities to derive the most accurate, final ground data. The development of analytical plotters adds a further extremely important factor. Modern analytical plotters permit sophisticated mathematical handling of data on line, whereas the operator retains complete control of the entire process. Thus various blunders and the need for repeated measurements are eliminated. Compared to stereocomparator measurements, determinations from single models on an analytical plotter are much faster and less tiring because they are carried out on oriented models. However, this does not prevent the inclusion of corrections in the determination of individual points.

The third requirement—the use of corrections for image deformation—is least known to practicing photogrammetrists.

Even the most carefully made photographs contain image deformation (19, 29, 30). We consider unpredictable deviations from the theoretical location of image points as such; deviations are caused by a number of factors, such as dimensional instability of photographic material, lack of film flatness at the moment of exposure, and unevenness of the upper surface of the photographic emulsion. Some causes of image deformation, alone or in combination, may have a systematic effect on the general pattern of the deformation, as for instance a specific pull of the film in the aerial camera during film transport and which causes the film deformation or delay in elastic recovery of the film. An image deformation, however, can be modified in any picture in a random manner by causes of accidental character, such as lack of flatness of the aerial film at the moment of exposure. This may happen if, for instance, an air bubble is trapped between the film and the suction plate.

Since image deformation, important from our point of view, is the one present in the photographic image used in the actual photogrammetric operation, any additional deformations introduced during the processing of aerial film and the production of glass or film diapositives must also be considered.

The only reliable way to control these deformations is to use the cameras equipped with the so-called réseau plate. The réseau plate carries at its surface, in contact or close to the image plane, small crosses arranged in a very precise, rectangular pattern spaced at 1 or 2 cm. These crosses, when centrally projected onto the film in the image plane, should form a similar pattern. Any deviation from the known location indicates an image deformation that can be established precisely by measurement of the crosses. Assuming a similar deformation of image points between particular réseau crosses (possibly by an interpolation process), relevant corrections for image deformation of any points can be derived and applied, and the overall accuracy of photogrammetric determinations can be significantly improved. Ziemann (19) established that the gain in accuracy can amount to about 20%.

By competently using réseau photographs under well-controlled conditions, the position of terrain points can be determined photogrammetrically with an accuracy of from ± 5 to ± 6 μm in the image plane.

Further increase of accuracy obviously can be achieved by using multiple photographic coverage of the area. Thus each point can be determined from a number of independent photographs. Using this approach, C. C. Slama of the National Ocean Survey in the United States, in an as yet unpublished report, quoted accuracies of up to 2 μm in the image plane.

PICTORIAL, GRAPHICAL, AND NUMERICAL PRESENTATION OF TERRAIN CONTENTS

General Considerations

Aerial photographs, particularly stereopairs of aerial photographs, contain a great deal of information on the terrain photographed. When using proper photogrammetric procedures, this information can be presented in a meaningful, metrically correct form, and significant accuracies can be achieved.

The three basic forms of presentation are

1. Pictorial
2. Graphical
3. Numerical

Any combination of these forms is possible and can be found in practice. All types of presentation are of interest in urban areas.

Pictorial Presentation

Prints of aerial photographs. In some applications, simple paper prints of aerial photographs or their enlargements are useful. Prime examples are all kinds of preliminary studies for which metric characteristics are of little importance. Photographs, particularly when observed stereoscopically, can provide information such as density and characteristics of buildings, land use, traffic patterns and characteristics, and topography of the terrain. When no other cartographic material is available, even uncontrolled mosaics of photographs can be helpful to the engineers and administrators of urban areas. These can never replace proper maps and other surveying and cartographic products, however, and must not be considered as a possible substitute for maps.

Rectified photographs. Compared to ordinary photographs, rectified photographs are a superior product. They contain the same information but have metric characteristics similar to maps; consequently, they can be used to determine distances and angles between points and directions on the terrain.

To perform rectification, control points that can be identified on aerial photographs are necessary. Field or photogrammetric methods can be used to establish control points. However, aerial triangulation methods are particularly suitable in view of the less stringent accuracy requirements in this particular application and the possibility of choosing control points from those well-defined *characteristic terrain features* (natural or man-made) that are visible on the photographs.

Assuming that the terrain is flat (not necessarily horizontal), classical rectification is a simple and rapid process, requiring relatively simple instruments, called rectifiers. In a well-organized production, rectification of a single photograph requires no more than 15 min, and much faster processes are known and have been used, as for instance, a highly automated system based on gyro indications defining the tilt and tip of photographs as developed by B. Dubuisson in France after World War II (31). The accuracy of rectified photographs of flat areas can be assumed to be

$$m_{\mathrm{p}} = \pm 0.3 \text{ to } \pm 0.4 \text{ mm}$$

as a root-mean-square error in position of any point within the rectified image. Truly flat areas are seldom found, however, and any difference in elevation over the reference plane results in a positioning error in radial direction, which for vertical photographs is defined by a simple formula

$$\Delta r = \Delta Z \frac{r}{f},$$

where ΔZ is the height difference, r is the radial distance of the point under consideration, and f is the focal distance of the camera (Fig. 8-10).

The formula permits the determination of radial shifts Δr that may occur in a given rectified photograph. At the same time, however, the formula suggests how

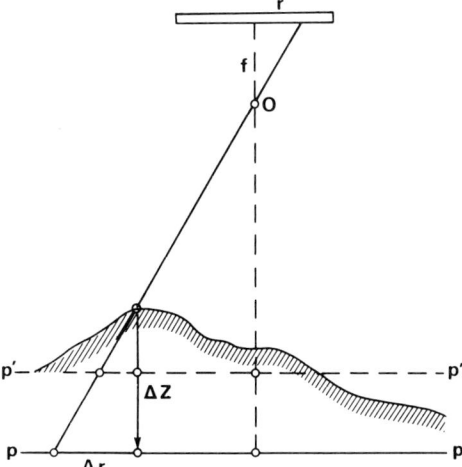

Figure 8-10. Effect of height differences ΔZ in terrain on the planimetric accuracy of rectified photographs.

the resulting shifts Δr can be minimized by proper measures, such as the following:

1. Reduction of the ΔZ value by the introduction of a suitable reference plane, *different* for each particular zone of allowable height difference. This consideration led to development of rectification procedures *by zones* which, however, lost importance once the differential rectification (orthophoto) was introduced.
2. Restriction of the radial distance r or use of central portion of each photograph only. This assumes that there is adequate longitudinal and side overlap between the adjacent photographs. Use of this measure is advisable in urban areas, because of *dead fields* created by the projection of elevated features such as buildings. Restriction of usable areas to the central portions of photographs, however, reduces the general efficiency of the rectification procedures and consequently is only a partial solution to the problem.
3. Size of radial shifts Δr and dead fields can be minimized by utilizing cameras of longer focal distances. One can see from Figure 8-11 that photographs of *identical size* and *scale* cover identical areas on the ground, independent of their focal distances. The selection of cameras with longer focal distances to reduce radial shifts and dead fields is a correct approach since cameras of narrower field of view usually provide a more uniform image definition. Some manufacturers of photogrammetric equipment offer cameras equipped with lenses of up to 60 cm focal distance. The possible drawback of this approach is the fact that most of the conventional stereoplotters do not accept this type of photograph. It should also be realized that the narrow-angle photographs with their smaller base–height ratio offer less accuracy in elevation determination than do wide-angle photographs.

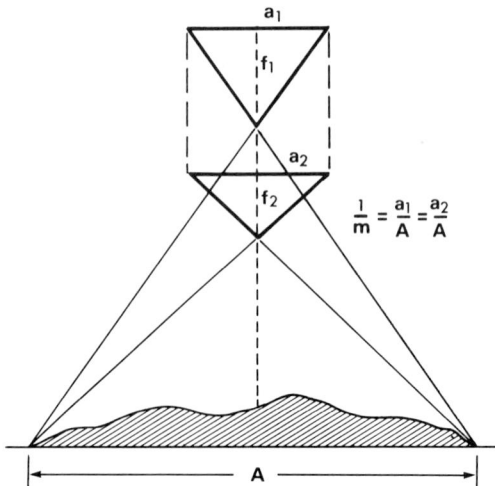

Figure 8-11. Ground coverage as the function of size and scale of the photograph.

Once the photographs are rectified and brought to a uniform scale, they can be put together in the form of a *controlled photo mosaic*, always a useful product in urban areas.

Orthophotos. The orthophoto technique offers the most advanced product in the pictorial-presentation category. Unlike the conventional rectification process, the orthophoto, or differential rectification technique, permits precise rectification of photographs taken over any type of terrain, including high and rugged mountains. Differentially rectified photographs, or *orthophotos*, are visually identical with conventionally rectified photographs; however, their geometric accuracy, if the terrain is not completely flat, is superior. Moreover, the orthophoto technique is relatively simple and is suitable for mass production and automation.

As the term "differential rectification" indicates, in the orthophoto technique, the photographic image is rectified differentially, or by applying rectification to elementary areas of photographs. The procedure can be easily explained when reference is made to Figure 8-12, which represents schematically an anaglyphic-type stereoplotter. Assuming that p_o represents an arbitrary projection plane, point P of the terrain model formed on the plotter will be projected onto projection plane in the point P''. On a correctly rectified photograph, however, point P should be depicted in position P^o as the result of an orthogonal projection of the point. This can easily be achieved by shifting vertically the projection plane p_o to position p_1. Assuming further that for each consecutive point of the terrain profile t–t, the height of the projection plane can be adjusted accordingly, all points of the profile will be depicted in a correct, maplike position. This procedure provides the basis for the orthophoto technique. Some orthophoto instruments follow this procedure precisely. For an anaglyphic plotter similar to

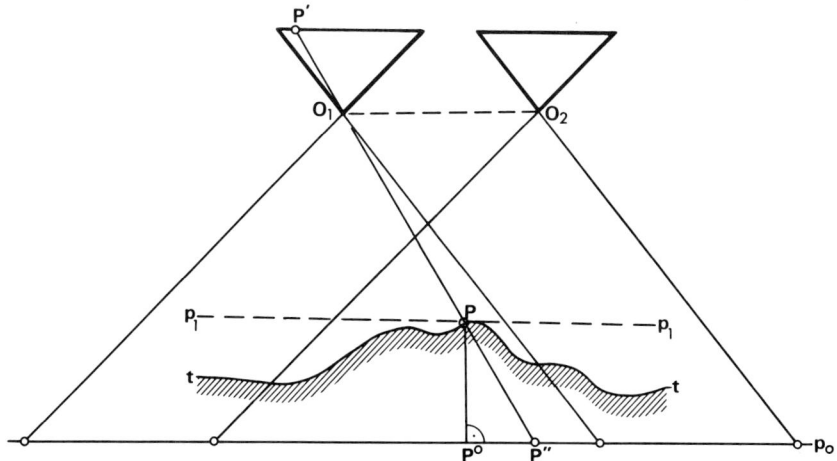

Figure 8-12. Basic geometry of the orthophoto process.

the Kelsh Plotter as an orthophoto-printer, for instance, the procedure is as follows. In the model space of the plotter, a vertically moveable, flat table is positioned. On this table is placed cartographic film sensitive to green light only. The film is covered by an opaque mat except for a small slit which can travel along one of the axes, for instance, the X-axis (Fig. 8-13). As the slit travels across the model, the operator keeps it at the level of the model surface by continuous adjustment of the height of the table. The film below the slit is illuminated by both projectors. However, since the film is sensitive to the green light only, only the terrain image projected by green light is recorded. One passage of the slit across the model produces a strip of a continuously rectified image of the terrain; the assembly of the parallel strips, covering the whole

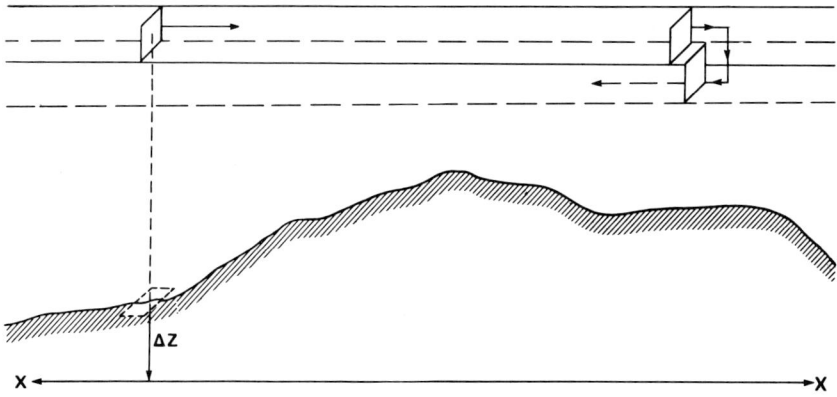

Figure 8-13. Movement of the slit in the model space.

Figure 8-14. T-64, example of early orthophoto equipment developed by the U.S.A. Geological Survey (courtesy of the U.S. Geological Survey).

photograph or part of it, constitutes an *orthophoto*. Depending upon the characteristics of the terrain, the size of the slit can be varied to achieve the required geometric accuracy. Specific measures ensure that there are no visible boundaries between individual strips and that the resultant orthophoto has the best possible uniform image quality. The first practical solution of the differential rectification was provided by R. K. Bean of the U.S. Geological Survey in 1956 (Fig. 8-14).

In other instruments, image transfer from the photograph to the projection plane may be effected by less direct means, but the basic geometry of the process is always identical. Because small areas (slits) are used in the rectification process of the photographic image rather than points in a mathematical sense, the differential rectification is generally an approximate procedure. By selecting sufficiently small areas, however, or by introducing secondary rectification of the elementary areas used in the process, the resulting errors can be significantly reduced.

For a comprehensive review of the great variety of orthophoto equipment, the reader is referred to Ref. (32).

"On-line" versus "off-line" production of orthophotos. The on-line operation is defined as a process in which the product is obtained simultaneously with the basic production operation. For instance, plotting of the map manuscript

directly on a stereoplotter is an on-line operation, whereas the complication of maps from the coordinates recorded on a stereoplotter constitutes an off-line process.

The initial orthophoto techniques were based on the on-line approach, as described in the example in the previous paragraph. In this method, the ortho-photos are produced simultaneously with the scanning of the stereomodel of the terrain along profiles. On-line techniques have the following characteristics: equipment which is simple and can be used for conventional plotting as well as for the production of orthophotos, and control over the entire process exercised by one person. Consequently, there are good arguments for those techniques in small projects with relatively small numbers of orthophotos to produce or in projects carried out by smaller companies or offices that cannot afford to purchase specialized photogrammetric equipment for each type of mapping operation.

On the other hand, there are certain distinct disadvantages to the on-line approach. For example, since the orthophotos are produced as the model is being scanned, any errors in profiling result in geometric inaccuracies of the orthophotos. Profiles must refer to the actual surface of the terrain (ground), which in built-up areas of cities or areas covered by trees may present difficul-ties. More precise results are achieved if the profiles needed for the production of orthophotos are recorded in a separate operation offering opportunity to eliminate profiling mistakes. There are also further operational considerations that support off-line solutions. The scanning of the model in the profiling method is a lengthy operation, particularly since short slits should be used to achieve required accuracy. This lowers the efficiency and the overall economy of the orthophoto technique. A possible solution to this problem is to record profiles at greater intervals than required and have them subsequently densified in the actual orthophoto production phase by a suitable automatic interpolation technique. By separating the orthophoto production phase from the acquisition of the terrain data (profiles), various advantages can be achieved. Profiles can be recorded on relatively simple plotting equipment (the accuracy of profiles does not need to be very high), and the more expensive and specialized ortho-photo instruments can be designated exclusively for fully automatic production of the orthophotos, a rapid process.

Moreover, it should be noted that precise information on terrain topography, usually in the form of contour lines, is available in some countries. It is a simple matter to record these contours numerically and use them in the construction of required profiles and the production of orthophotos. Generally speaking, time causes little change in the topography of the terrain; consequently, the existing terrain data can be used for automation of such processes as production of orthophotos.

The equipment used in an off-line production of orthophotos is more com-plex and expensive than that used in an on-line mode. Speed of production and other advantages already mentioned, however, justify the greater capital investment when a production of sufficient volume is assured.

Use of automatic image correlators in the production of orthophotos. The natural tendency to speed up the production processes found its ultimate expression in the development of image correlators that permit rapid identification of the corresponding points on photographs within a common overlap, and therefore can replace a human operator in recording topography of the terrain. In bare terrain, an image correlator can be used to locate terrain points along certain lines or to produce vertical profiles and contour lines.

This outstanding technology achievement (33, 34) has some limitations, particularly in the large-scale mapping of built-up areas. An automatic image correlator is not capable of recognizing whether the points of interest are ground points, points of buildings, or crowns of trees. Consequently, profiles of the terrain produced automatically on a stereoplotter with the help of an image correlator will include elevated points that should be disregarded. As referred to the orthophoto technique, this has two adverse effects in the large-scale mapping of urban areas:

> The orthophotos based on the automatically produced profiles can contain some planimetric errors owing to the fact that the derived reference level for a given elementary area does not coincide with the ground height.

> The ground elevation as recorded during automatic profiling or contouring of the terrain can be erroneous since it is affected by high-rise buildings or trees.

This latter fact is of particular importance. For this reason, when using orthophoto equipment based on an automatic correlation, the manual mode of operation should be used in built-up areas if the instrument offers such an option.

If precise vertical information on the terrain is available from previous mapping operations, it is logical that it should be used for the production of the orthophotos, thus providing a rapid solution and a superior uniform accuracy (disregarding the coverage of the terrain). Some off-line orthophoto equipment has characteristics suitable for adaptation of a general solution sketched above, and further development in this direction can be expected.

In orthophoto mapping from medium- and small-scale photographs, the orthophoto instruments based on automatic image correlation provide a technically and economically efficient solution.

In a rigidly controlled international experiment (35) with a 1:2500 orthophoto produced from 1:10 000 photographs, a consistent position accuracy $m_p = \sqrt{m_x^2 + m_y^2}$ of 35 μm in the image plane of the original photographs was obtained for higher-quality instruments, including fully automatic instruments based on the use of image correlators.

Stereoorthophotos. Available techniques permit production of orthophotos of excellent metric characteristics. However, the readability of single photographic images as compared to a stereoimage is limited. Moreover, single orthophotos cannot be used to measure elevations of terrain points or to present the topo-

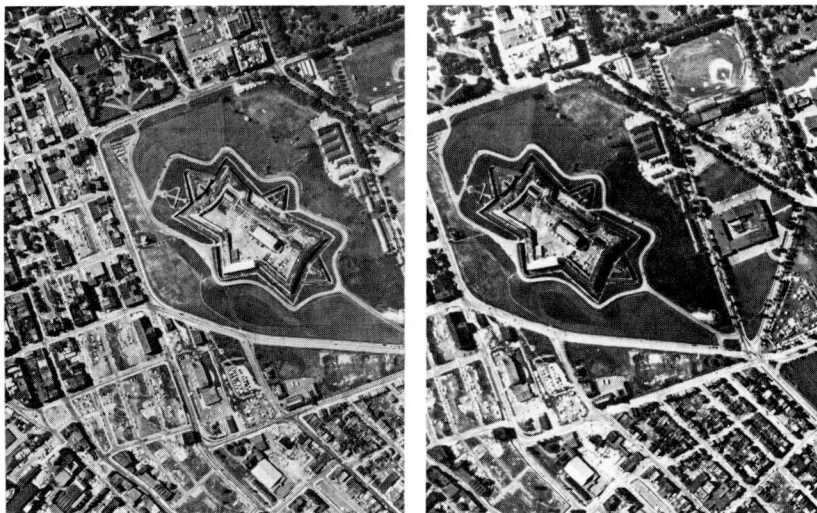

Figure 8-15.   A pair of stereoorthophotos.

graphy of the terrain by contour lines. To overcome these difficulties, Blachut (36) conceived the so-called stereoorthophoto system. By introducing horizontal parallaxes proportional to the elevation differences [according to Collins (37)] into the second orthophoto (called a stereomate) of a stereopair, a new pair of photographic images is created with the following characteristics:

> There are no vertical parallaxes between the corresponding points in the left and right image.
>
> When viewed stereoscopically, the left and right image present a geometrically correct, three-dimensional model of the terrain in which all three coordinates of any point can be measured.

Figure 8-15 gives an example of a stereoorthophoto pair that can be examined with a pocket stereoscope or by the naked eye of those accustomed to viewing stereopictures without benefit of a stereoscope.

The initial stereoorthophoto system consisted of the "Orthocartograph" (38), for the simultaneous production of the orthophoto and stereomate, and the "Stereocompiler," a very simple plotting instrument for the numerical and graphical evaluation of stereoorthophotos. In the meantime, Gestalt International Ltd. equipped its fully automatic Gestalt Photo Mapper with the capability of simultaneous production of stereoorthophotos. Whereas production of stereoorthophotos is a typical photogrammetric operation requiring a thorough understanding of photogrammetric principles and techniques on the part of the operators, the subsequent use of stereoorthophotos is uniquely simple; anyone possessing normal stereoperception should be able to operate the Stereocompiler. Figure 8-16 gives the general view of this simple plotting instrument.

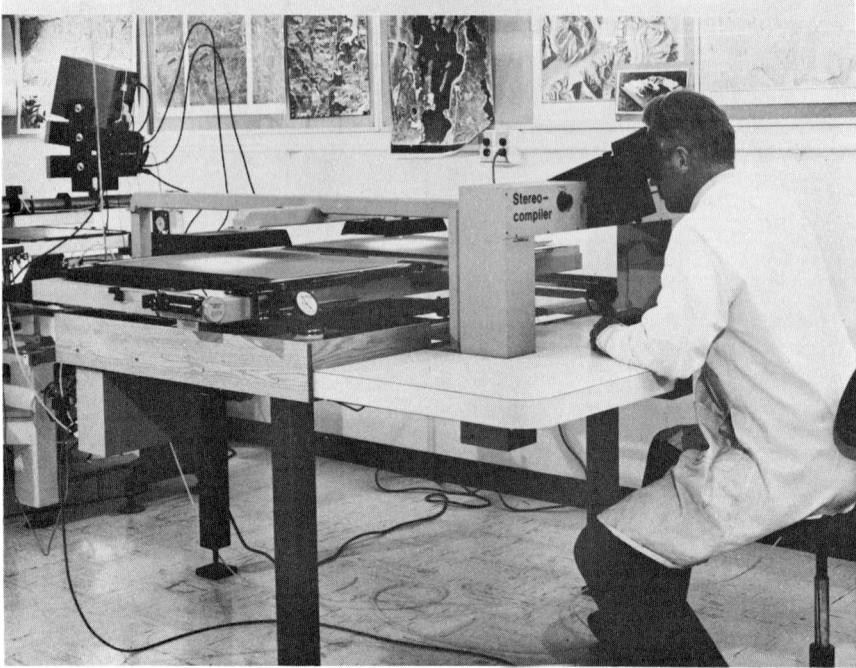

Figure 8-16. General view of the Stereocompiler.

Unlike conventional stereoplotters, plotting on the Stereocompiler is done on a plotting sheet overlaying the orthophoto image; therefore, the result of plotting is visible directly in the field of view together with the three-dimensional model of the terrain. This is an extremely important and convenient feature that makes plotting more efficient and eliminates the possibility of omissions. Moreover, a further most significant simplification can be achieved by using the "single" measuring mark only when plotting planimetry. Thus, the z-motion of the measuring mark is not involved during this part of the operation.

Because only *ground points* are presented in their correct geometric position on the orthophotos, roof or elevated wall points should not be used in plotting ground positions of buildings or other structures. This imposes certain limitations on the use of the orthophoto technique (including the stereoorthophotos) in built-up areas with high-rise buildings when optimum accuracy is required. Nevertheless, the stereoorthophoto technique, with its stereovision, speed, and economy, is most attractive for urban studies and projects and for planning purposes. The natural presentation of the terrain in orthophotos can be of particular value in studying and planning the development of the city's outlying areas or as up-to-date reference material.

Orthophoto plans and maps. It must be understood that a pictorial presentation of the terrain, i.e., an orthophoto, particularly of built-up areas with very complex, planimetric content, cannot replace a map, for several reasons. The most important is that numerous terrain details, natural and artificial, cannot be recognized in a single photographic picture. They must be identified otherwise and marked on the orthophotos by suitable symbols. This is accomplished by using stereophotogrammetric techniques, including the stereoorthophoto technique.

Orthophotos or orthophoto mosaics enhanced by additional symbolized, planimetric information are called *orthophoto plans*. An orthophoto plan with added contour lines is a modern mapping product, called an *orthophoto map*.

The orthophoto technique is relatively new, but its use is rapidly expanding; therefore, it may be useful to introduce a uniform terminology for its various products:

> *Orthophoto*: a differentially rectified photograph usually reproduced on photographic material;
>
> *Stereomate*: an orthophoto with horizontal parallaxes at each point (or elementary area), modified in such a way that when viewed stereoscopically with the orthophoto of another picture of the stereopair, a geometrically correct, three-dimensional model of the terrain is provided;
>
> *Stereoorthophoto pair*: an orthophoto and the corresponding stereomate;
>
> *Orthophoto mosaic*: a number of orthophotos joined together and possibly reproduced as a single image of a larger section of the terrain;
>
> *Orthophoto plan*: an orthophoto or orthophoto mosaic enhanced by additional information, including symbols of planimetric details, but excluding contour lines, which is usually reproduced by cartographic techniques;
>
> *Orthophoto contour map*: an orthophoto or orthophoto mosaic with contours added;
>
> *Orthophoto map*: orthophoto plan with contours;
>
> *Stereoorthophoto map*: a novel cartographic product based on the stereoorthophoto concept, which provides a true three-dimensional presentation of the terrain.

When producing orthophotoplans or orthophoto maps, two different approaches are possible. Usually, the terrain details, which must be symbolized on an orthophoto map, are plotted on a conventional stereoplotter. Subsequently, the plot, together with the orthophoto, is printed on a single sheet of paper. In this approach, the execution of the plot is divorced entirely from the actual orthophoto image with which the plot is eventually combined to form an orthophoto plan or map. Because of unavoidable geometric instrument and human errors, e.g., interpretation errors, certain geometric discrepancies between the orthophoto image and the plot result, which lead to possible confusion in presenting important terrain details. It may happen, for instance, that the boundary line of the plot does not coincide precisely with the boundary visible on the orthophoto. Therefore, when orthophoto plans or maps are produced, unless there is an important consideration such as precise numerical recording from which graphical plots are derived, the correct approach is to perform graphical plotting and symbolization using orthophotos. The *stereo-orthophoto technique* and the associated plotting equipment (Stereocompilers) provide a simple, versatile, and efficient solution.

Accuracy of orthophoto products: planimetric accuracy. A number of early results based on a large sample of *production data* indicated that, even in the early period of the orthophoto technique, an accuracy of from 0.2 to 0.3 mm in x- and y-coordinates of the orthophoto points was achieved, which is similar to the accuracy of conventional line-drawn plans. An international orthophoto experiment (35) proved that modern equipment can deliver a consistent accuracy of from 0.12 to 0.15 mm as standard position error, $m_p = \sqrt{m_x^2 + m_y^2}$, for 1:2500 orthophotos produced from 1:10 000 photographs or from 30 to 35 μm at the scale of original photographs.

The planimetric accuracy depends upon many factors, but primarily upon the orthophoto technique and equipment used plus the topography of the terrain. Consequently, more modern and sophisticated equipment is capable of providing products of superior quality. The results of a test orthophoto produced on the Orthophoto printer of the analytical plotter from photographs taken over the National Research Council of Canada test area are an example of advanced possibilities in this area. The terrain consists of bare rock formations of low but rather steep hills with slopes of up to 30°. The standard error in x- and y-coordinates of the orthophoto points amounted to ± 18 μm, expressed at the scale of original aerial photographs. Assuming a 5× magnification of the orthophotos, this would represent an error of only about 0.08 mm at the scale of the orthophoto, or below errors typical of line-drawn plans.

Accuracy of orthophoto products: height accuracy. The international orthophoto experiment mentioned also provided data on the accuracy of vertical information obtained in the orthophoto production process. The following modes of operation are of interest here:

> *A profiling mode typical to all manually operated equipment.* The vertical terrain profile data are recorded either numerically or graphically; they are subsequently used for an off-line production of orthophotos and computer-generated contour lines, or directly for an on-line production of orthophotos and simultaneous graphical generation of contours. In the latter case, the profile can also be recorded, and contours can be computed in a separate operation. It is obvious that, depending upon the equipment used and the care taken, the primary vertical information could be as precise as that obtained from any photogrammetric operation. Also, derived contours could be of high quality even if they lack the minute definition typical of the contours plotted directly on the stereoplotters. The standard elevation errors of contours in this class, as determined for the international orthophoto experiment, is of the order of 0.4 to 0.6 ‰H for the leading instruments.

An automatic correlation mode. More precisely, the "patch" mode as used on the Gestalt Photo Mapper. On this instrument, an elementary correlation area of about 7×7 mm in the plane of the original photograph is used. Within this area, about 1000 points are individually measured and rectified, and an optional number of them are numerically recorded. These data can be used to plot contours on-line or to compute contours in a separate process. However, the height information generated by an automatic image correlator is bound to be affected by elevated terrain details (buildings and trees) and therefore may not be suitable for the large-scale mapping of urban areas.

An altogether different option is offered by the stereoorthophoto approach. Using stereoorthophotos, elevation of any *discrete* points and the height of the elevated terrain details can be directly measured on the Stereocompiler. Similarly, the contours can be plotted with any degree of detail needed for presentation of minute terrain forms. The reported accuracy (39) of $0.3\%_{oo}H$ is adequate for many applications.

Graphical Presentation: Line-Drawn Maps and Plans

Early graphical presentations of terrain were line-drawn maps, and this type of presentation still constitutes the main product in the cartographic field, especially in the mapping of urban areas. The large number of man-made features in these areas calls for a graphical presentation in which each detail is clearly depicted by a symbol.

In some countries, there is a marked difference in terminology for the designation of different large-scale mapping products. In the present text, the following classification will be used:

> Large-scale maps or plans are at scales that permit the main features, such as streets and buildings, to be plotted according to their dimensions. The term is applicable to scales such as 1:5000, 1:2000, 1:1000, and up.
>
> Medium-scale maps are mapping products at a relatively large scale in which planimetric details are generalized or presented by symbols. Typical scales in this class range from 1:7500 to 1:15 000.
>
> Maps at 1:20 000 scale and smaller belong to the topographical maps category.

In well-organized work, the basic mapping effort of the city is concentrated on the *base city map*, usually at a scale of 1:1000 or 1:500 in large cities and 1:2000 in towns, because this map must provide the answer to the most demanding requirements. Most other plans and maps are derived from the base city map or from the field surveying or photogrammetric data, which are available from the production of the base map.

Preparatory work. Certain planning and preparation is required before the actual plotting work on stereoplotters is started. The aerial photographs should be examined, and suitable stereopairs (if 80% longitudinal overlap is used) should be selected for the plotting operation. It is also good practice to outline the approximate area that should be plotted from each stereopair, with particular attention to side-overlapping photographs. The main purpose of this planning work is to ensure the best possible insight into the plotted area for each stereopair. The presence of high buildings and trees, for instance, might suggest a shift of plotting limits in a given stereopair. The control points should be marked on the photographs.

For plotting, only dimensionally stable material (transparent plastic material) should be used. The sheets must be prepared in advance, the border lines and the grid crosses marked, and the horizontal control points carefully plotted.

Field identification should precede the actual plotting operation; suitable enlargements of aerial photographs are used, on which details not clearly identifiable on aerial photographs are marked (Fig. 8-17). J. Sima (11) recommends that the following terrain details should be identified in the field for 1:500 and 1:1000 maps:

> ground control points (including additional and auxiliary control points),
> location of bench marks and their elevation,
> land use within occupied land boundaries,
> fences, hedgerows, and walls,
> type of surface of road and parking areas, curblines, and entrances,
> highway signs, railroad crossing barriers, and billboards,
> elevated transmission lines, unless they are evident on the photograph,
> cliffs,
> delineation of natural and artificial slopes and break lines,
> differentiation of houses, annexes, garages, and sheds,
> all service and utility poles and posts, call-boxes, and transformers,
> fire hydrants, catchbasins, manholes, and valve chambers,
> less evident ditches or streams,
> names of streets, plazas, official buildings, etc.

For the 1:500 city map, it is recommended that the following be identified:

> less evident stairs,
> number and location of gasoline pumps,
> differentiation of verandas, porches, and balconies,
> differentiation of hydro, telephone, and other poles and lamp standards when required.

In precise work, field identification includes measurements of roof eaves, so that the photogrammetric plots of roofs can be reduced to the ground outlines of the buildings. The width of roof overhang is simply marked on the enlargements of aerial photographs.

Figure 8-17. Field-identified aerial photograph. In actual field work, information in white would appear in colored pencil (usually red).

Simple optical instruments permit the projection of the edge of the roof on the ground. The distance between the projected point and the wall is determined with a measuring tape (Fig. 8-18).

When property boundaries are to be plotted, special care must be taken that they are clearly identified when they follow features visible in photographs, such as fences, walls, and walls of buildings. Otherwise, their position must be defined precisely by auxiliary measurements with respect to neighboring well-defined features visible in the photographs. The auxiliary measurements are subsequently used to locate the property boundaries on the photogrammetric plot.

In some countries, only major features such as building and curb lines of a street are plotted, and no attempt is made to depict or classify secondary details. If this approach is taken, the identification work in the field can be almost completely dispensed with except for the identification and measurement of property boundaries. Restriction of the volume of plotted details to those of

Figure 8-18. Field instrument used for projecting roof overhangs (courtesy of Wild Heerbrugg Ltd.).

primary importance only, usually well visible and defined in aerial photographs, speeds up the plotting significantly and lowers its cost. It is disputable, however, whether this approach is the most economic and desirable in the long run. City administrations and various agencies must have data on the complete city fabric, and any fragmented data collection and mapping is always more expensive and less efficient than a unified approach. On the other hand, insistence on only complete and detailed mapping should not be a reason for delaying photogrammetric mapping projects of urban communities that are lacking even the most rudimentary plans and maps.

Differences in approach will be reflected in the final product—the map—as is shown in Figure 8-19, reproduced from Sima's study (11). Both maps were made from identical photographs and on the same plotting instrument. Map a was plotted without benefit of field identification; map b was supported by a field-identified photograph.

The actual plotting. Once the plotting sheets have been prepared and the supporting material and data have been secured, the actual plotting operation can

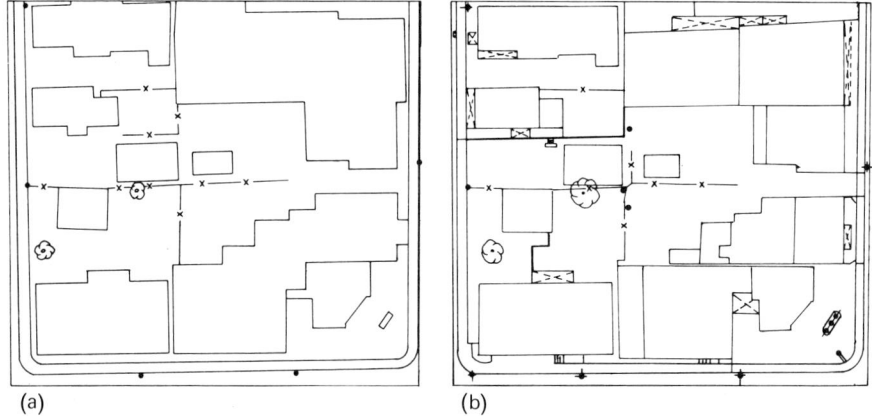

(a)   (b)

Figure 8-19. Comparison of a photogrammetric plot of an identical city block. Plot a without field identification; plot b with field identification.

start. There is no difference in basic procedure when plotting open areas and city areas. The complexity of plots in urban areas however, calls for particularly careful work. The operator must constantly check for the proper selection of stereopairs, and it may be necessary to modify the plotting limits assigned for individual stereopairs. If plotters with separate plotting tables are used, a draftsman is often employed at the plotting table to help the operator check the completeness of the plot, draw symbols, write spot elevations, number contour lines, and join boundary points, to give a few examples. The plotting of buildings is tedious work requiring much attention. Since the operator is supposed to plot the ground outline of buildings, he or she is forced to move the plotting mark frequently from the roof of the building to the ground whenever a fragment of the building at the ground level is visible. Therefore, plotting only the building corners is usually preferred. The procedure is not much simpler if the roof line is used for the plotting of buildings, particularly if the plots must be corrected by the width of eaves measured in the field. Thus the presence of a draftsman at the plotting table of conventional plotters is almost unavoidable in this type of work.

Usually, the planimetric contents are plotted first, which offers more freedom in correcting plotting errors and graphically enhancing the plot. Generally, contouring is a simpler operation than the plotting of planimetry, in the sense that the course of a contour line is determined by the operator, who relies upon stereoperception; an experienced operator seldom has to correct plotted contours. In densely built-up cities, however, difficulties of a different nature arise. The percentage of open areas that could be considered to represent the natural surface of the terrain may be very small and be restricted to backyards, streets, and occasional squares. When the streets are narrow and the block interiors are completely built-up, the usual contour lines might be reduced to infrequent, short sectors of contours that do not convey a coherent picture of the

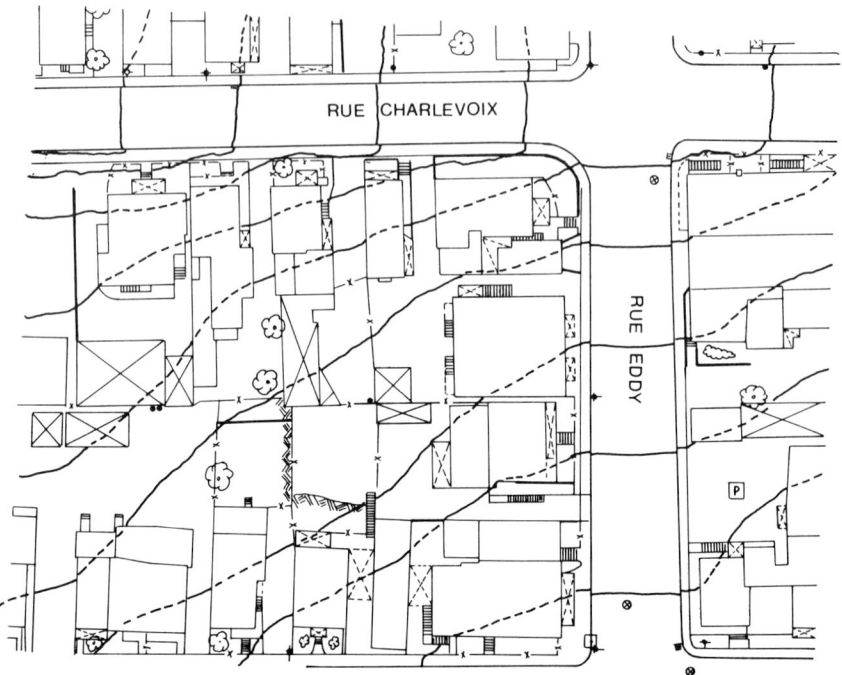

Figure 8-20. City plan with contours extended under the buildings and structures.

topography of the terrain. To overcome this difficulty and provide city engineers and planners with meaningful information on the general topography of the city, contours can be extended in *dotted lines* through all buildings and structures covering the terrain, assuming that the surface of the terrain has not been modified by buildings. In this approach, any cut or fill should be considered as part of the natural terrain (Fig. 8-20). To avoid further cluttering of the planimetric plans by contours, the contour plan should be presented on a separate sheet in the form of an overlay.

In contouring paved streets and squares, it should be remembered that the surface probably is smooth and, in well-built cities, maintains a theoretically designated shape. Consequently, the contours may be expected to have a predictable form or at least not to deviate drastically from specifications. Tonal differences in pictures may affect stereoperception and dictate an unlikely shape of the plotted contour, introducing microforms that are meaningless in view of the uncertainties accompanying photogrammetric measurements and plotting. Assuming that both the photographs and plotting equipment are of good quality, a vertical pointing accuracy (expressed at a terrain scale) of $m_z \cong \mu m \cdot m$, where m is the scale factor of photographs, can be expected (40). This represents a value of 2.4 cm for a scale of $1:6000$, with possible departures of up to 7.0 cm. Considering the other factors that affect pointing accuracy when a variety of points is involved, it must be concluded that departures of up to 10 cm are very

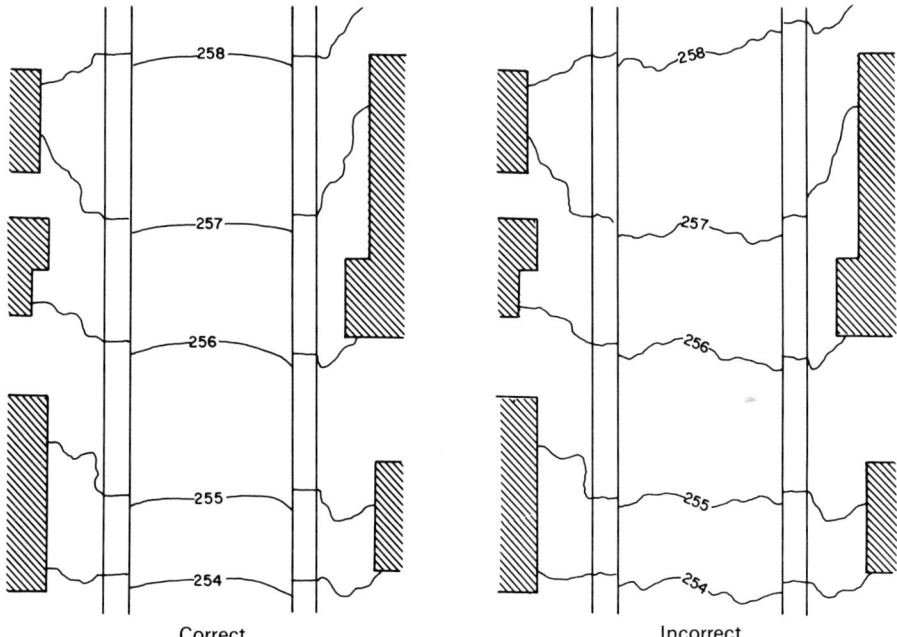

Correct   Incorrect

Figure 8-21. Contour lines on paved surfaces.

likely. This value must be further increased when considering a continuous contouring operation and the associated errors. It is, therefore, most difficult to dispute the correct course of any contour line within ± 15 cm, for the conditions specified above. The proper procedure is to smooth the contours within reasonable limits (Fig. 8-21).

For unpaved streets and squares, contours that are less smooth convey the character of the surface better. Any attempt to plot all microforms precisely is meaningless and should be avoided.

The accuracy of photogrammetrically determined heights (discrete points and contours) is, of course, lower than any vertical pointing accuracy, primarily because of model deformations. When *conventional procedures are used on a production scale*, the following accuracies can be expected, where m is the scale factor:

Height determination of discrete points

Analog instruments		Analytical instruments
$(10-15)\,\mu\text{m} \cdot m$	$= \quad m_z \quad =$	$(8-10)\,\mu\text{m} \cdot m$

Contouring accuracy

Analog instruments		Analytical plotter-type instruments
$(20-30)\,\mu\text{m} \cdot m$	$= \quad m_z \quad =$	$(16-20)\,\mu\text{m} \cdot m$

Table 8-6. Vertical Accuracy (in Centimeters) of Photogrammetrically
Determined Spot Elevations and Contour Lines

Scale of Photographs	1:4000		1:6000		1:8000		1:10 000	
Instrument	Analog	Analytic	Analog	Analytic	Analog	Analytic	Analog	Analytic
Spot elevations	4–6	3–4	6–9	5–6	8–12	6–8	10–15	8–10
Contour lines	8–14	6–8	12–21	10–12	16–28	12–16	20–35	16–20

For scales common to city surveying, Table 8-6 gives the expected accuracies expressed in centimeters. The accuracy of photogrammetrically derived quantities depends upon a number of factors; it is difficult to determine beforehand with precision. In Table 8-6, the expected range of standard errors is given; the reader must decide which figure is applicable to local conditions.

Because ground surface areas (paved or unpaved) in cities are smooth compared to the natural terrain, interpolation of contour lines from measured spot elevations will provide reliable contours. An alternative to continuous contouring on the plotter is therefore offered by the photogrammetric determination of sufficient numbers of spot elevations, with the subsequent manual interpolation of contours in which attention must be paid to planimetric features. This procedure offers good results, particularly if spot elevations are selected approximately along contour lines and at characteristic points of the terrain.

Another procedure, which is particularly fast and accurate when difficulties in continuous contouring are encountered, is as follows. The operator sets the measuring mark at the height of the contour to be plotted and moves the measuring mark, without changing its height, to intersect the terrain at the spot where the mark seems to touch the terrain surface. This position is quickly marked on the manuscript by a short line (made by lowering the plotting pencil). Then the operator moves away from the slope, so that the mark appears to float over the terrain surface, and proceeds to intersect the next contour point, as indicated in Figure 8-22.

In this procedure, each marked point is established with an accuracy close to that of the pointing of individual spots, and even minute terrain forms can be included; the contouring speed is only slightly slower than the usual continuous contouring. The operator should be assisted at the plotting table by a draftsman who joins the marked points to form a contour line.

Various engineering projects require higher elevation accuracy than that offered economically by current photogrammetric methods. Starting from bench marks, new bench marks, reference points, and longitudinal and transverse profiles of superior accuracy are determined by surveyors or engineers using field leveling techniques. However, the bulk of general vertical information

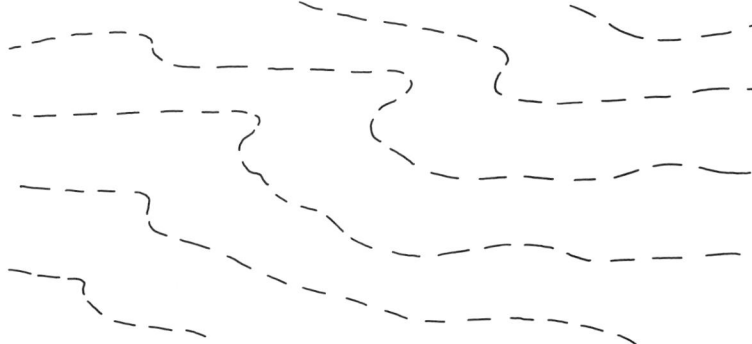

Figure 8-22. Contouring by marking short segments of contour lines.

needed in city areas can be determined more rapidly and economically by photogrammetry.

Numerical Recording with Subsequent Graphical Plotting: Automation in Plotting

When aiming for pictorial or graphical plots as the primary products, X-, Y-, and Z-coordinates of relevant terrain points are occasionally recorded in the process. Thus-generated numerical data can be used as digital terrain information or as numerical values defining the position of discrete points, from which also other quantities can be computed, e.g., coordinates of boundary points or size and form of land parcels. There is, however, a growing tendency to reverse the process and to use photogrammetry for the production of the primary numerical data, with subsequent automatic plotting in a line map form. Two reasons can be singled out for this latter approach: (1) Numerical accuracy is required, and (2) additional technical and economic advantages, such as ease of manipulation of computer-stored data and automatic drafting, are thought to result from this approach.

It should be obvious that in all applications in which accuracy is of primary importance, not only should numerical determinations be chosen but also all precautions should be taken to ensure that the required accuracy can be reached. We have in mind, for instance, optically and geometrically high-quality photographs, selection of a suitable scale of photographs, proper measuring or plotting equipment, and competent processing of data. The numerical accuracy to be expected can be found on pp. 251–252 and 280–281, the latter dealing with the problem of aerial triangulation. In specialized literature, there have been efforts to define the accuracy of photogrammetric determination as a function of various parameters. This may be a valid approach for organizations with a precise knowledge of the characteristics and quality of their work. It would seem more useful, and sufficient, to accept that in large-scale work referred to well-defined points, a coordinate accuracy m_x and m_y of the order of from 6 to 8 μm in the photographic plane is *operationally* achievable. Introduction of analytical plotters should permit the improvement of these latter figures.

Numerical data are usually converted into maps eventually. To avoid misconceptions, the basic techniques involved will be explained.

When using automatic image correlators, contour lines, profiles, or simply spot elevations of the terrain can be determined and recorded automatically or with the occasional intervention of a human operator. Image correlators are optional parts of some analytical plotters and are used also on some orthophoto equipment. Assuming suitable terrain, that is, terrain void of prominent vegetation or buildings, the topography of the terrain can be recorded graphically or numerically in an automatic mode of operation.

The plotting from aerial photographs of planimetry in the conventional symbolized form, however, cannot be done automatically, because such plotting requires intelligent interpretation and identification of the plotted details, and this cannot be accomplished at present by any instrument.

In urban areas, the amount of terrain details to be identified and symbolized on the plot is enormous. During conventional plotting operations, the operator follows terrain details with the floating mark, and the map contents are marked in pencil or scribed on the manuscript. Since manual plotting of well-defined details such as buildings, sidewalks, and railways is difficult, usually the fair drawing in ink or in scribing is performed for the final reproduction.

Fair drawing is a lengthy and costly operation. In addition, some countries encounter a shortage of persons trained to work as cartographic draftsmen. Numerical recording of all details during the plotting operation, and their subsequent drafting by so-called automatic plotting tables, offers an alternative. This has become an important trend in several countries, and there is little doubt that in future it will be an area of intense development. This trend will be particularly strengthened once the second generation of automated drafting systems reaches the market. Even now, the performance of some systems is quite impressive. For instance, there are automatic drafting tables with drawing speeds exceeding 50 cm/s.

Probably the strongest argument for the use of an automated drafting system results from additional operational considerations, such as difficulties in updating maps. This is particularly true in the large-scale mapping of urban areas, in which there are frequent physical changes that require modification of the contents of the base map. Because of the high rate at which changes occur, the map must be revised either continuously or at frequent periods. In the automatic drafting system, map contents are numerically stored on magnetic tape, and any alteration can be effected relatively easily. New maps of the whole city cannot be drawn every day, but a periodic automatic redrawing of a map sheet is a practical proposition. Moreover, it is a relatively simple matter to have a small section of a map automatically drawn at a moment's notice. In addition, with modern techniques, any portion of the stored contents (or any section of the map) can be instantaneously displayed on a cathode-ray tube, and a copy can be made. This visual and graphical presentation lacks the rigorous accuracy and dimensional stability of a conventional map, but this may not be of concern in many applications.

The use of automatic drafting tables is a logical extension of the introduction of computers and data banks. Computers have reached a high level of performance and reliability and will be used increasingly for storage, manipulation, and automatic drafting of information.

Automatic drafting replaces, to a certain degree, only the manual work of the draftsman and not that of a photogrammetric operator, who must still numerically record the basic manuscript. The operation outlined above should not be confused with "automatic mapping," which cannot be achieved in a conventional form.

In automatic drafting, the plot content is numerically recorded on electro-magnetic tape. The operator can record certain features in an "automatic mode"

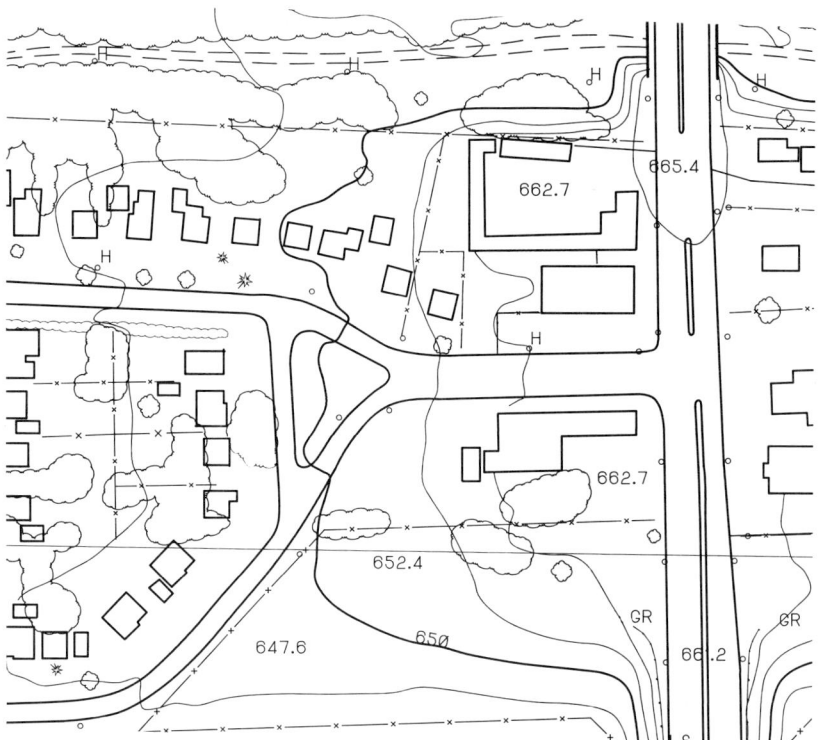

Figure 8-23. Sample of computer drawn map of an urban area, scale 1:2400. Analytical-type plotters are particularly suitable for this purpose since an essential characteristic of these systems is the capability of rapid recording, storing, and even on-line manipulation of recorded data (courtesy of the Ministry of Transportation and Communication of Ontario, Canada).

or without personally activating the recording system at selected points. This is achieved by setting either a time interval or geometric spacing between the consecutive points, as the basis for the frequency of the recording of points of the terrain detail under consideration. This mode of operation is particularly suitable when recording contour lines, profiles, courses of rivers, or other continuous "natural" lines.

When recording geometrically regular lines, recording of a discrete characteristic point is required. For example, when recording buildings, the x and y machine coordinates of corner points are registered in addition to a code which classifies the point and tells the computer controlling the automatic drafting operations how the points must be connected. Depending upon the program used, and within a specific category, the instruction code for recording is usually introduced before the actual recording of points starts. The recording of individual points, however, requires an individual command (pressing of a button or foot pedal) from the operator.

Photogrammetric numerical records present valuable data, which are comparable to field surveying data. Since the recording must be done with care and accuracy, a good stereoplotter is necessary to ensure the required readability and pointing and measuring accuracy. There are companies that specialize in modern recording hardware that can be connected to conventional photogrammetric stereoplotters with coordinate measuring capability. An example of a map plotted automatically from suitable records is shown in Figure 8-23.

PHOTOGRAMMETRIC CADASTRE IN URBAN AREAS

Introductory Remarks

The possibility of using photogrammetry as a surveying and mapping technique in establishing cadastre in urban areas depends upon the type of the cadastre to be established, the accuracy required, the physical characteristics of the community (difficulty of ground surveying and visibility of planimetric details from the air), and the resources available, including qualified manpower.

Most modern cities are the least suitable areas for establishing cadastre by photogrammetric means, because of limited insight from the air and the difficulty of efficient targeting of boundary points, which limits the resulting accuracy. Also, the use of photogrammetry is not practical for the continued updating of cadastral systems; therefore, direct field surveying methods are to be preferred.

Photogrammetric techniques can be considered, however, if a cadastral system is used in which a precise determination of boundary coordinates (direct or indirect) is not a requirement or in a multipurpose approach combined with field procedures. Numerous urban communities do not have a coherent cadastral or mapping system, nor do they possess the knowledge and resources necessary to establish a more precise cadastre. In this situation, photogrammetry can provide a relatively rapid and economic solution that may depart from traditional forms and yet satisfy pressing needs.

Assuming that photogrammetry has been chosen as a satisfactory technique, it is important that the following points be realized:

1. The basic difficulty in using photogrammetry for cadastral purposes, or at least some phases of it in city areas, is an economic, rather than an accuracy, problem. In many studies and operational projects, it has been found that photogrammetry can deliver a higher and more uniform general accuracy than is obtainable in conventional field procedures. This requires targeting on the ground, which may not be practical in a city area. In addition, sophisticated photographic missions, including very precise, "pointed" aerial photographs, careful ground preparatory and complementary work, and precise coordination of all operations would have to be considered.
2. No cadastral work can be carried out solely from the air. Various cadastral information and data can be gathered only on the ground; consequently, when a "photogrammetric cadastre" is discussed, it is in reality a combined field–photogrammetric operation.

Establishment of Additional Control Points

One area of cadastral work where photogrammetry can be used is the establishment of additional control points, particularly in areas surrounding a city.

Depending upon local conditions and requirements, one of the procedures described on pp. 266–283 could be used. However, since the accuracy requirements are most demanding and probably will increase with the extension of the city, the most precise procedure, described on pp. 282–283, should be considered first. The use of réseau photographs processed analytically with suitable corrections could be considered, for example. If block triangulation is used, sufficient density of peripheral ground control points must be provided.

The best possible accuracy can be expected when the densification of control points is carried out within single models (each controlled by at least five known terrain points). By also using the analytical approach, the root-mean-square coordinate error of established points should not exceed ± 6 to ± 8 μm in the plane of the photographic image. All points must be suitably targeted.

It can be assumed that the triangulation carried out on analog instruments will provide results 20 to 25% less accurate than those quoted above, partly because analog plotters are less precise than analytical-type plotters or comparators. It is also difficult to apply corrections for image deformations, even when they are known from suitable fiducial marks or from réseau photographs. This, and any "identification" mistakes, may occasionally lead to gross errors.

To eliminate such excessive errors, in some countries, triangulation is carried out twice—on two *independent* sets of photographs. In Austria (41, 42), three independent flights are used. The area is covered by two identical but independent flights at 1:16 000 scale. Each "double" stereopair is controlled by five ground points—four in the corner of the model and one in the center. In each single model, the ground control points and the points to be established are read and transformed to the ground coordinate system. The third flight at the 1:8000 scale is arranged so that the 1:16 000 models are covered symmetrically by four models at large scale. In these models, all points are reread, and the machine coordinates are transformed into the ground system by using *mean values* of all points determined from 1:16 000 photographs. Apparently, consistent and uniform results are obtained from this approach. Standard coordinate errors from 1:16 000 "double" models amount to ± 10 cm on the ground (or $\pm 6\ \mu$m in the plane of photographs), which is further reduced to ± 7 cm by introducing 1:8000 photographs.

Proper monumentation and targeting of all points involved is an essential part of the operation. Well-defined images must be produced by the targets in the photographic pictures. The size of the targets for a given scale of photographs depends to a great extent upon the contrast. In some countries, the area surrounding the target is either covered or painted with a black, antireflecting substance or paint. If, in the location of the target, the terrain is too loose or is covered by grass which could soon overgrow the target, the terrain surface around the point can be consolidated by a thin layer of concrete. This layer is painted matte black, and a target is placed on top.

In deciding the size of targets, it must be remembered that the image of the target should be slightly larger than that of the measuring mark projected on the photographic plate (p. 265).

Because the initial investment in any photogrammetric project, such as the cost of photogrammetric instruments and photographic flights, is substantial, it is essential that the projects undertaken are of a certain size and complexity. This means that a sufficiently dense set of supplementary control points, such as 10 per square kilometer over larger areas, should be attempted. The major expense is the monumentation and targeting of points; the cost of photogrammetric determination of an additional number of points is negligible.

Numerical Determination of Property Boundaries

The procedures mentioned on the previous pages can be used to determine the coordinates of any points identifiable in aerial photographs. From the coordinates, other quantities, such as distances between the points (length of the property boundaries) or size of the individual lots, can be computed.

Two basic cases should be considered: (1) the boundaries are monumented or otherwise precisely defined; high accuracy in mathematical definition of boundaries is expected; (2) the "general" boundaries concept is accepted by the cadastral system in operation or reduced accuracy in mathematical presentation is sufficient.

High accuracy requirements in the determination of property boundaries. If only the highest accuracy satisfies the requirements of the cadastral system, the procedures mentioned above should be used. Low-altitude photography taken at scales such as 1:4000 to 1:6000 allows a uniform coordinate accuracy of the order of very few centimeters on the ground, which is better than results obtainable from current field surveyings. Comparative studies carried out in various countries have proved this fact conclusively. This very excellent accuracy is only possible, as has been repeatedly stressed, if the points are carefully targeted before the flight, which may not be practical in built-up parts of urban communities.

It is also occasionally argued that the geometrical form and size of a lot is of primary importance in cadastre and not parcel's location as referenced to the coordinate system. Open, accessible boundaries of short length can doubtlessly be measured more easily directly in the field than photogrammetrically; an example would be the boundary lines that follow the faces of buildings along the streets. If, however, consecutive boundary points are separated by building or other construction or obstacles and there is no intervisibility between them, the situation may become extremely complex and difficult for field measurement, and the final accuracy of field determination by current field methods may be questionable. It is not so much the accuracy consideration that restricts the use of photogrammetry for cadastre in urban areas, but very real difficulties imposed by the limited visibility of boundary points, particularly in densely built-up, modern cities.

General boundary concept. The use of photogrammetry is facilitated if the cadastral system in question operates with the general boundary concept for recording and map display purposes. In this approach, the cadastre does not accept responsibility for the precise location of boundaries, but records and guarantees property rights within general boundary limits as described and otherwise defined. In this system, natural features such as hedges and trenches or man-made features such as fences and walls of buildings define the location of boundaries. To assess the accuracy of field procedures compared to photogrammetry in locating "natural" boundaries, studies were arranged in which the results of two or more independent field surveys were compared with similar results of photogrammetric determinations. These operational-type experiments show that photogrammetry offers more consistent results. It has been proven, for example, that it is simpler to determine the location of the median line of a larger and irregular natural feature from aerial photographs than in the field.

If the boundaries do not follow natural or man-made features, they must be tied into the nearby features that are visible on the photographs, and therefore, supporting or complementary field surveying work cannot be avoided. Since the boundary coordinates are not expected to be correct within a few centimeters, the whole procedure is significantly simplified. In particular, this approach eliminates the need for very careful and extensive targeting, a cumbersome and expensive operation that is the chief difficulty in any photogrammetric project requiring it.

Neighboring accuracy in photogrammetric cadastre. One characteristic of the field surveying procedure is that the relative position of close-by points can be established with higher accuracy than that of distant points. This is particularly true if as "relative position," the distance between respective points is meant, such as widths of buildings along the streets that represent sides of individual properties or length measurements referencing property corners (or any other survey points) to well-defined permanent features. Under favorable terrain conditions, distances within the surveying-tape length can be measured with an accuracy of ± 1 cm or better, which is significantly greater than the photogrammetric accuracy that could be economically obtained in this kind of application.

If, however, ground points are separated by physical obstacles such as buildings, even when the points are relatively close to each other, determination of their relative position may require involved field surveys, with a resultant drop in accuracy in establishing the respective distances. When considering, for example, a continuous row of buildings, it is a simple matter to determine with great accuracy the distances between consecutive boundary points along the street and along the boundary line on the other side of the row buildings. The correlation of respective boundary points belonging to the same property and separated by the buildings is much more difficult, however, and the distances derived from careful field surveys may easily vary by from ± 3 to ± 5 cm, particularly if significant elevation differences and other terrain difficulties occur.

In contrast, photogrammetric accuracy is most uniform, and coordinate accuracies of the order ± 3 to ± 4 cm have been reported on various occasions. The photogrammetric accuracy in length determination is of the same order, following the formula

$$m_d = m_p = \sqrt{m_x{}^2 + m_y{}^2}.$$

The main difficulty is targeting and obstructed visibility of points concerned. The proper approach is to combine the remarkably uniform photogrammetric accuracy over large areas with precise short length measurements in the field in order to provide supplementary evidence in cadastre surveys.

Graphical plots in cadastre. As the universal multipurpose type of cadastre becomes widely accepted, a cadastral map that contains the most important terrain features, man-made or natural, is gaining in importance. Photogrammetric techniques are particularly well suited for this approach, since the cost of photogrammetric plotting operations constitutes only a relatively small part of the total expenses. There is a strong, economic justification for extending the photogrammetric plots beyond the requirements of property cadastre (basically property boundaries, buildings, and soil classifications), thus creating valuable maps of general use (43, 44).

The comments on photogrammetric plotting in previous paragraphs are applicable here also. Since simultaneous numerical recording of basic data, such as the coordinates of control and boundary points, may be mandatory for the establishment, continuation, and operation of land records, the analytical-type plotter could provide the highest efficiency and operational flexibility as well as the best possible accuracy.

Field identification of boundaries and the acquisition of other information that constitutes part of the cadastral data of the system in question are an absolute must, and suitable enlargements of aerial photographs are used for this purpose. The degree of enlargement depends primarily upon the density and characteristics of the detail to be marked and recorded in the field. To ensure the proper understanding of requirements and difficulties in the operator's work and a good coordination with the field operations, it is good practice to give field workers an opportunity to spend enough time on the plotters to familiarize themselves with photogrammetric plotting procedures. Similarly, operators should be sent to the field occasionally. This significantly increases their capability in the correct interpretation of various details and helps them to formulate properly their requirements concerning field work. An example of cadastral map produced by photogrammetric means is shown in Figure 8-24.

Use of orthophotos in cadastral work. The extent to which orthophoto technique can be used in cadastral work depends upon the type of cadastre and its specifications. If numerical accuracy is prescribed, as is expected in the cadastre of urban areas, the orthophotos are relegated to the role of supporting pictorial material displaying the boundaries of individual land parcels and their numbers (Fig. 8-25). In this example, no particular effort is made to draw the boundary lines in their precise location. The abundance of information, high graphical accuracy, and speed and relatively low cost of orthophoto production make it attractive as supporting pictorial material.

When a graphical type of cadastre is used and thought to be acceptable in the city area, orthophotos can be used as the basis of a cadastral map. To minimize "dead areas" in the orthophotos, either a narrow-angle camera (long focal distance) should be used or only the central portions of the usual photographs should be considered. The property boundaries must be carefully identified, either on the orthophotos or on enlarged photographs in the field and subsequently plotted on orthophotos.

To determine the geographical location of individual land parcels and compute their sizes and so on, orthophoto coordinates of boundary points can be measured and transferred to the ground coordinate system via suitable control points that have been established by photogrammetric means. The accuracy of boundary coordinates is not sufficient for locating or staking out the boundaries on the ground, however, and in this case, more precise evidence, such as suitable monumentation and referencing of boundary points, must be used. Nevertheless, the graphical accuracy of coordinates should be sufficient for the computation of property taxes and for most administrative and technical operations, particularly in the planning sector.

In certain countries (45), orthophoto techniques are used as the basis for a modern countrywide cadastre that includes computerized data banks and automatic updating features as integral parts of the system. This interesting

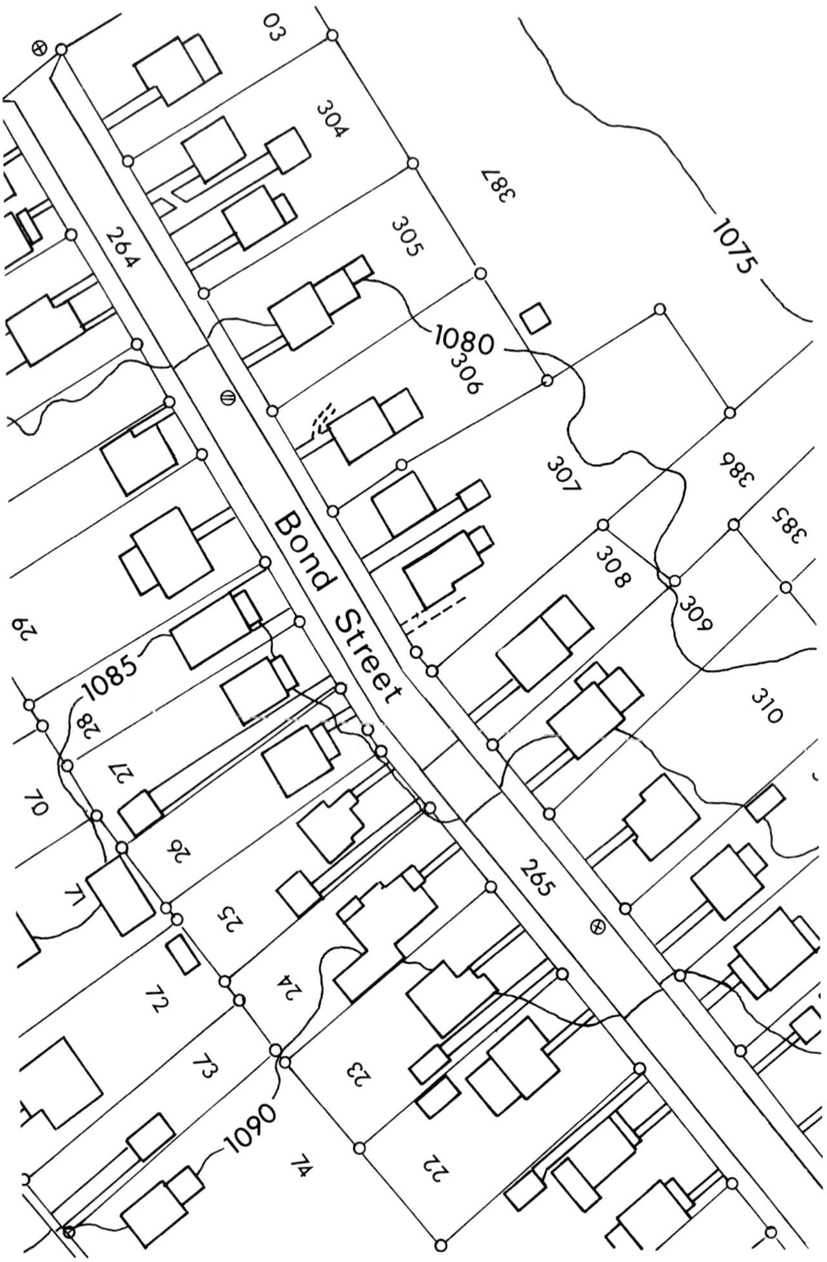

Figure 8-24. Cadastral map, at scale 1:1000, produced photogrammetrically.

Figure 8-25. Cadastral orthophoto map at a scale of 1:1000.

approach is not very practical for densely built-up urban areas, but it could be used for the rapid establishment of a cadastral system in outlying areas of the city. To this end, the use of the stereoorthophoto technique is suggested (46) since it offers a particularly simple, complete, and flexible system—one that also permits the integration of cadastral operations into the general surveying and mapping scheme. Obviously, restrictive features of orthophoto techniques in densely built-up areas should be noted.

SPECIAL APPLICATIONS OF PHOTOGRAMMETRIC METHODS IN URBAN AREAS

Regional Land Inventory

Cities are complex, dynamic organisms, whose growth is extremely difficult to control. Even in ancient times, the population of principal cities could reach several hundred thousand or even a million. The rapid increase in world population in recent years, combined with the marked development of industrialization and accompanying phenomena, further strengthened the general trend toward urbanization, with the result that, in some countries, over 80% of the population is living in cities of various sizes.

It is clear that the city and the surrounding region are structurally connected. Water and often an important quantity of fresh food, manpower, recreation areas, and space for expansion are provided by the surrounding area, which, depending upon the size and characteristics of the city and country, may extend over 100 km from the city.

To secure the necessary resources and space for the healthy functioning and development of the city and to study the effect of the growing city on the surrounding region (e.g., pollution), a suitable land inventory of the whole region and all pertinent changes should be established and kept up to date. In more advanced and well-organized countries, this basic information is provided by existing surveying and mapping systems, such as cadastre (43). If such a system does not exist or is not satisfactory, the required information should be provided by photogrammetric mapping with overlays depicting geological content, soil categories, agriculture activities, vegetation, and the climatic, social, and economic data needed for long-range planning and monitoring of changes, particularly those detrimental to the future development of the region.

It is disturbing to see the disparity that exists between the recognized need for this kind of approach and the actual measures taken by responsible authorities toward implementing some semblance of intelligent planning and management of the growth of urban communities. As a result, fertile land, which is becoming increasingly scarce everywhere in the world, is used primarily for housing and other construction; water is mercilessly contaminated, making it unfit for human and animal consumption over significant areas outside of the city; terrain ideally suited for parks and recreational areas is converted into ugly industrial sites or garbage dumps. Only after the natural beauty of the landscape has been destroyed, the ambient conditions haphazardly altered, the land put to wrong use, and the city developed into an awkward and unmanageable conglomeration of all types of unsightly and unfunctional structures are expensive, and often impossible, corrective measures contemplated. Thus lack of foresight and incompetent management lead to the irrevocable destruction of ambient conditions and, consequently, to huge rehabilitation expenses later.

This could be easily prevented if more thought was put into the planning and development of urban communities, including outlying areas, from the outset. This work does not have to be unduly complex, and the needed basic information contained in the land inventory can be built up gradually as requirements increase in order to accommodate the growing size and complexity of the city. An early start is essential, however, not only to prevent undesirable development but also as an educational measure. Members of urban communities, as well as managing authorities, should learn to manage city affairs looking constantly into the future and considering the harmonious development of the whole region, of which the city is an organic part.

The use of photogrammetry, with its various techniques, particularly photo-interpretation, permits a relatively rapid establishment of the needed land-inventory data, the extent of which should be dictated by local conditions and requirements. A map constitutes the basis of such a land-inventory system. If the country does not have a satisfactory topographical mapping program, orthophoto maps should be considered as an efficient solution. In addition to their metric correctness, they contain a wealth of information of primary importance (pp. 286–295).

Orthophoto maps, like conventional maps, must be based on a reliable control net. Since we are referring to relatively small scales, such as 1:25 000, 1:50 000, or even 1:100 000, the relatively sparse primary control points are satisfactory. The remaining control points, as required for plotting operations or for the production of orthophotos, can be established photogrammetrically.

Application of Photogrammetry to Architectural Surveys

Photogrammetry is a versatile and powerful indirect measuring method that can be used for the precise recording of buildings and their structural and ornamental features. A multitude of precise measurements is required, particularly if conservation work on buildings and monuments of historical value is considered. In this work, usually both the exterior and interior are included with all the important details. In some countries, building facades along whole streets or squares of particular interest are carefully recorded by photogrammetric means.

Photogrammetric recording can be carried out in form of (a) pictorial presentation (rectified photographs), (b) graphical presentation (plots in the form of drawings), or (c) numerical recordings (X-, Y-, and Z-coordinates of discrete points). In more complete inventory work, all three forms of presentation are used at the same time since they complement each other.

   a. Rectified photographs. This is a very simple and rapid product, particularly when recorded buildings or structures contain a limited number of plane surfaces. Ideally, each photograph should be taken with the camera axis perpendicular to the specific surface of the building in order to simplify the rectification process and provide the best possible photographic image of the building. Since the magnification range of commercial rectifiers is quite large (0.5 to 7.0 ×), the required scales of rectified photographs can easily be met. If a larger building or a number of adjacent buildings should be covered, a mosaic of rectified photographs can be produced. Figure 8-26 illustrates a rectified photograph of a building.

   When speaking about rectified photography we have also in mind the ortho-photographs and stereo-orthophotographs that, depending upon the circum-stances, may offer important technical and operational advantages.

   b. Graphical plots. These are drawings of buildings produced from stereo-grams in a manner similar to any photogrammetric plotting operation. The drawings are produced in an orthogonal projection on vertical planes as is

Figure 8-26. Rectified photograph (courtesy of the Indian and Northern Affairs Dept., Ottawa).

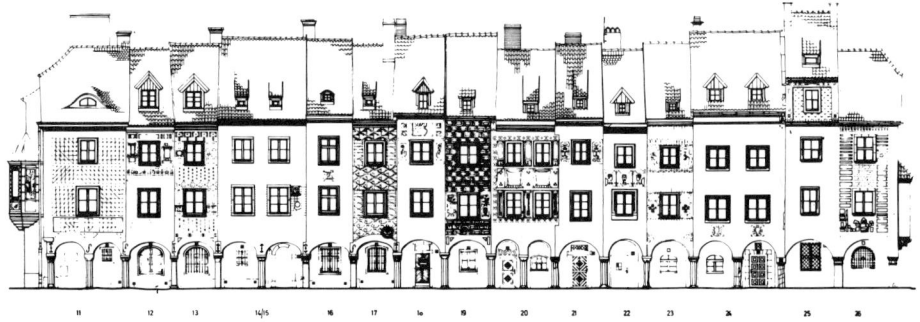

DOMKI BUDNICZE

Figure 8-27. Example of a stereoplot of a row of buildings at a strongly reduced scale (courtesy of the City Survey Office of Warsaw).

customary in architecture. For the same reasons mentioned in the discussion of rectified photographs, the stereophotographs should be parallel to the main surfaces of recorded buildings if conventional plotters are used.

As a rule, only stereophotogrammetry provides a simple and complete solution to the many surveying and measurement problems encountered in civil engineering and architecture. The efficiency of stereophotogrammetry has been further improved by the recent introduction of analytical plotters, which allow the processing of arbitrary stereophotographs and the selection of the most convenient projection plane in plotting operations. This is possible by the use of any arbitrarily oriented, reference coordinate system on the plotter.

An important difference between a rectified photograph and a stereo-plot is in content. The rectified photograph contains all the features of the recorded building, but they must first be *identified* before any further use of the rectified photographs can be made. The stereoplot contains only those features that have been identified during the plotting process. Since interpretability of stereograms is superior to the interpretability of single photographic images, the stereoplots must be regarded as more complete and reliable products (Fig. 8-27).

c. Numerical recordings. In more specialized studies or operations requiring the highest possible accuracies (Fig. 8-28), numerically recorded X-, Y-, and Z-coordinates of selected points are used. Numerical recordings are also used to supplement graphical plots or to determine supplementary control or reference points, as well as to plot particularly complex features.

Any numerical determination of X-, Y-, and Z-coordinates requires a stereo approach. At the measuring end, the work is particularly efficient if an analytical-type plotter is used that allows measurements to be performed *on an oriented model* with all the rigor of the analytical approach and with the on-line inclusion of any corrections required to provide the best possible accuracy.

Figure 8-28. Ceiling of a chapel in Ottawa, which had to be taken apart and reconstructed in another location. Numerical records provided most of the data in this project.

Accuracy considerations. Some confusion exists concerning accuracy requirements and the realistic possibility of meeting exaggerated accuracy specification. This is due to the following factors:

> It is generally recognized that higher accuracy, such as of the order of a few millimeters, is meaningful only if referred to well-defined or specially targeted points. The usual building features that are the subject of measurements *are seldom defined* with this order of accuracy.
>
> There is a substantial difference between the determination of a single dimension (distance, length) and the determination of a position in space expressed by the X-, Y-, and Z-coordinates of a point. For example, the width or height of a window can be determined quite easily by a steel tape to within an accuracy of a few millimeters. However, the determination of the relative position of two window corners located along different sides of the building requires a most complex measurement procedure and an accuracy of a few millimeters can be achieved only by using extraordinary care.
>
> Buildings are not usually constructed to an accuracy of a few millimeters; historical buildings do not have the type of geometrically defined finish that would justify an exaggerated order of accuracy in their general recordings.

For these reasons, it is suggested that an accuracy better than 1 cm (as standard error) should be reserved for establishment of control and reference points, measurements of single dimensions or ornamental details of lesser dimensions, and deformation and other special studies, such as the effect of geometric characteristics of surfaces on the acoustic properties of interiors.

A coordinate accuracy, defined as the standard error $\sigma = m_x = m_y = m_z$, of 1–2 cm is quite satisfactory in current architectural recordings. Assuming that the graphical plots are at scale 1 : 100, the above standard error corresponds to the graphical drawing accuracy of 0.1 to 0.2 mm.

The highest accuracy in the determination of discrete building points is achieved by angular measurements from two or more field stations. It must be realized, however, that this is a lengthy procedure only suitable for the determination of a limited number of points. If a large number of points is involved, a photogrammetric technique must be used.

Choice of photographic equipment. If precise measurements and graphical plots are intended, only proper photogrammetric cameras should be used.

For rapid photographic stereorecording of elements of limited size that are typical of close-range photogrammetry, use of the so-called stereometric cameras could be considered. They are particularly convenient for photographing interiors, and any external features of buildings or monuments that should and can be photographed from a short distance. The equipment consists of two cameras rigidly connected by a metal tube or frame so that their relative orientation is maintained. The optical axes of the cameras are parallel, but the whole unit can be inclined by preset angles or rotated. The central exposure-control mechanism ensures simultaneous exposure. These cameras are designed

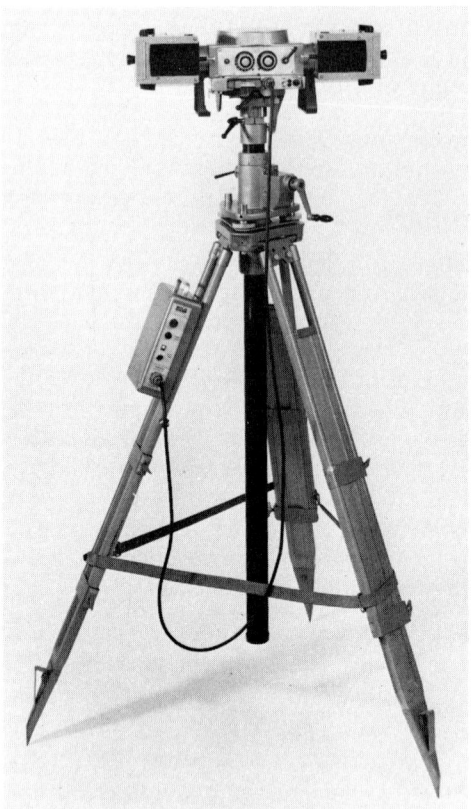

Figure 8-29. Wild Stereometric Camera C-40 (courtesy of Wild Heerbrugg Ltd.).

for glass plates, ranging from 60 × 90 mm to 130 × 180 mm. In Figure 8-29, the Wild C-40 Stereocamera (base, 40 cm) is depicted.

Single cameras are used when photographing the whole of the building exterior because the geometry of the photogrammetric technique requires a larger camera separation than is provided by stereometric cameras. They are usually equipped with devices that permit precise orientation of the camera, whereby pairs of parallel photographs can be taken from respective stations. Figure 8-30 shows a Carl Zeiss Oberkochen TMK camera.

When selecting a suitable camera, factors such as optical performance and stability, which are of primary importance, should be considered. The camera body must have a certain rigidity to secure unchangeable inner orientation. Modest weight and ease of manipulation, along with sufficient range in various settings, are other considerations. Flexibility in the use of these cameras can be increased by introducing either continuous or stepwise focusing capability.

Compatibility between the intended camera and the available plotting equipment is another basic consideration. This also includes an analysis of the field

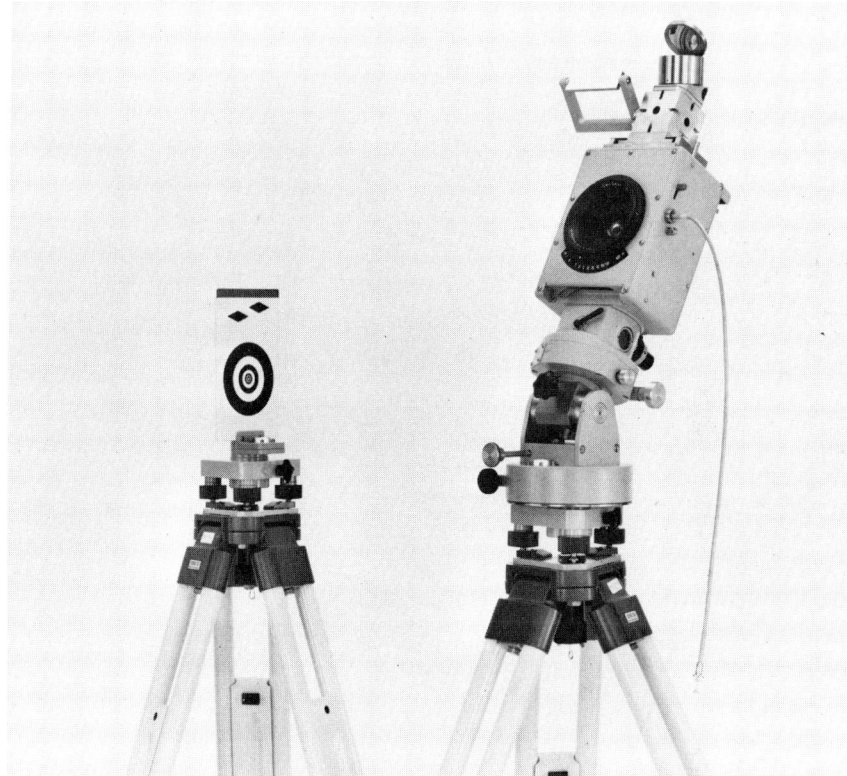

Figure 8-30. A Carl Zeiss Oberkochen TMK Camera (courtesy of Carl Zeiss Oberkochen).

procedures contemplated, to ensure that the proposed mode of operation is within the working range of the photogrammetric plotting equipment.

Field work. The procedures used must guarantee good matching between the exterior and interior structure of the building. This can be achieved best by the establishment of a control net that will be adequate for measurement of both external and internal features of the building. The geometric shape and characteristics of such a net are determined primarily by the configuration of the building to be recorded and the limitations imposed by local conditions on the site. The control net must provide an independent and adequate basis for the assessment of the total geometry of the recorded structure; under no circumstances should assumptions concerning such characteristics as rectangularity of external and internal elements of buildings or uniform thickness and verticality of walls be made in advance.

To keep the amount of field work to a minimum, the points of the primary control net should be suitably located to serve as photogrammetric stations as well. Usually, a closed polygon is established around the building (Fig. 8-31). All supplementary photogrammetric stations are tied to this traverse. From the

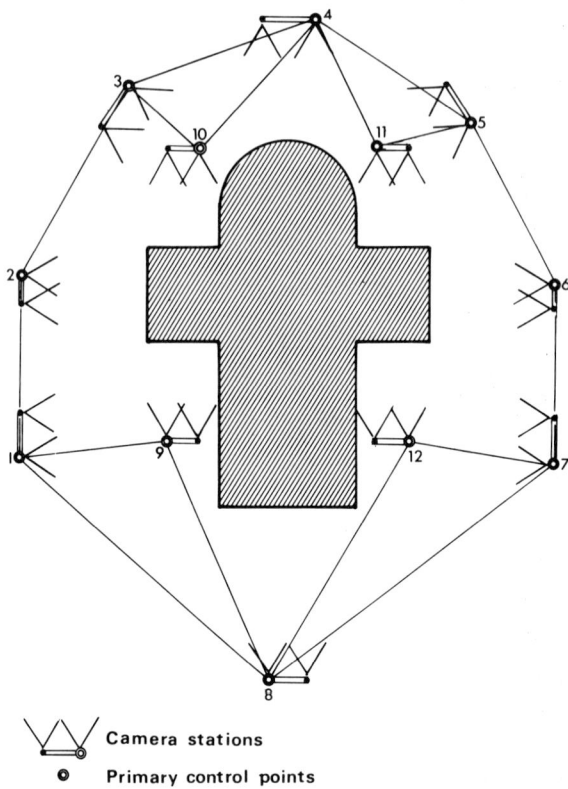

Camera stations

○   Primary control points

Figure 8-31. Example of a control net and photogrammetric stations.

traverse and photogrammetric stations, control points, including those affixed to the building, are determined by the intersection technique. At the same time, control lines and points that are used for the recording of courts or building interiors must be established and marked for further use.

The next step in field operations is the taking of photographs. The guiding principle is to cover the building with a minimum number of stereograms and keep, insofar as is possible, a frontal view of the building. Ideally, the axis of the photographic camera should be perpendicular to the dominant surface of the building being photographed and, at the same time, perpendicular to the photographic base. This is not always possible, because of limited space around the building, overall dimensions of the building, or topography of the terrain. Consequently, stereopairs of parallel photographs, not perpendicular to the photographic bases or convergent photographs, can be used in addition to "normal" photographs. Similarly, photographs with the cameras tilted are often used to cover the whole vertical extension of the building. One should note that use of an analytical plotter permits much wider latitude in planning photographic coverage and, therefore, greatly simplifies both field and office work.

The base–distance ratio should provide adequate stereocoverage and should not be too small in view of the accuracy requirements, since this affects depth determination. For conventional cameras, this ratio varies from 1:4 to 1:20. Convergent photographs offer a more favorable base ratio, with stereocoverage close to 100%; however, viewing convergent photographs in conventional plotters or measuring instruments is less comfortable than viewing parallel photographs. The degree of deviation from normal photography acceptable by the available plotting equipment must also be established in advance.

Several factors concerning the selection of the scale of the photographs have to be considered. The most important is accuracy, which, assuming that correct procedures and high quality photogrammetric equipment are used, is a function of the photographic scale and the base ratio. Peczek (47) quotes standard errors for the x- and y-coordinates of the determined points in the image plane as varying from 30 to 60 μm and values for the depth as varying from 120 to 240 μm. Rinner and Burkhardt (16) give somewhat different values defined as relative errors of distances measured in the image plane of

$$\frac{\sigma_s}{y} = \pm 0.07\%$$

and in depth

$$\frac{\sigma_{\Delta y}}{y} = \pm 0.17\%.$$

Assuming that 45 μm is the standard error in x- and y-coordinates in the image plane and ± 2 cm the expected accuracy, the required scale of photographs is derived:

$$\frac{45}{20\,000} = \frac{1}{444}.$$

In reality, a compromise solution may be imposed because of local terrain conditions. Again, magnification range of the plotting and rectifying equipment must be taken into account when planning the photographic field work.

For the recording of very large or complex structures such as cathedrals, helicopters have been used to obtain adequate photographic coverage.

Photogrammetric office work. General rules are applied in photogrammetric office work concerned with applications to architecture. Stereoplotting provides not only very detailed plots but also numerical data (X-, Y-, and Z-coordinates) that can be used for measurements of even the most complex buildings and richly ornamented structures, including precise plots of interiors. An impressive example is provided by a plot of ancient organs carried out by the City Survey Office in Warsaw (Fig. 8-32).

Figure 8-32. Photogrammetric plot of ancient organs carried out by the City Survey Office in Warsaw (courtesy of the City Survey Office of Warsaw).

Figure 8-33. Rectified photographs of three wall sections (courtesy of the Indian and Northern Affairs Dept., Ottawa).

Part of the elevation of the Parliament Building in Ottawa is reproduced in Figure 8-33 as an example of rectification work. Three parallel wall sections were involved. Each was rectified separately and then mounted into a single image through the use of suitably distributed control points. The scale of rectified photographs was 1:24, and the scale of original photographs was 1:60.

Use of Photogrammetry in City Traffic Studies

The term city traffic in this context means the movement of population and goods by both natural and artificial means. The complexity of problems in this area is increasing, and one of the main purposes of modern planning is to provide an efficient solution to traffic problems within city areas. It is not always realized that photogrammetry can offer an excellent means for studying traffic characteristics and for determination of its most important parameters. From aerial photographs taken at an appropriate time of day, the general traffic pattern (48) showing both distribution and density, as well as available ancillary facilities such as parking, can be determined. Furthermore, if aerial photographs are taken with short time intervals between exposures, the traffic flow and its characteristics can be studied. Since metric requirements in these applications are not too stringent, both surveying cameras and reconnaissance-type cameras can be used.

Ordinary aerial survey cameras have a relatively long minimum time interval (about 4 s) between successive photographs, which is somewhat restrictive in the application in question.

In contrast, various reconnaissance cameras, which are smaller and lighter than survey cameras and which permit photographs to be taken every second, are most useful if a study of the dynamic characteristics of traffic flow is intended.

The scale of photographs can vary in this kind of work, and it appears that scales of up to 1:15 000 can be used for studying motorized traffic.

Ideally, photographs should be taken over specific circulation arteries at a desirable time. If variations in traffic pattern and intensity during the day are to be determined, photographic coverage at suitable moments must be obtained.

Some traffic parameters that could be established by photogrammetric technique are as follows.

Spacing between vehicles. Each photograph is an instantaneous record of the position of vehicles. If the photographs are rectified, the distance between vehicles can simply be measured on the rectified picture. The determination can be even more precise if the coordinates of the vehicles are measured on a suitable instrument. For this purpose, it is important that the same point of a vehicle, such as "front-center," be used.

Velocity calculations. The position of vehicles in successive photographs is transferred to one of the pictures; thus the relative position of vehicles at known time intervals is obtained. Rectified photographs at a suitably enlarged scale should be used for this purpose. From the distances between vehicles, referred

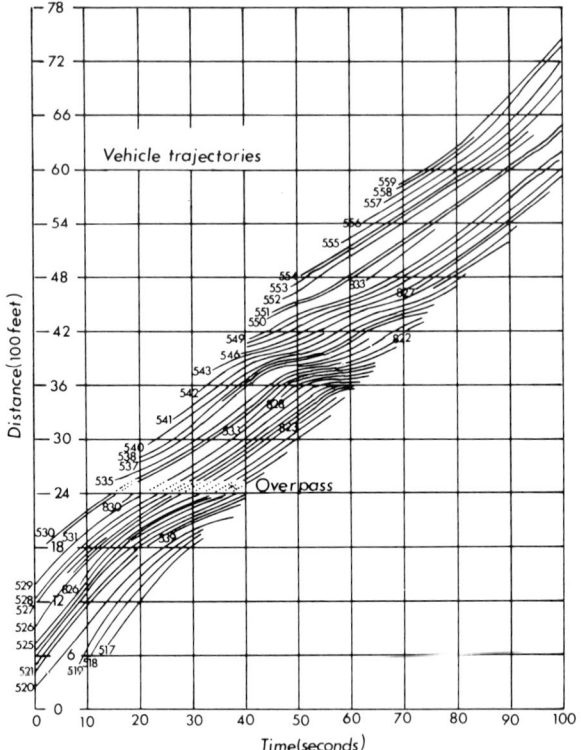

Figure 8-34. Time–distance diagram depicting vehicle trajectories [after Taylor (48)].

to front-center points and the known time intervals between the corresponding exposures, the speed of vehicles, expressed in kilometers per hour, can be calculated. When the time interval between successive photographs is not known, the ground speed of the aircraft can be used to determine the time interval between exposures.

These simple computations can also be made by using coordinates of the vehicle, recorded on stereoplotters or other coordinate measuring instruments. By using an appropriate program, the speed computations can be performed automatically on an electronic computer.

From the accumulative distances covered by each vehicle and the corresponding times, "time–distance diagrams" can be constructed (48), which are useful when studying traffic characteristics at any given moment or at any point along the traffic artery (Fig. 8-34). The position of a vehicle at any time is plotted in a coordinate system, in which one coordinate represents cumulative distance along the roadway and another the corresponding time and the slope of projectories defines the speed. Consequently, not only the trajectories of all vehicles are visible from the diagrams but also the speed fluctuation.

Dubuisson and Burger (49) proposed some definitions and suggested how the characteristic patterns of traffic flow can be calculated. First, they determine the average speed of vehicles involved and, based on this quantity, they calculate the following: deviations from the average speed for individual vehicles, positive and negative acceleration, variation of acceleration, changes of sign of acceleration, and traffic load.

In addition to the basic data on traffic pattern and its characteristics, aerial photographs are an excellent means for rapid and reliable gathering of other information needed in studies of various factors connected with urban traffic and transport. In particular, information on open parking facilities, parking along the streets and squares, and the condition of the traffic markings on the pavement, can be determined easily from suitable aerial photographs.

References

1. Pichlik, V., and Roule, M. Planimetric accuracy of plotting in built-up areas. Paper presented at the XI Congress of ISP, Lausanne, 1968.
2. Sima, J. Contribution to the Experimental Research of Photogrammetric Mapping of Densely Built-Up Areas (Czech.), *Geodeticky a Kartograficky Obzor*, No. 8, 1964.
3. Brock, G. C. *Physical Aspects of Air Photography*, Longmans Green and Co., London, 1952.
4. Carman, P. D. Camera vibration measurements, *The Canadian Surveyor*, September 1973.
5. Carman, P. D., and Brown, H. Resolution of four films in a survey camera, *The Canadian Surveyor*, December 1970.
6. Carman, P. D. Cameras, Films and Camera Mounts. Paper presented at the 2nd Seminar on Air Photo Interpretation in the Development of Canada, March 1967.
7. Blachut, T. J. Should réseau photographs be considered for improving photogrammetric accuracy?, *Photogrammetria*, No. 21, 1966.
8. Umbach, M. J. Color for metric photogrammetry, *Photogrammetric Engineering*, March 1968.
9. Schallock, G. W. Metric tests of color photography, *Photogrammetric Engineering*, October 1968.
10. Mega System Design Ltd., 1766 Midland Ave., Scarborough, Ontario, Canada, M1P 3C2.
11. Sima, J. Compilation Photogrammétrique de Cartes Urbaines aux Échelles de 1 : 1,000 et 1 : 500, *The Canadian Surveyor*, No. 5, 1971.
12. Blachut, T. J. Analytical aerial triangulation based on the use of a point transfer device and monocular measurements, *Bollettino di Geodesia e Scienze Affini*, No. 3, 1961.
13. Blachut, T. J. Monomeasurements in photogrammetric operations. Paper No. 49 at the Conference of Commonwealth Survey Officers, 1963.
14. Young, M. E. H. Block triangulation on the NRC Monocomparator, *The Canadian Surveyor*, No. 4, 1966.
15. American Society of Photogrammetry, *Manual of Photogrammetry*, 1966.
16. Rinner, K., and Burkhardt, R. *Photogrammetrie*, Vol. III a/1, III a/2, and III a/3 of Jordan, Eggert, Kneissl, *Handbuch der Vermessungskunde*, Stuttgart, 1972.

17. Schwidefsky, K., and Ackermann, F. *Photogrammetrie*, B. G. Teubner, Rinner, Stuttgart, 1976.
18. Blachut, T. J. L'Aerotriangulation sur A6. Paper presented at the VI Int. Congress of Photogrammetry, The Hague, 1948.
19. Ziemann, H. Image deformation and methods for its correction, *The Canadian Surveyor*, No. 4, 1971.
20. Schut, G. H. *An Introduction to Analytical Strip Triangulation with a FORTRAN Program*, Publication P-Pr 44, Division of Physics, National Research Council of Canada, 1973.
21. Schut, G. H. Block adjustment by polynomial transformations, *Photogrammetric Engineering*, Vol. 33, No. 9, 1967.
22. Schut, G. H. Review of Strip and Block Adjustment During the Period 1964–67, *Photogrammetric Engineering*, Vol. 34, No. 4, 1968.
23. Anderson, J. M., and Ramey, E. H. Analytic block adjustment; final summary of ISP Commission III Working Group reports 1968–72, *Photogrammetric Engineering*, Vol. 39, No. 10, 1973.
24. Ackermann, F. Experience with applications of block adjustment for large scale surveys, *The Photogrammetric Record*, April 1973.
25. Ackermann, F. H. Photogrammetric densification of trigonometric networks—the project Appensweier, Bull., 6, 1974.
26. Kupfer, G. On accuracy achieved by different triangulation procedures. Paper presented at the ISP XIII Congress, Helsinki, 1976.
27. Brown, D. C. Densification of Urban Geodetic Nets. Paper presented at Am. Soc. of Photogrammetry meeting, Seattle, Wash., September 1976.
28. Kratky, V. Normes de Précision pour les Points de Contrôle Photogrammétriques au Sol, *The Canadian Surveyor*, No. 5, 1971.
29. Ziemann, H. Sources of image deformation, *Photogrammetric Engineering*, Vol. 37, No. 12, 1971.
30. Ziemann, H. Economics of image deformation correction, *Photogrammetric Engineering*, Vol. 38, No. 2, 1972.
31. Dubuisson, B. La photographie aérienne et la photogrammétrie dans l'établissement des plans de reconstruction et d'urbanisme en France et quelques travaux originaux connexes, *Archives Internationales de Photogrammétrie*, Tome X, 1950.
32. Blachut, T. J. Methods and instruments for production and processing of orthophotos. Paper presented at the International Congress of Photogrammetry, Ottawa, 1972.
33. Hobrough, G. L. Automatic stereo plotting, *Photogrammetric Engineering*, Vol. 25, No. 5, 1959.
34. Blachut, T. J., and Helava, U. V. Automatic stereo plotting in small and large scale mapping. Paper presented at the International Congress of Photogrammetry, London, 1960.
35. Blachut, T. J., and van Wijk, M. C. Results of the international orthophoto experiment 1972–76, *Photogrammetric Engineering and Remote Sensing*, Vol. 42, No. 12, 1976.
36. Blachut, T. J. *Mapping and Photointerpretation System Based on Stereo-Orthophotos*, National Research Council of Canada, Publication No. 12281, Ottawa, 1971.
37. Collins, S. H. Stereoscopic orthophoto maps, *The Canadian Surveyor*, No. 1, 1968.
38. Blachut, T. J., et al. Stereocompiler, Canad. Pat. 072475, January 1970, and Orthocartograph, Canad. Patent 074470, February 1970.
39. Collins, S. H., and van Wijk, M. C. Production and accuracy of simultaneously scanned stereo-orthophotos, *Photogrammetric Engineering and Remote Sensing*, Vol. 42, No. 12, 1976.

40. Blachut, T. J. An experiment on photogrammetric contouring of very flat and feature-less terrain, *Nachrichten aus dem Karten- und Vermessungswesen*, Reihe V, Sonder-heft No. 9, 1965.
41. Peters, K. Tauglichkeit von photogrammetrischen EP-Netzen für Katastermessungen, *Österreichische Zeitschrift für Vermessungswesen*, 1966.
42. Kovarik, J. Wird die Genauigkeit von Einzelmodellauswertungen durch die rech-nerische Transformation der Maschinenkoordinaten beeinträchtigt? *Österreichische Zeitschrift für Vermessungswesen*, 1966.
43. Blachut, T. J. *Cadastre as a Basis of a General Land Inventory of the Country*, National Research Council of Canada, 1973.
44. UN Report. Medium-scale and large-scale surveying and mapping. Cadastral Sur-veying and Mapping. *Cadastre*, PAIGH publication, National Research Council of Canada, 1974.
45. Garcia Conzalez, J. A. The use of orthophoto techniques in a modern cadastre in-cluding a data bank, *Cadastre*, PAIGH publication, National Research Council of Canada, 1974.
46. **Blachut, T. J. The Stereo-orthophoto Technique in Cadastral and General Mapping.** *Photogrammetric Engineering and Remote Sensing*, Vol. 42, No. 12, 1976.
47. Pęczek, L. *Application of Photogrammetry to Architectural Surveys*, National Research Council of Canada, 1972.
48. Taylor, J. I. Photogrammetric Determination of Traffic Flow Parameters, *International Archives of Photogrammetry*, Vol. 17, Part 10, Lausanne, 1968.
49. Dubuisson, B. L. Y., and Burger, A. A. J. Etude de la Circulation par Interprétation de Photographies Aériennes, *Photogrammetria*, XVI, No. 4, 1959/60.
50. Slama, C. C. High Precision Analytical Photogrammetry Using a Special Réseau Geodetic Lens Cone. Paper presented at the Intern. Soc. of Photogrammetry Com-mission III Symposium, Moscow, July 1978.

Additional Readings

Bonneval, H. *Photogrammétrie Général*, Eyrolles, Paris, 1972.
Dubuisson, B. *Pratique de la Photogrammétrie et des Moyens Cartographiques Dérivés des Ordinateurs*, Eyrolles, Paris, 1975.
Hallert, B. *Photogrammetry, Basic Principles and General Survey*, McGraw-Hill, New York, 1960.

Chapter 9

City Maps

INTRODUCTION

In cities there is always great demand for a variety of maps. City administration, technical services, planners, investors, and those concerned with all facets of city life cannot operate properly without suitable maps. These maps may vary in scale, content, or form of presentation. General planning requires maps at scales such as 1:5000, and 1:10 000, but the execution of some projects may require plans at a scale of 1:1000 or larger. For administration, land use, cadastre, or detailed planning, maps at scales of 1:500 to 1:2000 are usually used. There is also a need for a variety of special or thematic maps, particularly at smaller scales. These maps usually depict specific information; the scale is selected to permit the coverage of the whole city by a single map sheet.

To systematize the subject, we shall subdivide city maps into the following three categories:

1. The *city base map* is the original map of the city prepared from actual field surveying or photogrammetric plots. It contains complete planimetric and vertical information and may consist of several overlays.
2. *Derived maps* are basically similar in content to the city base map but are usually at smaller scales; they are derived from the base map and use a certain generalization in presentation. Reduction to required scales, such as 1:5000 and 1:10 000, may be done photographically or by other, cartographic means.
3. *Thematic maps* include maps not listed in category 1 or 2. Usually, they are single-factor maps providing, on a reference background, quantitative or qualitative information on the phenomenon in question.

With the advancement of the techniques of computer storage of map content and automatic map drawing, some changes may be expected in the approach to the general question of city maps. Computers provide almost unlimited flexibility in handling the stored data and can translate data into a graphical product—a map manuscript—at any desired scale. As a result, it is likely that the traditional classification of maps into "original" or "derived" categories may lose its significance in future. Maps will be at different scales; each map will have its own characteristics, but all will be plotted from the original and continuously up-dated field or photogrammetric data. Even then, the concept of a "base city map" will probably continue to be used in reference to the most detailed map, which depicts the ground situation with maximum fidelity.

The production of a specific map at any scale covering the whole city is a major and costly undertaking. Understandably, many map users, preoccupied with their own responsibilities, insist on having their particular requirements satisfied. This could result in a very costly proliferation of all kinds of maps at various scales, none of them properly maintained or kept up to date. By making appropriate decisions on the selection of map scales for the city in advance, the number of maps can not only be reduced to a minimum but the general quality can also be improved. The availability of maps at definite scales usually induces the users to tailor their demands accordingly, with a resultant uniformity of scales.

Two distinct steps are involved in the production of a map: (1) compilation of the map manuscript and (2) final drafting, cartographic enhancement, and map reproduction.

The compilation of a map manuscript for the base map and the derived map is the responsibility of the city surveying department or the commercial or other survey organization that has been contracted for the job. The actual carto-graphic processing of the manuscripts, including the final printing, can also be done by specialized printing houses on a contractual basis. In larger cities, however, the volume and variety of cartographic products is such that the establishment of cartographic printing facilities within the city-survey depart-ment should be considered. The availability of cartographic in-house printing facilities permits prompt attention to urgent needs and better integration of cartographic work within the framework of the surveying and photogrammetric activities of the city.

BASE MAP

Selection of Scale

Several factors have to be considered in deciding upon the scale of a base map. The most important are (a) degree of detail presentation, (b) accuracy that can be derived from the map, (c) production time, (d) production cost, and (e) number of map sheets involved.

a. Degree of detail presentation. The main feature of a base map is the detailed presentation of the terrain. With the exception of very small objects, for which symbols are used, the planimetric information must be presented in its true form and dimensions. This requires a scale sufficiently large to permit the clear presentation on the map of the details surveyed in the field. The scale of the base map must also be suitable for cadastral purposes. In the highly complex and densely built-up areas of city centers, the scale of 1:500 would satisfy most of these requirements. It should also be kept in mind that, at the detailed planning stage, engineers require plans at a scale sufficiently large to permit the drawing of the projects with all the important details.

b. Accuracy. The accuracy of graphical plotting, when conventional manual plotting procedures are used, is accepted as 0.2 mm (standard coordinate error). On the ground, this represents a 10-cm error at the 1:500 mapping scale and a 20-cm error at the 1:1000 scale. Since important points are measured in the field or photogrammetrically with an accuracy of better than ± 10 cm, an argument is raised occasionally on behalf of 1:500 scale. Similarly, measurements taken from 1:500 maps are more consistent with the true dimensions on the ground. The accuracy of plotting is increasing with the progress being made in the mechanization of plotting processes; at present, the standard coordinate plotting error for good-quality map manuscripts of 0.1 mm appears to be a realistic figure. Any further increase in graphical plotting accuracy—which is technically possible—may prove of little value because of limitations imposed by dimensional instability of the manuscript sheets, definition of conventional graphical presentation, and the resulting difficulty in extracting much higher accuracy from maps in practical operations.

c. Production time. The larger the map scale is, the richer the map content. This means that a larger number of details must be surveyed in the field or photogrammetrically. Consequently, more time is needed for the production of a large-scale map. In addition, there is an increase of cartographic work due to the fact that as the map scale increases, the volume of drawing and the number of map sheets grows.

d. Production cost. In an interesting study (1), W. Kłopociński gives a detailed account of comparative costs involved for a map sheet covering 48 ha when produced at 1:1000 and 1:500 scales. According to the author, the base map of a large city, particularly of its central part, must be at a scale of either 1:1000 or 1:500, with the *supporting map*, derived through suitable enlargement or reduction, at the alternative scale. Consequently, the production cost of the supporting map must also be considered. The cost given by this author (1) covers complete field surveying of both planimetry and elevation (contour lines and spot elevations) and further processing of the data, with the inclusion of printing plates. Because of difficulties in expressing the purchasing value in foreign currency, the actual figures are not quoted. Instead, the respective costs in relative numbers are given (Table 9-1) assuming that the cost of production of the base map at a scale of 1:1000 equals 1000.

Table 9-1. Comparative Cost of Producing Base
Maps at Scale 1:1000 and 1:500

	Scale of base map	
	1:1000	1:500
Cost of producing the base map	1000	1349
Change: 1:1000 → 1:500 scale	324	
Cost of 1:1000 map + 1:500 map	1324	
Change: 1:500 → 1:1000		18
Cost of 1:500 map + 1:1000 map		1367
Total cost	1324	1367

From these figures, it can be seen that the cost of producing a 1:500 map is approximately 35% higher than that of a similar map at a 1:1000 scale. However, assuming that both scales are required, one of the maps being the original base map and the second a derived supporting map, the total cost is practically identical. In conclusion, Kłopociński suggests that the centers of large cities should be mapped at 1:500, with the remaining areas mapped at 1:1000. In this solution, for the center of the city, the 1:1000 scale would be derived from the original 1:500 maps; in the remaining areas having a 1:1000 base map, the derivation of a supporting map at 1:500 might be necessary only occasionally.

When reducing a 1:500 map to the 1:1000 scale, all the minute details of the former map are retained. On the other hand, when enlarging a 1:1000 map to the 1:500 scale, certain details expected to be present at the latter scale are missing. Therefore, 1:500 maps derived through an enlargement must be completed by additional field measurements, which is the main reason that enlargement of a map costs more than reduction to a smaller scale. This refers, of course, to maps with a well-specified content at a given scale and not to a simple photographic enlargement or reduction. The enlargement of the base map is frequently used if, for various reasons, a larger scale is needed without any change in the content of the map.

e. Number of map sheets. The number of map sheets is an important considera-tion because of the time and cost of producing the map and the practical use to which the map can be put. The larger the scale, the greater the probability that a specific fragment of the city will be depicted on two or more adjacent map sheets. This is an inconvenience from the user's point of view.

The decision concerning the scale of the base map is a fundamental one that has important financial and operational consequences. It can be made only after very exhaustive studies of local conditions and requirements. Considera-tions should center on the above points and should also include two basic rules: The smallest acceptable scale should be selected for the base map (e.g., 1:1000 instead of 1:500 if 1:1000 is acceptable), and only one original field or photo-grammetric survey should be necessary to provide a set of maps at the different scales required by the city.

Map Content

The fundamental premise that one field or photogrammetric survey should suffice to establish all city maps is valid and serves to determine the content of the base map. This map must contain all the man-made and natural terrain details that are important to the administration and management of complex city areas in the planning and monitoring of various projects and the operation of multiple technical and other services. Obviously, some engineering projects still require very specific maps, such as street maps at the 1:250 scale, with longitudinal and transversal profiles. In spite of these additional requirements, the city base map at the scale of 1:1000 or 1:500 must contain as much infor-mation on the city fabric as possible; otherwise, unsatisfied users would produce their own maps which would jeopardize the very purpose of the base map.

The use of computer-stored survey data or automated map drafting does not necessarily change this basic approach. The demand for base maps in a modern city is such that they should be available at a moment's notice. Also, the cost of a specifically produced map sheet from the computer-stored data is prohibitive at present for a single user.

Since the base map must be quite complete, it would lose its readability if all the information was placed on a single sheet. Therefore, the base map may consist of several overlays. As an example, the base map may consist of the following sheets: planimetry (including cadastral information), topography (contour lines and spot elevations), overhead facilities, and underground facilities.

If the original map manuscripts are on transparent, dimensionally stable material, working copies of any combination of the overlays can be obtained by using common copying techniques. Under usual conditions, the map should be north-oriented, and the border should parallel the grid line system.

The following details and information should be depicted:

reference grid,

horizontal and vertical control points,

administrative and property boundaries, parcel numbers, walls and fences,

all buildings and permanent structures such as elevated highways, bridges, and monuments,

streets and squares, with curb lines, trees, lampposts, hydrants, manholes, and catchbasins,

land use and vegetation,

railroads and other transportation facilities,

transmission and communication facilities such as masts, towers, and cableways.

underground utilities,

terrain relief features, contour lines, spot elevations, fills, and embankments,

drainage features, shorelines, canals, and ditches.

With the exception of small features such as manholes, lampposts, and catch basins, which are represented by symbols, all other objects are represented in their true form and size at ground level. The size of symbols at a given scale, according to Ref. (2), should be as close as possible to the true size of represented features. This approach permits photographic enlargement or reduction of the map to another scale without the necessity of drawing the symbols anew. Therefore, the size of symbols used in the base map should be carefully selected, with the designer bearing in mind the other maps in which they will be used. There is greater flexibility with the line thickness on the base map, and selection of line thicknesses may depend upon local customs and drafting and reproduction facilities.

Further decisions concerning the content of a base map can be aided by the list of symbols compiled by J. Sima (3) for city maps at scales of 1:1000 and 1:500. These symbols may have different size in the actual maps.

Symbols for City Base Maps

1. *Ground control points*

   Triangulation point   △ **171**

   City survey monument   ☐ **2373**

   Bench mark

2. *Boundary lines*

   Visible land-use boundary

   Wall   less than 30cm   retaining wall
                                                  more

   Fence

   Hedge

   Outline of wooded area

3. *Land use and vegetation*

Green area

Orchard

Scrub

Individual tree   1:1 000

Bush

Sand

Swamp

Cemetery

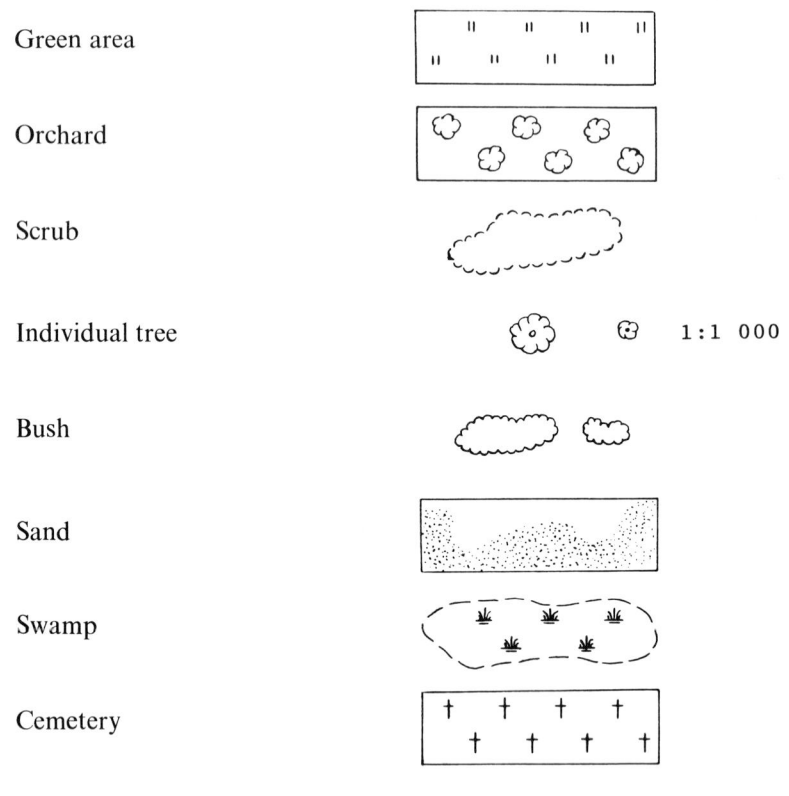

4. *Roads and related features*

Highway, main road (paved)

Unpaved road

Private entrance

Footpath

Culvert

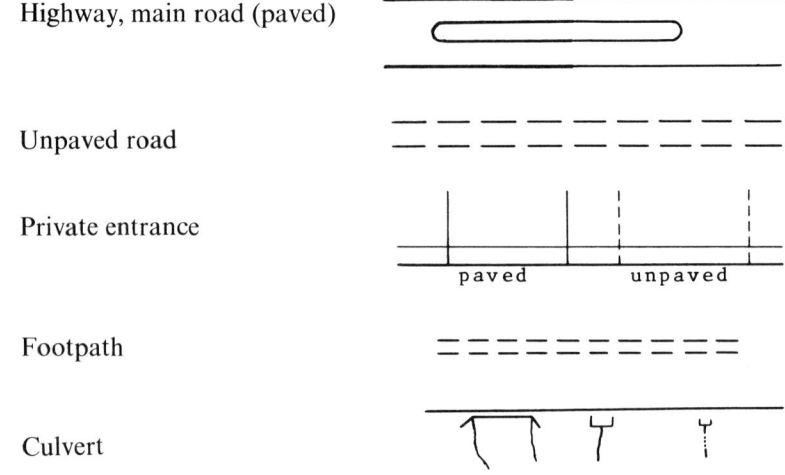

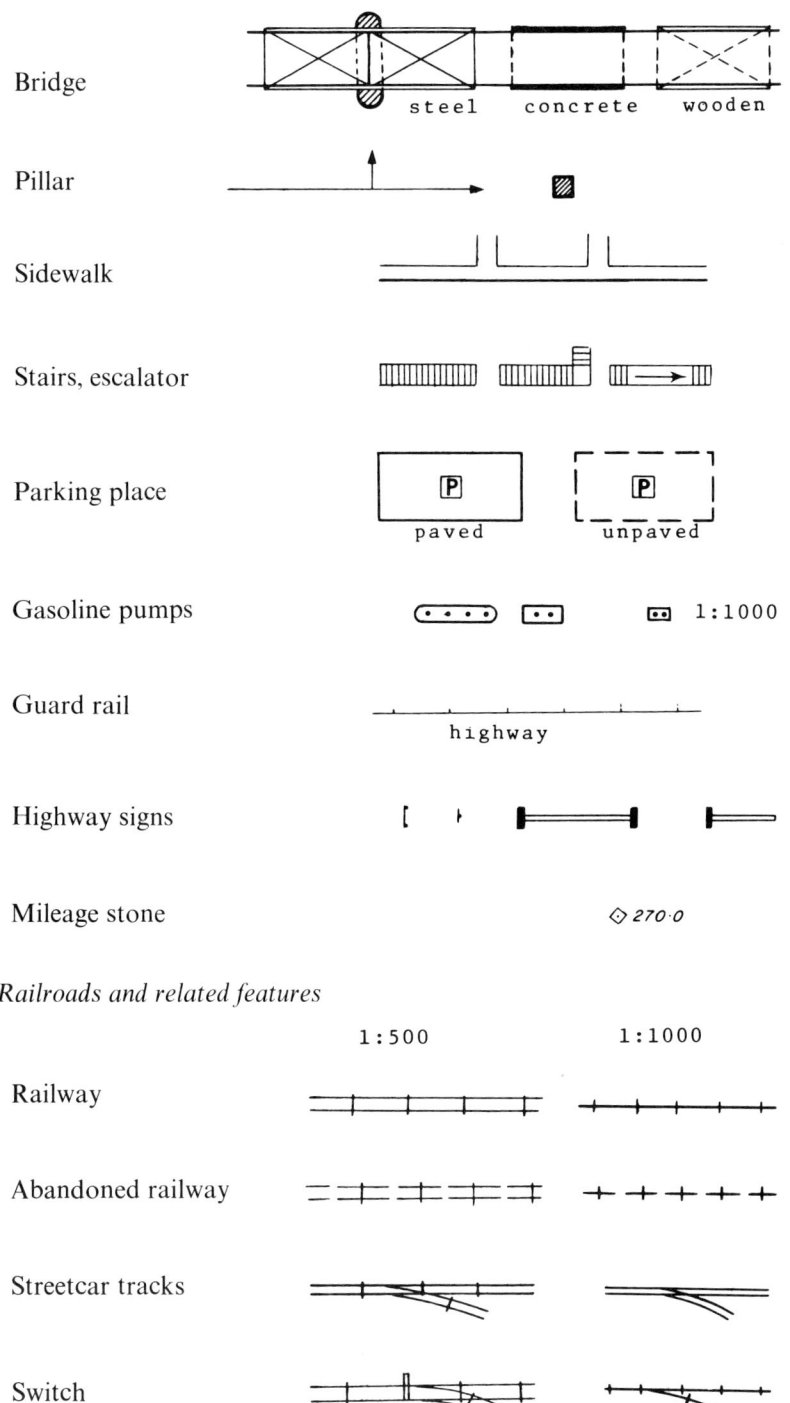

Bridge   steel   concrete   wooden

Pillar

Sidewalk

Stairs, escalator

Parking place   paved   unpaved

Gasoline pumps   1:1000

Guard rail   highway

Highway signs

Mileage stone   ◇ 270·0

5. *Railroads and related features*

	1:500	1:1000

Railway

Abandoned railway

Streetcar tracks

Switch

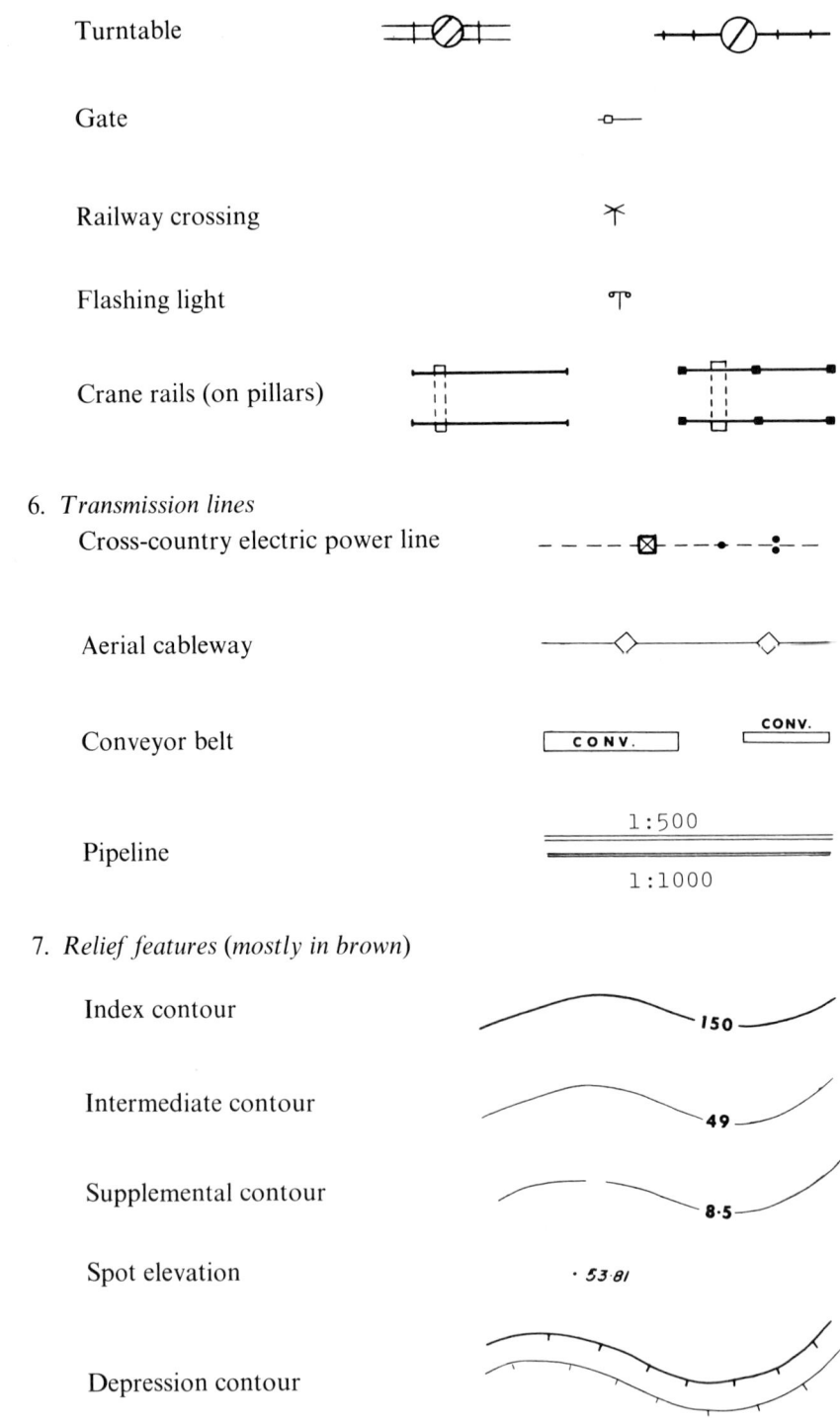

Turntable

Gate

Railway crossing

Flashing light

Crane rails (on pillars)

6. *Transmission lines*
   Cross-country electric power line

   Aerial cableway

   Conveyor belt

   Pipeline

   1:500

   1:1000

7. *Relief features* (*mostly in brown*)

   Index contour

   150

   Intermediate contour

   49

   Supplemental contour

   8·5

   Spot elevation

   · 53·81

   Depression contour

   CONV.

   CONV.

Rock, cliff

Fill, embankment

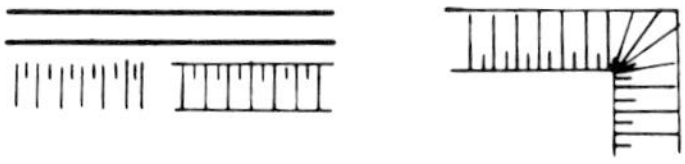

Cut

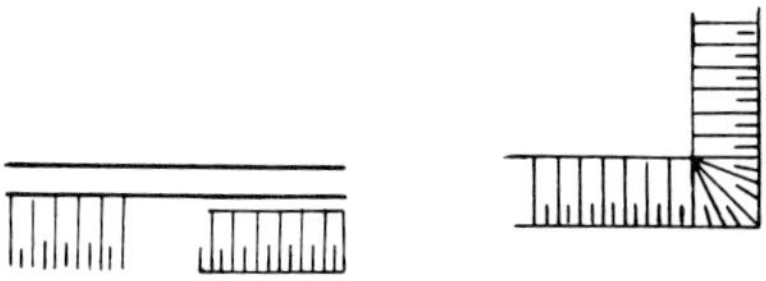

Break line, escarpment

8. *Buildings and constructions*

Building foundations

Building roof

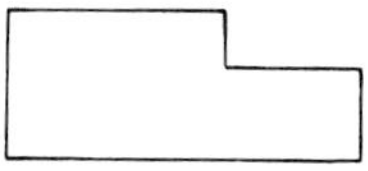

Annex, garage, shed

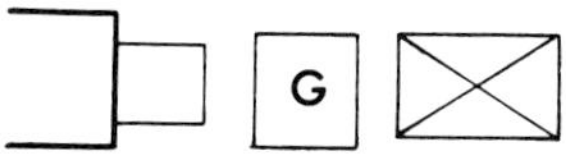

Verandah

Porch, balcony, canopy

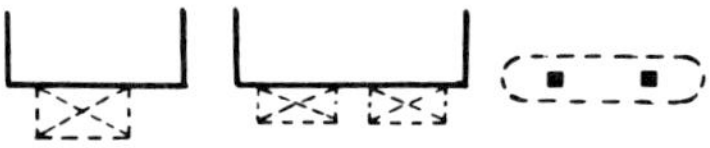

1:500 only

Building under construction

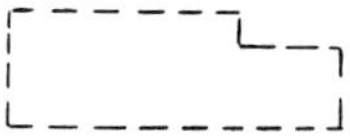

Ruin

Monument

Separate chimney

Fountain

9. *Services and utilities*

Radio or TV mast		
Hydro tower		⊠ 1:1000
Hydro pole	♦	
Telephone pole	–•–	
Lamp standard	↓ ↓ ∨	
Other poles and posts	•	∘
Signal light	┿	┿ (suspended)
Transformer		
Call-box	☐T	
Well	◖	∘ 1:1000
Fire hydrant	⊤	
Water conduit	⊖	
Sanitary manhole	⊗	
Gas manhole	⊕	
Catch basin	◫ ▥	

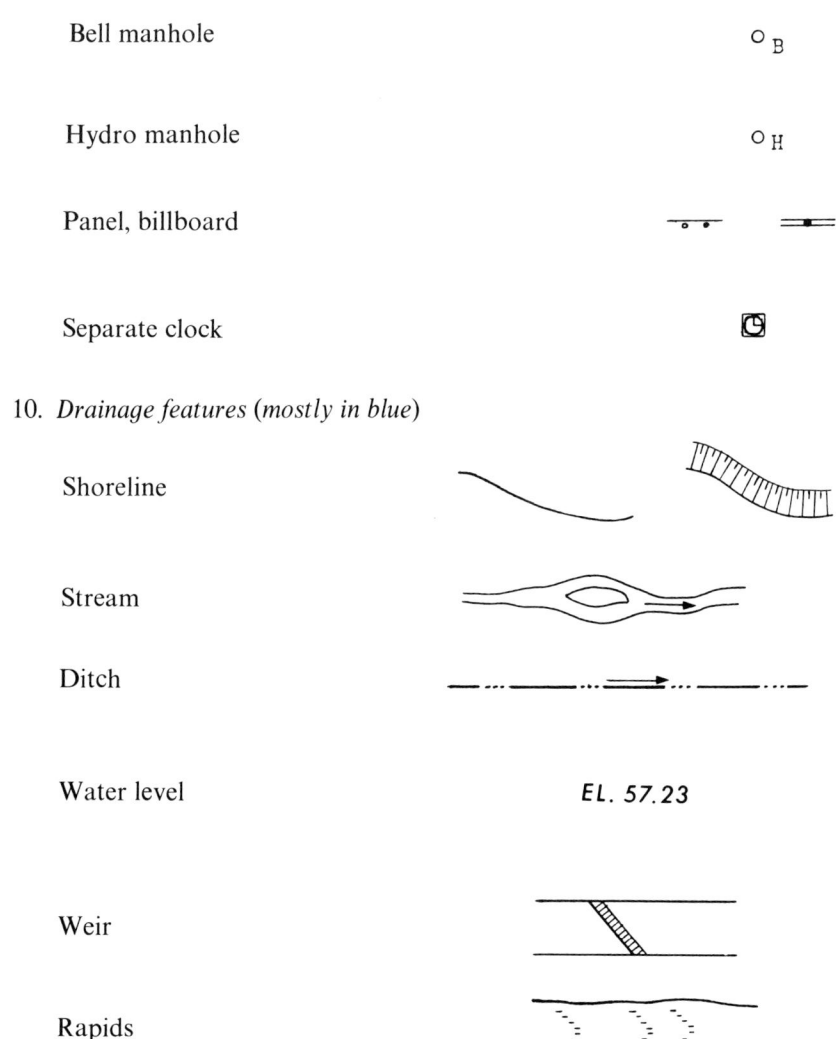

Bell manhole

Hydro manhole

Panel, billboard

Separate clock

10. *Drainage features (mostly in blue)*

Shoreline

Stream

Ditch

Water level   *EL. 57.23*

Weir

Rapids

Compilation of a Base Map

In this chapter, we refer to base maps that are compiled from field survey data. Readers interested in the photogrammetric approach are referred to Chapter 8.

To ensure good "internal" accuracy of each map sheet, it is recommended that grid crosses and basic horizontal control points, including traverse points, be plotted during the same operational step, regardless of whether manual or automatic coordinatographs are used. Dimensionally stable transparent plastic sheets are used for the purpose at present. Once the horizontal control points

Figure 9-1. Transportable Haag-Streit coordinatograph. Plotting area 50 × 40 cm.

have been plotted, the draftsman can proceed with the plotting of the actual map content by following, point by point, the sequence of the field surveying operation. The plotting can be done by using extremely simple drafting instruments such as triangles, ruler, compass, protractor, and pencil. However, portable coordinatographs, either rectangular or polar and of various sizes, may still be available on the market. They provide good accuracy in a relatively rapid and smooth manual operation. In contrast to the plots in which primitive means are used, only the points of the map content are marked by coordinatograph. In consequence, the resulting drawings are very clean, and a plotting accuracy of from ±0.05 to ±0.02 mm can be obtained. In Figure 9-1, the Haag-Streit portable coordinatograph, which permits plotting of surveying results over a sheet area of 50 × 40 cm is shown. The coordinatograph is equipped with a microscope and exchangeable numeration tapes.

Portable coordinatographs, particularly the larger ones, can also be used for plotting points determined numerically by photogrammetric means.

Polar coordinatographs are used when field surveying is carried out by a polar method. Aristo used to offer two models of these coordinatographs, both 40 cm in diameter. The more sophisticated model is equipped with a reading dial for distances; in the simple model, readings must be done with the help of a vernier. The accuracy in plotting distance is ±0.02 and ±0.05 mm, respectively. The vernier reading of angles in both models allows a resolution of 30″ or 1ᶜ (centesimal minute).

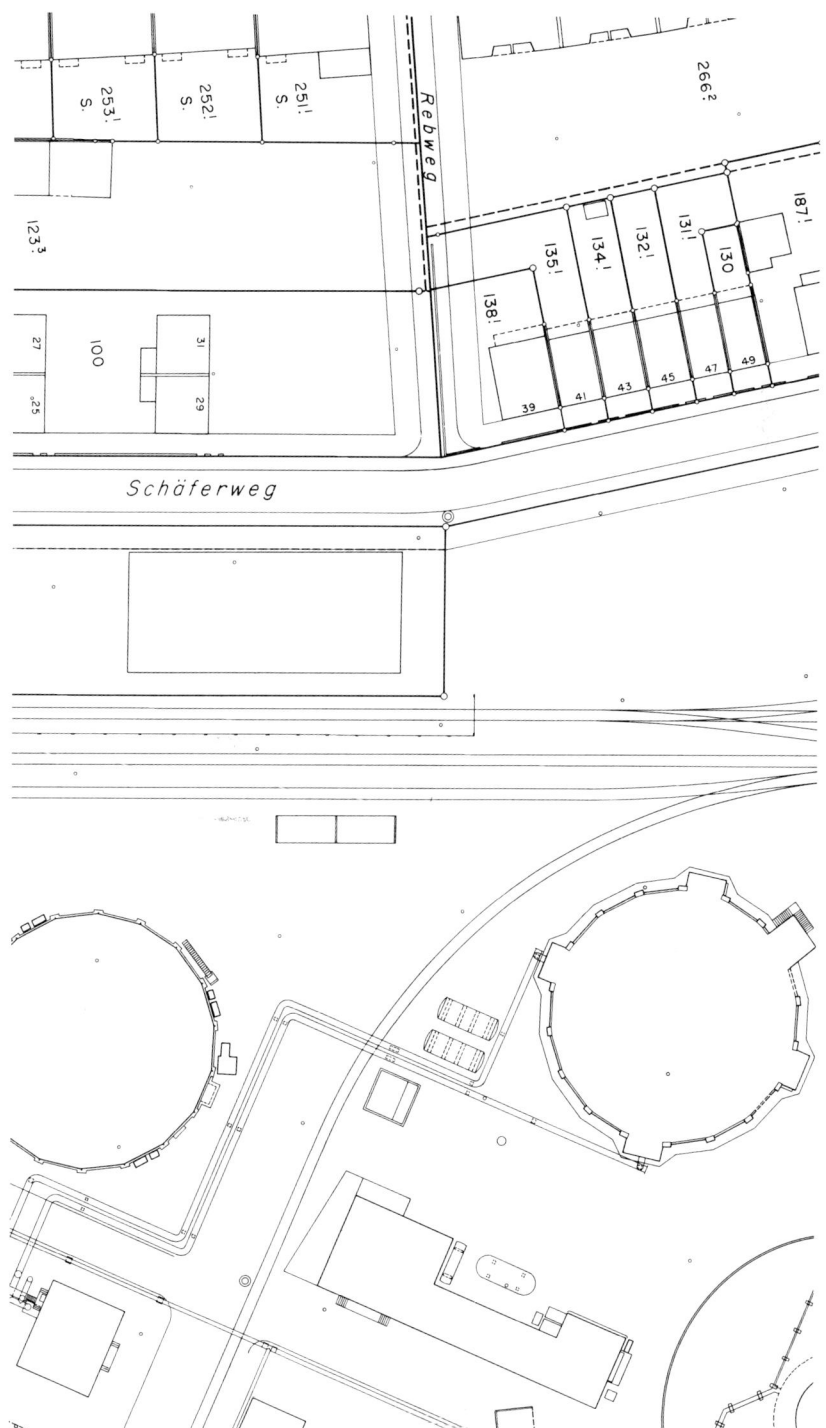

Figure 9-2. Example of a city base map at the scale 1:1000 (courtesy of the City Survey Office of Basel).

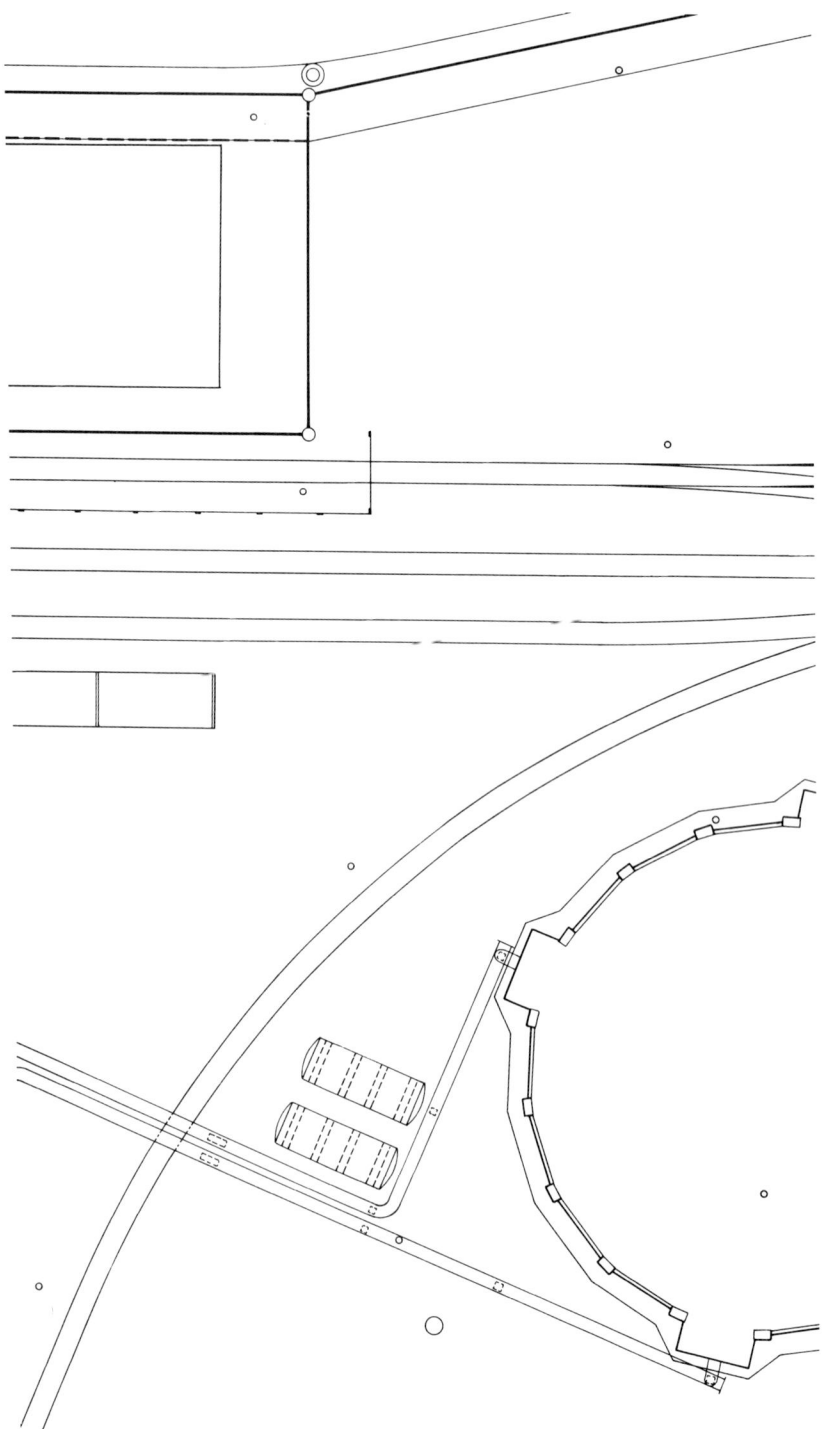

Figure 9-3.   City map at the scale 1:500 (courtesy of the City Survey Office of Basel).

The compilation of a map manuscript is an excellent verification of survey work and of the ensuing computations. Discrepancies encountered during plotting must be clarified by checking in the field. The map, therefore, is not only the ultimate purpose of field survey operations but also the final control phase.

As an example, an identical section of a base map at the scales of 1:1000 and 1:500 is shown in Figures 9-2 and 9-3.

DERIVED CITY MAPS

A carefully compiled and properly maintained base map permits the derivation of any other map, particularly if computer-stored map information from field or photogrammetric surveying is considered an integral part of a modern mapping technique. In this advanced approach, there is no limit to the scales at which either the whole or selected parts of the city can be presented in map form. This possibility should be kept in mind by those responsible for city surveying and mapping. Most urban communities, however, do not yet have access to this advanced technology and therefore must use more conventional means in meeting multiple mapping requirements. It is also true that the simplest and the least expensive way of changing over from one scale to another is by direct photographic reproduction. Unfortunately, direct photographic enlargement or reduction can be used only within a limited range of scale differences, mainly for the following reasons: Photographic enlargement also amplifies the graphical inaccuracies of the original map, more detailed presentation is expected at a larger scale or an expanded map content, and less detailed information is expected from the map at a smaller scale or a reduced map content.

Therefore, when changing from a smaller to a larger scale, the question of accuracy of the derived map, as well as the problem of map content, must be considered. If the derived map is at a smaller scale than the original, only generalization of the map content imposes problems. By clearly setting priorities and technical procedures in advance, some of the problems can be overcome or alleviated. For instance, when using relatively simple portable coordinatographs for the plotting of the base map manuscript (see the preceding section), the accuracy of the plot can be very high, and therefore a map enlargement may still have very acceptable metric characteristics. It should also be remembered that large-scale maps, such as 1:500 or 1:250, are needed not so much because of accuracy considerations as simply because there is a requirement for more "paper space" to record graphically the expanded map content encountered in central city areas or to draw details of new projects. Consequently, the specifications for the base map should be drawn up in a form that permits a changeover to other scales in the most economic and rapid fashion, without jeopardizing the essential characteristics of the derived maps.

Numerous municipal departments and services require maps at various scales derived from the base map. The most important of these are listed in the table

Table 9-2. Recommended Map Scales for Different Uses

Department	Conceptual projects	Realization studies, pre-projects	Final projects execution plans
Urbanistic planning	1:10 000–1:5000	1:1000	1:500–1:250
Roads and streets	1:10 000–1:5000	1:1000	1:500–1:250
Water and sewage	1:10 000–1:5000	1:1000	1:500–1:250
Power and heating	1:10 000–1:5000	1:5000–1:1000	1:1000–1:250
Parks and recreation	1:10 000–1:5000	1:1000	1:500–1:250

above. It must be understood, however, that usage may vary from country to country; the table provides a general orientation only. There is also a need for topographical maps at much smaller scales, such as 1:25 000, 1:50 000 and 1:100 000, but we assume that these maps are provided by the national mapping program of the country.

Table 9-2 indicates the main map scales to be considered for different uses.

The 1:250 *map.* This is basically a street map derived from the base map by suitable enlargements of a section covering a required street and a strip of terrain on both sides of the street about 10 m wide. Of the details expected to be shown on the street map, as many should be included in the initial field survey work for the base map as is practical; this will avoid the occurrence of independent field surveys. Missing details (microarchitectural details) should be measured and plotted additionally.

The street maps should contain not only the usual planimetric and topographic information but also the information needed to meet the requirements of various engineering projects. It is useful to have information on the type of street surface, underground utilities (with the respective surface features such as manholes and catch basins), overhead utilities, traffic signs, spot elevations of characteristic points, soil characteristics, ground water level, and so on. An example of a street map at a scale of 1:200 is shown in Figure 9-4.

The 1: 500 *map (derived from the* 1: 1000 *base map).* In large cities, and particularly in their densely built-up central parts, the 1:500 scale could be considered for the base map. In such a case, general rules for base maps apply. Otherwise, it is a supporting map obtained by an enlargement of the base map. Unlike the previously discussed 1:250 map, the 1:500 map usually consistently covers the designated area by map sheets that follow the map projection grid. Specific details not present in the base map must be obtained from additional

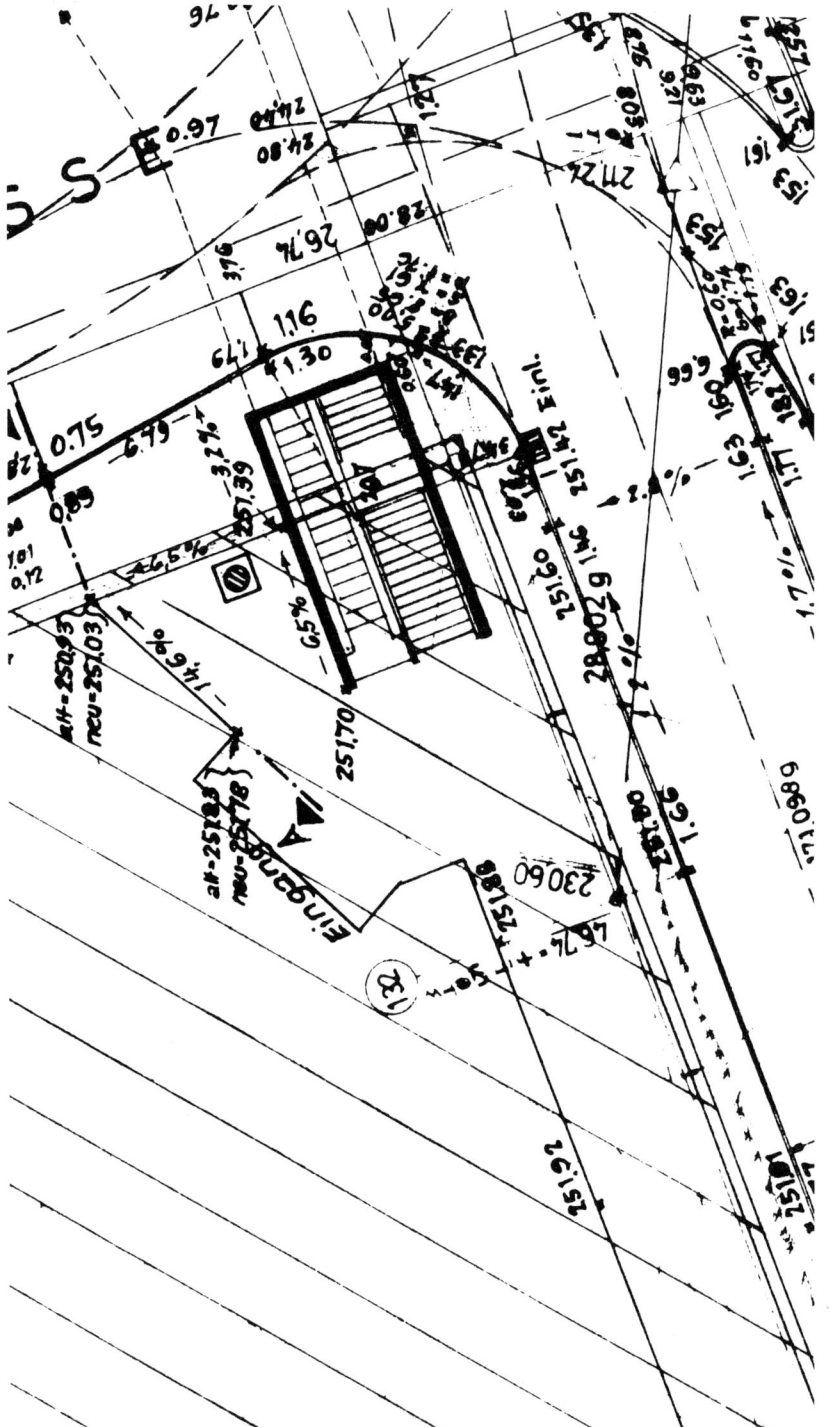

Figure 9-4.   Street map at a scale of 1:200 (courtesy of the City Survey Office, Stuttgart).

field survey. If suitable symbols are selected, a direct photographic enlargement of the 1:1000 map provides acceptable results.

The 1:1000 *map* (*derived from the* 1:500 *base map*). The 1:1000 scale is considered indispensable in city areas. Each section combines a sufficiently detailed and precise presentation of the terrain content with coverage sufficiently large to be meaningful and useful in many important applications. If the 1:500 scale is selected for the base map of densely built-up centers of the city, then the 1:1000 map is produced by photographic reduction, so that the whole city area is covered by uniform 1:1000 maps.

The 1:2000 *map.* In some countries, a 1:2000 map is produced by photographic reduction of the base map. Unless there is a very good reason for using it, a map at this scale should be avoided. The guiding principle should be production of as few maps at different scales as possible, and those that are produced should be kept in good condition. In some countries, the 1:2000 scale is used for base maps of smaller towns and villages.

The 1:5000 *map.* The next map of importance is the city map at the 1:5000 scale, which is derived from the base city map by suitable cartographic reduction. Compared to the 1:1000 map, the single sheet covers an area 5^2, or 25, times larger. At the same time, the scale is still large enough to include single buildings, property boundaries, regulation lines, and other features of importance at the initial planning stage of various larger projects. As a result, the 1:5000 map is considered a most useful and necessary general information map and can be used as a base for studies and planning. Often, this map is published in multicolor editions. If this is not possible because of financial considerations, the printing of at least contour lines in a suitable brown color greatly enhances the map presentation. An example of a 1:5000 map is shown in Figure 9-5.

Because of significant scale reduction between the base map and the 1:5000 map, the content of the latter must be generalized, and the symbols must be drawn anew. There is, therefore, an appreciable amount of compilation and editorial work involved in the preparation of this map.

The 1:10 000 *map.* The scale 1:5000 is too large to cover the whole city by a number of sheets convenient to form a handy general city plan for both studies of the future development of the city and conceptual elaboration of various projects. There is also a need for a suitable cartographic base for a multitude of special maps. In these applications, the mutual dependence and the interplay of various factors must be clearly depicted for the whole city area; therefore, smaller-scale maps are required. This function is usually fulfilled by a map at the scale of 1:10 000, which is derived from the previous map at 1:5000. There is, however, a distinct difference between those two maps. The 1:5000 map is a relatively detailed general information map; the content of the 1:10 000 map is restricted to the main physical features of the city. An example of a 1:10 000 map is shown in Figure 9-6 (see insert opposite page 247).

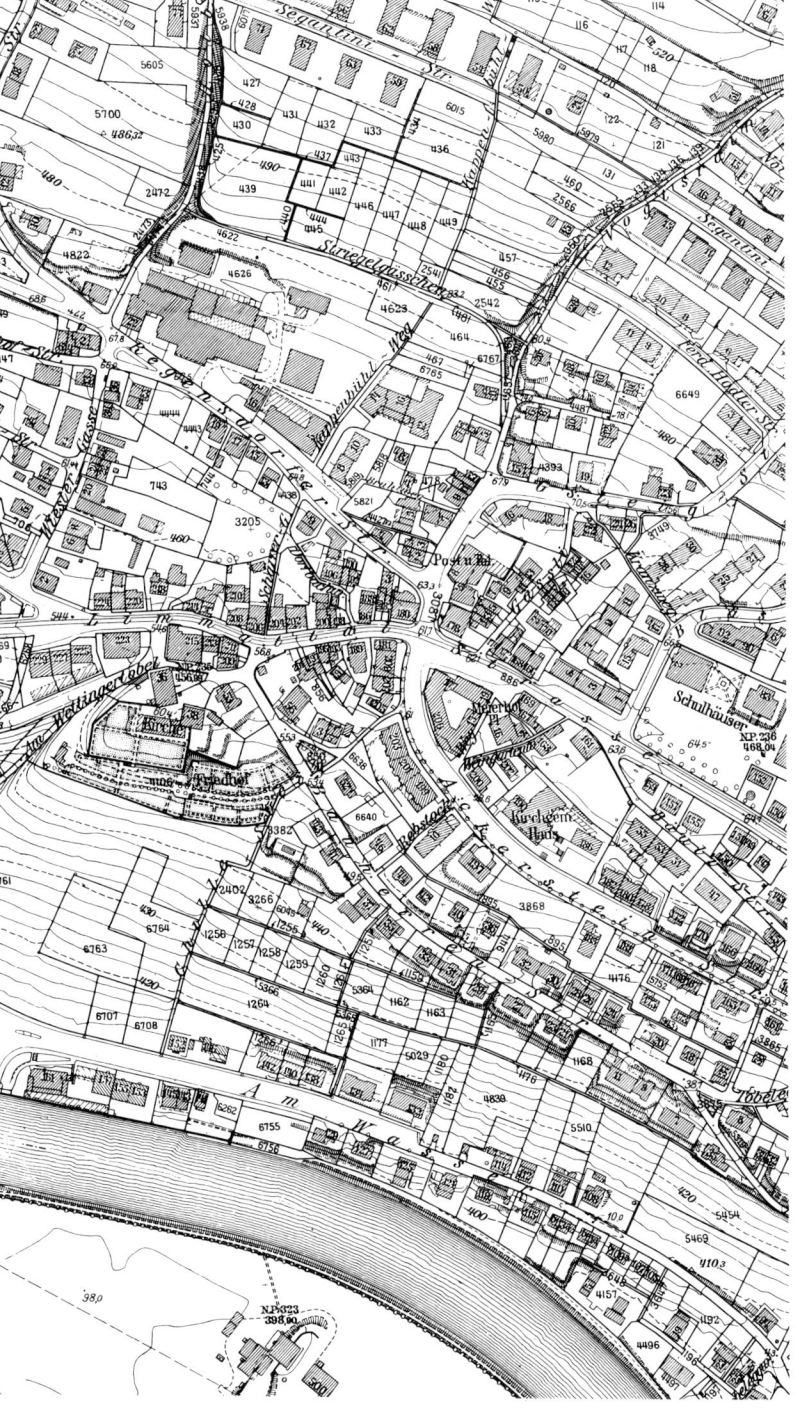

Figure 9-5.   Example of a 1:5000 map (courtesy of the City Survey Office of Zurich).

1:25 000, 1:50 000, 1:100 000, AND SMALLER-SCALE TOPOGRAPHICAL MAPS

These are not city maps but the general topographical maps produced and maintained by state authority. They are used extensively by city and regional authorities, particularly in studies and projects involving the city and the surrounding regions. In larger cities, they are also used as the base for special maps.

MAINTENANCE OF CITY MAPS

"Maintenance" we understand to be not so much the keeping of city maps in good condition to preserve their initial metric characteristic and its graphical quality as their continual updating so that they represent, at any given moment, the actual situation on the ground. This is a most complex and important matter since, unless it is solved in a satisfactory manner, full advantage cannot be taken of the information provided by city surveying and mapping. The difficulties are twofold: technical difficulties and organizational difficulties.

The technical difficulties are due to the fact that the introduction of changes into existing maps is always a delicate proposition. Once a number of copies are printed from the original map manuscript, it is not possible to make alterations on all copies. Also, the number of changes occurring on the ground every day can amount to several hundreds or thousands in a large city. If these changes are not promptly marked on the map, they accumulate very rapidly, and the usefulness of the maps and of the whole survey system can be jeopardized. Therefore, in well-organized city survey departments, each physical alteration on the ground or change in possession is immediately marked on the original map manuscript or a copy of this map used as a "service map." These changes are usually made in a distinct way, so that they can be easily recognized. It is understood that an alteration on the map is only made when the actual physical change has taken place or the transfer of ownership rights has become legally valid. Consequently, precise updating procedures must be laid down and observed by all associated departments of city administration. For example, a copy of a building permit issued by the associated department must be sent to the city survey department. The builder may proceed with the building only after the city survey department has certified that the foundation follows precisely the approved regulation line. At this point, the outline of the new building should be marked tentatively on the map, and the alteration should be completed once the construction work is finished. Similarly, notification of any transfer of ownership rights, creation of new land parcels, or other modifications that would cause changes in property boundaries or reference numbers on city maps must be immediately passed to the city survey office for alteration on the map.

The proper and efficient organization of transfer of pertinent information among various authorities and parties involved is a major difficulty in the updating process of city maps because it requires the rigorous adherence of all concerned to the operational scheme and regulations. This is usually most difficult to achieve unless specific bylaws are passed and strictly enforced.

Since printed copies of the maps cannot be modified to provide the numerous users with complete and up-to-date maps, the actual practice involves a different solution. The individual map sheets (of the base map) are republished as soon as a certain volume of modifications on the terrain has been exceeded. Meanwhile, the users are offered copies of the updated original map (or its duplicate, the so-called service map).

For the derived maps at smaller scales, such as 1:5000 and 1:10 000, either a similar approach is taken or a complete reediting of the map is done at suitable time intervals depending upon the frequency of the changes that occur. A period of from 5 to 10 years is the time interval usually considered.

The alternative to continuous maintenance of city survey data and maps is wasteful and expensive duplication, lack of foundation for a rational administration and planning, and general chaos. It is also apparent that computer storage of survey data and computer mapping provide an efficient new approach that will eventually dominate this important activity. Although there is operational complexity in maintaining city maps in an up-to-date condition, the experience of cities in some European countries proves that an efficient and practical solution can be achieved, with obvious economic benefits.

THEMATIC MAPS

The last category of city maps is thematic maps, or single-factor maps. Unlike the base map and the derived map, thematic maps are usually initiated and compiled by departments other than the city survey department. Their purpose is not to present the planimetry and topography of the terrain occupied by the city but, rather, to present selected information against the background of the city's planimetry and general topography. The geometric accuracy of thematic maps is therefore of secondary importance. Of primary interest is presentation of selected data or phenomena from which an immediate and meaningful conclusion can be drawn. The variety of thematic maps is practically unlimited since most data or phenomena can be presented with advantage in this form. A well-organized city must produce and maintain a limited number of these maps. Otherwise, no use can be made of the variety of information that is indispensable for city administration and planning, such as location of fire stations and their respective territories, distribution of school children of various age groups, and density and pattern of city traffic. Figure 9-7 gives some examples of thematic maps.

Thematic maps are frequently produced as transparent overlays for use in conjunction with regular maps at suitable scales. This approach allows the use of a number of overlays superimposed on each other to provide a block of correlated information.

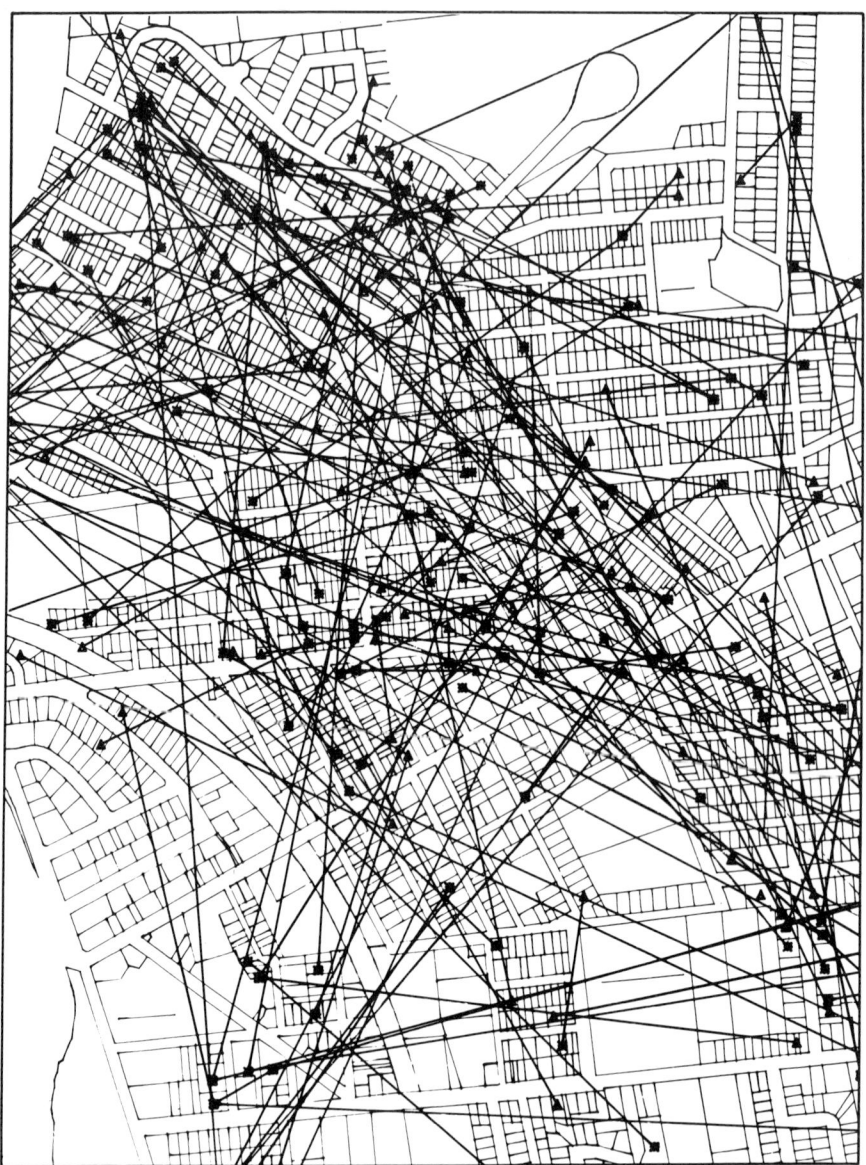

Figure 9-7a. Interior population migration for the municipality of Vanier during the year 1971. (Courtesy National Capital Commission, Ottawa.)

Figure 9-7b. Percentage change of population (growth and decline) in a section of Ottawa for the period 1971–76. (Courtesy National Capital Commission, Ottawa.)

COMPUTER-SUPPORTED MAPPING SYSTEMS

Until recently, map drafting was an exclusively manual operation—the field surveying records or photogrammetric plots, including names, numbers, and symbols, were converted to the final map manuscript by hand. The introduction of scribing and the use of special sets of symbols, numbers, and letters that can be added to the map manuscripts did not change the manual nature of the work.

However, the progress made in computers and ancillary technology has altered this situation drastically. Computers permit not only the storage and retrieval of a large volume of data, both numerical and descriptive, but also extensive handling of these data and their display on CRT (cathode ray tube, or video tube), or in a map manuscript form if a numerically controlled plotting table is used.

There are several reasons for the growing interest in computer-drafted maps. The underlying general factors are general expediency and the trend toward automation, particularly the elimination of painstaking, manual drafting operations. Despite the high-unemployment situation, it is becoming increasingly difficult in many parts of the world to find persons trained as cartographic draftsmen and willing to work as such. From the purely technical point of view, however, the automation of map drafting is only a logical extension of increasing automation in the initial phases of surveying and mapping processes. For instance, the use of some electromagnetic distance-measuring devices permits the recording in the field of final ground coordinates of surveyed points in a form suitable for direct storage and use in a computer. Similarly, modern photogrammetric analytical plotters provide *final ground coordinates* of measured points in an on-line operation. Since an electromagnetic tape or equivalent record of surveying information (field or photogrammetric) is becoming the standard output of the initial surveying phase, not to use these computer-compatible data for further automatic processing would appear, technically and economically, to be a backward approach.

Moreover, automatic drafting offers important features: It is rapid and of very uniform quality, both graphically and metrically. It is also certain that further spectacular progress in this field can be expected.

The strong argument, however, for computer-supported, versus conventional, mapping is the ability of this technique to overcome the inherent technical difficulties in the continual updating of maps and the derivation of maps at different scales. Particularly in city areas, with their complex physical content, the question of continual updating of relevant surveying information, and its speedy presentation in a map form at a certain scale, is an on-going problem that requires a more satisfactory solution.

In computer-supported integrated mapping systems, the information that constitutes the base map content or that may be considered for thematic mapping is stored in computer-based files and subfiles. The information is usually arranged thematically, which means that each information class, such as buildings, property boundaries, and contour lines, is stored separately. This facilitates automatic plotting according to specific algorithms devised for the purpose. Relevant changes that have occurred on the ground or otherwise (e.g., in ownership conditions) since the original data were compiled can be immediately introduced into the corresponding files.

In spite of the relative rapidity with which automatic plotting can be carried out, it is not yet practical to consider the stored data as a "numerical map" that could be converted at a moment's notice into a graphical product. Therefore, the general approach to city maps, as outlined in the previous paragraphs, remains valid: Uniform city maps at a few selected scales must be published even when automated drafting is considered, and they are used until their new publication in an updated form can be undertaken. Nevertheless, a computer-supported city mapping system offers some important features and, from an operational and even conceptual point of view, brings about the following changes: a unique flexibility in producing maps of specific content, up-to-date maps over limited areas that can be produced on short notice, and loss of validity of the present distinction between "original" and "derived" maps if the derived map is produced in an independent plotting operation from information stored in the base file.

It must be noted, however, that the apparent simplicity of computer-supported mapping is deceiving. In reality, it is an operation of considerable complexity, requiring sophisticated and expensive hardware such as computers, interactive display and editing systems, and automatic plotting tables, all supported by complex software. For a better appreciation of the whole operation, the typical phases of a computer-supported mapping system are outlined.

1. Mapping data is collected in a computer-compatible form. These data can be from either field surveying or photogrammetric plotting operations. In a properly designed, integrated system, suitable procedures and instruments should be used in this initial phase of the work. Careful planning of the whole mapping process from the very beginning, considering even the most minute details, is an absolute prerequisite and constitutes one of the greatest difficulties in setting up a computer-supported mapping system.

2. The raw data stored in computer-compatible form is checked for all kinds of errors, duplications, gaps, and mutual compatibility and is arranged for the automatic drafting stage. Thus edited data are transferred to a base data file that will also be used to meet further requirements, such as the provision of specific terrain information, derived from the digital terrain model. The editing process carried out on a computer requires an extensive software package covering editing, storage, and manipulation of data.

   Changes that have occurred on the ground that affect the mapping content must be introduced into the base data file, so that it represents the actual situation on the terrain.

3. As a result of the editing process, a plot tape is generated. Preparation of these plot tapes requires special subroutines to optimize the computer processing of data. Depending upon the organization of the system, interactive display hardware can be used in editing and checking processes. These systems contain a cathode ray tube (CRT) on which selected information can be instantly displayed and modified. With such systems, not only can rapid visual checking be exercised over modification of the magnetic tape content, but also a "hard copy" of the actual situation within a limited area on the ground can be produced, without going into much more rigorous, time-consuming, and expensive automatic plotting.

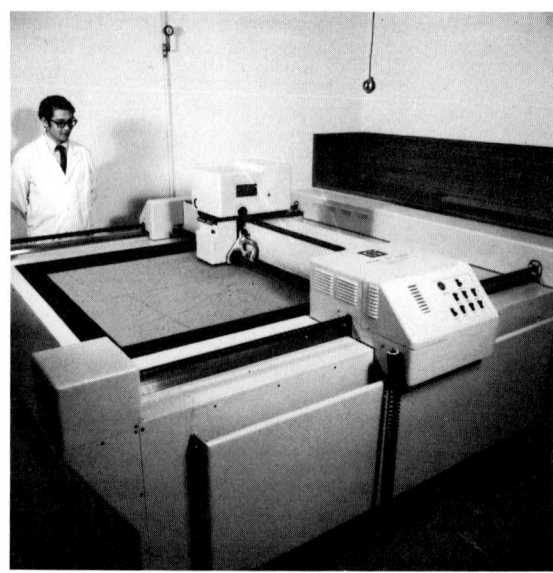

Figure 9-8. Gerber Automatic Plotting System (courtesy of the Ministry of Transportation and Communication of Ontario, Canada).

4. The plot tape is loaded into the automatic plotting system (Fig. 9-8), usually an electromechanical plotter. There are a number of these devices on the market, varying in price, size, precision, and speed of plotting operation. There are drum plotters and precision flatbed plotters. They can be equipped with ink pens, scribe tools, or light heads; the latter provide excellent quality drawing, but unfortunately the light-sensitive film used for plotting does not tolerate general exposure. As a result, the plotting results cannot be inspected during the plotting operation. Complex city maps generally require the use of precise flatbed plotters. An example of a complex map plotted on a flatbed table is depicted in Figure 9-9. However, for many immediate engineering and administrative applications, copies produced on much less expensive plotters, such as some of the drum plotters, can be used.

The general structure of a computer-supported mapping system can be presented in schematic form as in Figure 9-10. The actual structure depends upon the scope of the system and the degree of sophistication intended. According to some authors (4), the cost of maps produced in a computer-supported system is about the same as that of manually produced maps. As mentioned at the outset, however, the main reason for introducing automation into the city mapping process is the speed and flexibility of the computer-supported techniques and their ability to solve most of the difficult problems of map updating. Nevertheless, the initial investment is considerable, and the overall complexity should not be

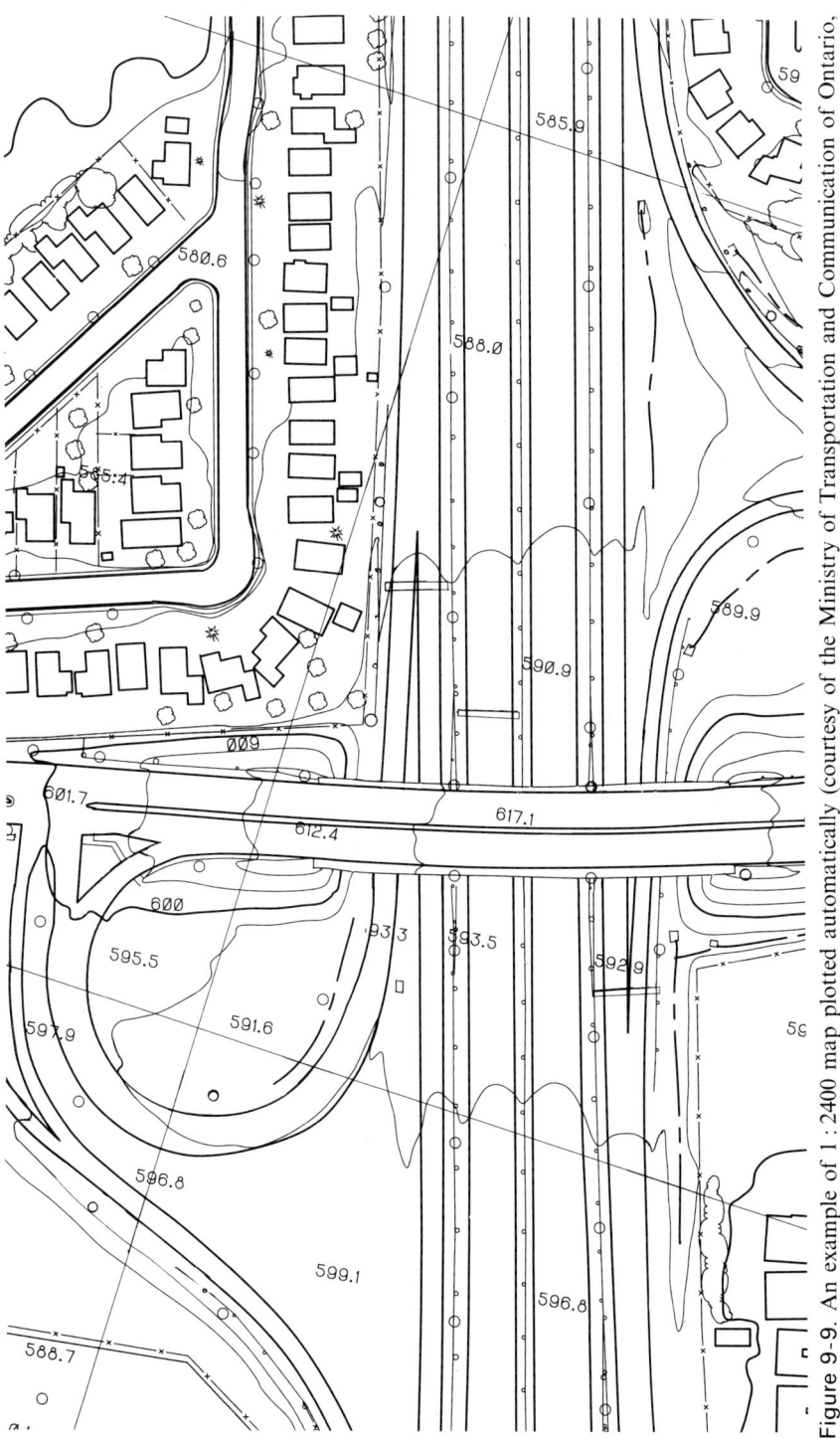

Figure 9-9. An example of 1 : 2400 map plotted automatically (courtesy of the Ministry of Transportation and Communication of Ontario, Canada).

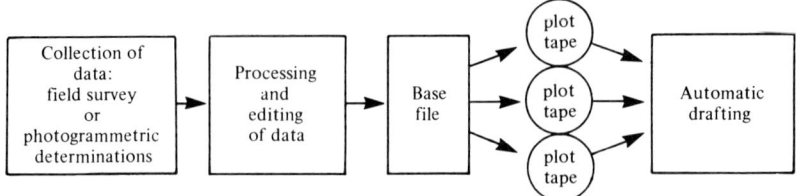

Figure 9-10. General structure of a computer-supported mapping system.

underestimated. Experts in the field warn against premature decision to implement automated map drafting. In any event, such decisions should be preceded by extensive consultations with those centers and cities that have had experience with this technology. An interesting study of the general problems to be considered can be found in the report recently prepared under the direction of A. Hamilton (5).

A map is the goal and the final product of surveying. To take full advantage of automation based on modern computer data processing, the whole surveying and mapping operation must be designed as a totally integrated entity.

References

1. Kłopociński, W. Content and the Scale of the Base Map of a Large City, *Proceedings of a National Consultation on City Base Maps* (in Polish), Polish Geodetic Society, Warsaw, 1967.
2. *Cartography at Large Scale in the Life of the Nation* (in Polish), Polish Geodetic Society, Poznań, 1968.
3. Sima, J. Compilation Photogrammétrique de Cartes Urbaines aux Échelles de 1 : 1,000 et 1 : 500, *The Canadian Surveyor*, No. 5, 1971.
4. Howell, T. F. Automated mapping system implementation, *Photogrammetric Engineering*, No. 12, 1974.
5. Hamilton, A. *The Principal Concepts for a Long-Term Mapping Program in the Maritime Provinces*, Department of Surveying Engineering, University of New Brunswick, 1976.

Chapter 10

Principles of Surveying and Mapping Data Banks in Cities

by R. A. SMITH

Automation of data gathering by using electronic measuring and recording instruments in the field and office, provides raw surveying and mapping data in a computer format. Computer programs and interactive systems enable the manipulation, analysis, and editing of the raw data, which results in files of location-oriented image data. The use of a common geometric (coordinate) reference system facilitates integration of data from various missions or projects. The resultant data base is a scale-free digital map. It may be retrieved and presented in graphic or numeric form.

The traditional approach to the development of a survey project is to study the user's problem, define the output or map needed, design the system that will produce the end product, and, finally, implement the system by gathering and synthesizing the data and drawing the map. The map may be a conventional map in *hard copy* form or a numerical 1:1 scale "map" expressing areas, volumes, and relationships in terrain dimensions.

The development of a graphic base-map system requires a cost/benefit analysis. What map data, at what accuracy, timeliness, scale, orientation, sheet format, etc., are required by several users? Are there economies to be realized by producing base maps? Will the costs and maintenance exceed the benefits? What are the costs to the user of restructuring the maps to suit the project requirements? What are the costs to the user of gathering and adding the special-purpose data that are required? Does the availability of map data facilitate decision making and contribute to the organization's fulfillment of its programs, goals, and objectives? Would a digital map data base provide benefits over and above the conventional mapping approach?

A map data base is only part of the urban data resource. The recognition that map data are an integral part of urban data as a whole leads to a data-base philosophy. Data are to help the whole urban community fulfill its goals and objectives; therefore, they are a resource to be managed. The data base will be used to produce information, and the information will be used to make decisions. There need be *no direct relationship* between the input to the data base and the problem-solving output, since input is only added to a shared data pool.

Consequently, a data base is not the integration of various functional data bases: It is the integration of the data. The retrieval process must provide for direct access by the user and use of the data base as an *ad hoc* inquiry facility. In addition, there should be a reduction in the number of scheduled reports and maps.

The shared pool of data in a data base may be considered a fourth basic resource along with land, people, and capital.

A municipality's main responsibility is for goods and services, so any data base it develops must respond to these areas. The data base is to provide data for creating information and subsequently making decisions. The result is that the data base must also provide indicators (information) to assess the success of these decisions. In the private sector, indicators of customer satisfaction are easy to identify. In the public sector, satisfaction is difficult to identify, but inadequate policies or poor decisions usually result in dissatisfaction in one of the following general areas:

Area	Dissatisfaction indicator	Items
Land	Depletion	Of open space, wildlife, flora, minerals, agriculture, hills, trees, lakeshores, wetlands
	Pollution	Air, water, noise, visual
People	Individual waste	Welfare and unemployment. Detriment to health, education, culture, social life, physical fitness
	Social failure	Crime, welfare, accidents, social disorder.
Fixed capital	Breakdowns	In utilities, communication networks, transportation, blighted areas, fires
	Shortages	In dwellings, utilities, energy, communication networks, transportation, parking, industry
Finance	High costs	Taxes, municipal debt, costs of alternate services, costs due to maintenance costs, tax burden, lack of services.

A municipality requires a data base that will allow it to respond to change, identify indicators, and plan and execute programs. The data base must not be so overwhelming that it cannot be managed. Data gathering and maintenance must be defined in realistic and attainable terms. The output must not drown the user with either too much data or data presented in an unmanageable format. On the other hand, there should not be information deficiencies.

The following illustrates the type of information the surveying and mapping sector can provide to the shared data base pool:

Land	People	Fixed capital	Finance
Location, topography, inventory of natural features	Location of events, living and working places, crime and welfare	Location, inventory of man-made features, boundaries	Location of events, taxpayers, assessment of political areas

Location is a key element in a data base, for it provides the common link for the integration of all data. The data are image data, event data, or attribute data that happened at, or can be associated with, a location. Location, therefore, allows not only for the integration of data but also for the manageable display of data. It is well accepted that thematic maps aid users in quickly and intelligently comprehending and integrating large masses of information. Location also enables data to be stored in a detailed disaggregated form at the parcel level. Although the parcel identifiers need only be unique, they must be cross-referenced to a location or geometric identifier to enable the aggregation of data for all or any part of the city.

Two ways of expressing geographic location are by reference to a geographic name or by reference to a geometric location. Geographic names include country, province, district, region, county, township, municipality, ward, division, block, plan, lot, street, block face, street number, parcel, and subparcel. Also included are planning districts, census areas, economic regions, railway subdivisions and mileages, freight areas, postal routes, and postal codes. Geometric locations, on the other hand, are specific, and all locations are mathematically compatible. Geometric identifiers include geographical coordinates (longitude and latitude), geographical grid areas, plane coordinate zones, and points or grid areas. The surveyor provides the geometric location system. By placing map data in the data base (i.e., the reference grid and the images it relates to), the surveyor can provide the user, on an *ad hoc* basis, a means to assign geometric identifiers, aggregate and integrate data, and display masses of information in meaningful ways.

A detailed examination of the data elements of the base map is required to determine the extent of detail, accuracy, and timeliness of the data. Certainly, the political boundaries of the city and the general patterns of the community (streets, parks, developed areas) are required. A general view would not require parcel boundaries in the data base. Any study of several contiguous parcels, however, would be enriched and made more manageable if various map elements, including the parcel boundaries, are included. Each of us carries a mental spatial image of our community. It is from our own vantage point that we see and perceive the world around us. A data base that capitalizes on our ability to use spatial images would lead to better decisions.

To be more specific, the inclusion of boundary data would enable the calculation of areas and other spatial data. If boundary coordinates of sufficient accuracy and stability are included, any boundary change could be precalculated and displayed, both in map form and in numerical form for field stake-out. Manmade physical features such as buildings, roadway, and utilities have a life expectancy of from 30 to 70 years, and sometimes longer. City planners and engineers continually need details concerning their features and would use terrain models to design improvements. Topographic data are used on an ongoing basis by engineers as well as by parks and conservation staff. In addition, by including boundary data, it would be possible to compare mapping data elements to various social-dissatisfaction indicators.

Map data may be at four basic levels in a city data base:

1. Precise: Boundaries. Consists of calculations and stake-out of geometric reference system, land boundaries, engineering works. These data are usually merged with level 2.
2. Detailed: Structures. Displays building parcels, boundaries, etc., for engineering design, maintenance, etc. Usually at a large scale (1:500 or 1:1000).
3. Pictorial: Structures. Displays major boundaries, buildings (as pictograms), etc., for planning and operations. Usually at a medium scale (1:5000 or 1:10 000).
4. Detailed: City. Displays political boundaries, street patterns, etc., for planning and general administration. Usually at a small scale (1:50 000 or 1:100 000).

Although the above map images could be aggregated from the parcel-level data, they could also be stored as separate data elements, updated as changes are recorded.

The responsibility for recording changes in a data base rests with the functional users, which thus ensures the accuracy and timeliness of the data. A data-base administrator would have the sole responsibility of managing the resource of data. This person would be responsible for the integrity of the whole to ensure that it truly expresses the parts and that all the parts are there. He or she would be the data-base comptroller, and would decide what to include in the data base and what data are of specific interest to a functional group and should be maintained in a separate data base. In addition, the data-base administration would arrange a data dictionary that includes a description of each data element and defines the standards to be applied and would be responsible for data gathering, accuracy, and timeliness.

City surveying and mapping systems may start as functional data bases but, because they provide the fundamental location building block and a means of displaying data in an understandable spatial form, they will become a part of the city's shared data base. Awareness of these implications must guide the development of standards and programs for mapping.

There are long-term benefits in having map data in digital rather than graphic form. A computer-supported mapping system may not be cost-effective as a functional application. However, having map data in digital form as the building block of a shared data base may be a cost-effective way of fulfilling the goals and objectives of the whole city organization.

A shared data-base system requires the adoption of an overall data-handling approach. Once data are stored in the shared data base, all systems must access it through a data-base management system (DBMS). A user-oriented language is used to request information. The DBMS uses report writers, query language, and retrieval programs to access, manipulate, and report the required information. Data of specific interest to a functional group could be in a file format that is compatible with the DBMS. The DBMS retrieval programs could then access data directly from the shared data base and the functional file. When a functional file is not designed to be compatible with the DBMS, it may still be possible to prepare functional access programs that access data both from the functional file and through the DBMS from the shared data base (see Figure 10-1).

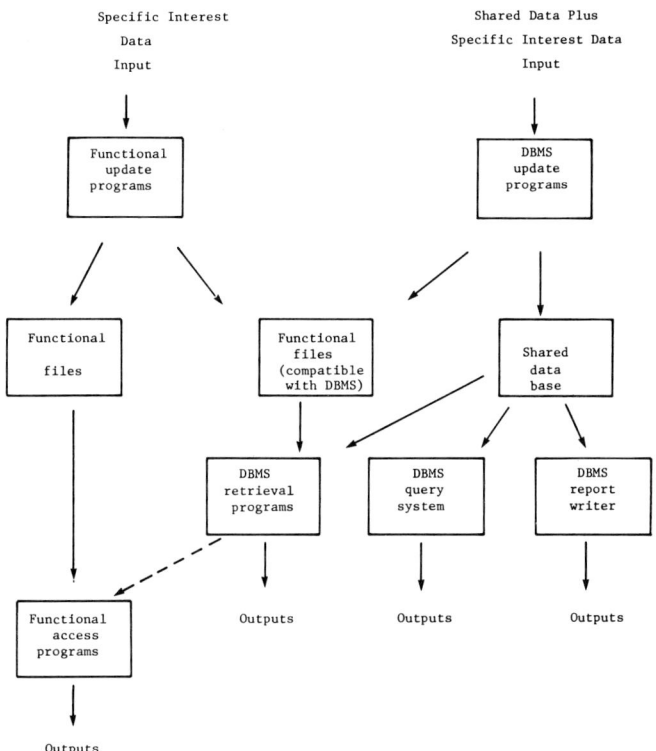

Figure 10-1. General organization of an urban data bank.

When selecting a computer-supported mapping system or when structuring functional files for surveying and mapping application, careful consideration should be given to such factors as language and data structure. Eventually, most surveying and mapping data will become part of the shared data base or will need to be merged with the shared data base to produce meaningful outputs. Not all computer-supported drafting systems will be compatible with the DBMS, and changing to a shared data-base system may prove highly impractical.

Index

Accuracy   57, 133, 134, 161
  neighboring   309
  of adjusted network   128–131
  of aerial photographs   251, 252
  of angle measurements   98–102
  of detailed surveys   201–204
  of EDM   84–90
  of horizontal control   47–49
  of utility surveys   222
  of vertical control   140–142
  photogrammetric   304, 310, 319
Accuracy preanalysis   57–69, 101,
    102
Ackermann, F. H.   281
Adjustment
  of horizontal control   121–128
    preparation of data for   122,
    123
  of vertical control   149–156
Aerial photographs, scales of
  large   261
  medium   260
  small   260
Aerial photography   247–266
  aimed   263

atmospheric conditions in   253,
    254
black and white   248, 249
characteristic curve   249
color   248, 251, 252
cost of   253
dimensional stability   251
exposure latitude   250
geometric accuracy   251, 252
image motion in   248
longitudinal overlap   260–262
range of exposure   249
resolving power   248, 251, 259
side overlap in   260–262
use of helicopters in   263
Aerial triangulation   266–283
  analogue   272
  analytical   273–283
    accuracy of   280–283
    adjustments   276–280
    maximum accuracy of   282, 283
    of models and strips   274
  analytical plotters in   270–272
  monocomparators in   270
  stereocomparators in   269

Alidade   189
Anaplot   270
Angle book   97
Angle method   97, 98
Angles, horizontal
  measurement of   93–110, 185–191
  observation equation   62, 63, 124
  reduction to the ellipsoid   112
  reduction to the projection plane
    32–35
Angles, vertical (*See* Leveling,
  trigonometric)
Angular units   46, 47
Architectural surveys,
  photogrammetric   315
Arc-to-chord correction   32–35
Area, units of   46
Automated drafting system   304–306
Automatic drafting   354
Automatic image correlators   290,
  295, 304
Automatic mapping   305
Automation in plotting   302, 304,
  356
Auto-reducing tacheometer   182, 185
Azimuth   27, 40, 53, 185
  geodetic, grid, conversion of   28,
    29, 32–35
  observation equation   62, 124

Base data file   355
Base/height ratio   260
Base map (*See* City base map)
Base number   211–213
Bauernfeind prism   186, 187
Bean, R. K.   288
Bearing angle   185, 186
Bench mark   144–146 (*See also*
  Monumentation)
Blachut, T. J.   291
Blunders (*See* Gross errors)
Bomford, G.   140
Brown, D. C.   281
Burger, A. A. J.   327
Burkhardt, R.   323

Cadastral maps   236, 243
Cadastral surveys   235–245
  basic functions of   236
  boundary definition of   8, 236
  definition of land parcels   235
  field work   235
  integration of   8, 9
  operational requirements   236, 237
  purpose of   235
Cadastre
  completeness of   236, 242
  computational   240, 242
  exaggerated accuracy of   242
  finality of   237, 242
  graphical   239, 243
  numerical   239, 243
  of utilities   222, 226–232
    idealized system of   227
  orthophotos in   311
  photogrammetric   306–314
  stereo-orthophotos in   314
  updating of   237, 242
Carman, P. D.   249
Cassini's solution (resection)   172,
  173
Catenary (*See* Sag)
Centering   95, 96, 98, 99, 190
  on towers   95, 104
Centesimal system   46
Central meridian   15
  choice of   38, 39, 41
  UTM   25
Central scale factor   26
  application of   39
Characteristic curve   249
Chi-square test   128, 129
Chord length   112
City base map   295, 330–345, 348,
  359
  accuracy of   332
  compilation of   341, 342
  content of   334
  detail presentation   332
  production cost   332
  production time   332

selection of scale   331
symbols for   335–341
City maps (*See also* City base map)
   cadastral   236, 243
   computer-drafted   354
   derived   330, 345–349, 351
   large-scale   295
   medium-scale   295
   street   227, 231, 346
   thematic   330, 351–353
   updating of   354
City survey office   2–4
   cadastral operations in   8, 9
   structural and functional
      characteristics   5
   suggested organizational models
      9–11
Clarke 1866 Spheroid   16, 17
Coefficient of refraction   113, 157,
   158
Coefficient of thermal expansion   177
Collins, S. H.   291
Collins' solution (resection)   171, 172
Computer-drafted maps   354
Computer-supported mapping system
   364
Cone angle   88
Confidence interval   128, 129
Confidence level   57, 58
Convergence, grid   27–29
Coordinatographs, portable   342,
   345
   polar   342
   rectangular   342
Covariance   57
Curvature correction, EDM, long
   distance   112–115

Dangerous circle   173, 174
Data banks   359–364
   geographic location in   361
   geometric location in   361
Data base   359
Dead fields, in photogrammetry
   262, 285

Deflection of the vertical   44
Deviation of the vertical   44
Differential rectification   286
Direction correction   32–35
Direction group   96, 103
Direction method   96, 97
Direction set   96, 97
Directions, horizontal (*See* Angles)
Direction theodolite   190
Dissatisfaction indicator   360
Distance measurements (*See also*
   Length measurements)
   eccentric   117, 118
   electromagnetic   77–93
      corrections, first-order   112–118
      corrections, lower-order   119
   observation equation   61, 123
   optical   179–185
      sources of error   184, 185
Drafting, automatic   304–306, 354
Dubuisson, B.   284, 327
Dynamic height   150

Eccentric observations
   in distance measurement   117, 118
   in horizontal control   104,
      110–112
   in polar method   207, 208
EDM instruments   77–79
   choice of   90–93
   principle of operation   80, 81
Electromagnetic distance
   measurement   77–93
   accuracy analysis   84–90
   basic principles and classification
      77–81
   choice of instruments   90–93
   determination of corrections
      82–84
Electronic data processing, of ground
   surveys   218
Ellipsoid (*See* Reference ellipsoid)
English units of length   45
Error curve   57

Error ellipse 58, 59
Error propagation 59–65, 84–87
Eternal books 231
Extension method 162–164, 192, 193, 200, 201, 205

False easting 25
False northing 25
Field identification, photogrammetric 296, 297, 311
Field records 97, 193–200
Film diapositives 251
Focusing error 100
Foot, international, in meters 46

Gauss, Carl Friedrich 14
General boundary concept 309
General information map 348
Geodetic datum 17, 39, 40
Geodetic network, national 50, 133, 134
Geographical coordinates 17
  conversion into Transverse Mercator coordinates 22, 23
  precision of 22
Geoid 43, 44
Geoid chart 44
Geometric identifiers 361
Geopotential numbers 150
Geopotential surface 149, 150
Gestalt Photo Mapper 291, 295
Glass-plate diapositives 250
Goulier prism 187
Graphical plots 316
Grid, local, establishment of 38–41
Grid convergence 27–29
Grid coordinates 25
  convention for notation 12
  transformation into another system 26
Grid north 27
Grid tables 14

Gross errors
  in control net 129
  in taping 179
Ground swing 88–90

Hamilton, A. 358
Hansen's problem 174
Hayford's ellipsoid 16
Heat island 56
Heights, systems of 149–151
Heuvelink, H. J. 98
Horizontal control
  first-order 48, 50, 51, 56, 90, 91, 93–96
  fourth-order 53
  maintenance and record keeping 134, 135
  new developments in 135
  preliminary design 56
  second-order 48, 51–53, 91–96
  third-order 48, 53, 91–96
Horizontal stadia surveying 209

Image correlators, automatic 290, 295, 304
Image deformation 282, 283
Inertial surveying 135
Infrastructure 221, 230 (See also Utilities)
Instability, of towers, tripods 101
Interactive editing system 355
International ellipsoid 16, 17
Intersection of points
  by angle measurements 162, 168, 169
  by distance measurements 162, 163
  of fourth order 53
Invar subtense bar 183, 184

Kaestner's solution (resection) 169–171

Katona, S.   225
Kłopociński, W.   332, 333
Kratky, V.   281

Lagrange, Joseph Louis   14
Lambert, Johann Heinrich   14
Land inventory, regional   314, 315
Latitude, geodetic   17, 22
Least-squares adjustment, parametric
    method of   123–128
Length measurements
    allowable tolerances of   203–205
    mechanical   175–179
    notation of   194
    optical   179–185
    (See also Distance measurements)
Leveling (See Vertical control)
Leveling error, of theodolite   99
Leveling instruments   146–148
Line-drawn maps   295
Longitude, geodetic   17, 22

Maintenance
    of city maps   350
    of horizontal control   134, 135
Map data base   360
Mapping, automatic   305
Map projection
    conformal   14
    urban, general criteria of   13, 14
Meade, B. K.   114
Mean square error   57
Metric Convention, the   45
Metric units   45, 46
Midlatitude formulas   20–22, 31, 32
Modified TM coordinates   25, 26
Modulation-frequency errors   86
Monumentation   69–77 (See also
    Bench mark)
Mountains   39, 112, 114, 115, 117

Nassar, M.   151
Natural boundaries   309

Neighboring accuracy   309
New York City   49
Normal equations   125
North American Datum 1927   17
Numbering of points   195, 211–215
Numerical map   354, 359
Numerical processing, of ground
    surveys   216–218
Numerical recordings   317

Observation equation   123, 124
Obstacles, on survey line   176, 177
Optical plumb   96, 98, 99, 190
Optical plummet   95
Optimization of network   66–69
Organizational schemes   1–11
    Central Europe   5
    Eastern Europe   5
    West Germany   5
Orthocartograph   291
Orthogonal method   162, 165, 166,
    192, 193, 200–206
Orthometric height   150
Orthophoto contour map   293
Orthophoto map   293, 315
Orthophoto mosaic   293
Orthophoto plan   293
Orthophotos   286–295, 311
    accuracy of   294
    automatic production of   290
    definition of   293
    off-line   288
    on-line   288
Overhead utilities (See Utilities)

Parallactic triangle   180
Parallel light printer   251, 282
Pęczek, L.   323
Pedal curve   58, 59
Pentagon prism   186, 187
Phase-difference error   87–90
Photogrammetric accuracy   304, 310

Photogrammetric cadastre   306–314
  establishment of additional control
    307
Photogrammetric plotting   298–306
  of contour lines   299
  of planimetry   299
  vertical accuracy in   301, 302
Photographic missions   258–266
  choice of camera   258
Photography, aerial (*See* Aerial
    photography)
Photo mosaic, controlled   286
Plane azimuth   32
Plane coordinates (*See* Grid
    coordinates)
Plate signals   94
Plot tape   355
Plotting
  automation in   302, 304, 356
  of ground surveys   215, 216
Pointing error   99, 100
Polar method   162, 167, 168, 192,
    193, 201–203, 206–211
Precision reduction tacheometer
    182, 183
Prisms   186–189
Projected grid azimuth (*See* Azimuth)
Projection corrections
  practical application of   36, 37
  Transverse Mercator   32–36
Projection zone   15, 25, 26
  crossing the boundary of   131–133
Prorated observations   108

Radian, in seconds of arc   19
Raster system   231
Reading error, of theodolite   100
Reconnaissance   53–56
Rectified photographs   284, 316
Reduction of observations   110–121
Reference ellipsoid
  definition   43
  dimensions   16
  length of meridional arc   17–22
  radii of curvature   18

Refraction, atmospheric   55, 56, 135
Refractive index   80–83
Refractive-index errors   86, 87
Reiteration theodolite   190
Repeating theodolite   190
Réseau camera   282, 283
Resection   162, 169–174
Rinner, K.   323
Rooftop marker   54, 56, 76, 77
Root mean square error   57

Sag, of tape, correction for   178
Scale correction   35, 36
Scale factor
  definition   14
  of modified systems   30
  practical computation of   31, 32
  Transverse Mercator   29, 30
Schriever, H.   5
Sea-level arc, reduction to the
    ellipsoid   117
Sea level, mean   43
  reduction to   116, 119
Service map   350, 351
Sexagesimal system   46
Sima, J.   296, 298, 335
Slama, C. C.   281, 283
Slope correction   115, 116, 119, 175,
    177, 178
Slope length, geodetic   112
Speed of light, in vacuo   80, 85
Spheroid (*See* Reference ellipsoid)
Stadia surveying   208, 209
Standard deviation   57, 58
Standard error   57
State Plane Coordinate System   40
Station adjustment
  of complete sets   104–107
  of incomplete sets   108–110
Stereocompiler   291, 292, 295
  plotting on   292
Stereographic projection   14, 15
Stereomate   293

Stereo-orthophoto map   293
Stereo-orthophoto pair   293
Stereo-orthophotos   290, 314
  vertical accuracy from   295
Stereo-orthophoto system   291
Street map   227, 231, 346
Strip maps   227, 232
Subdivided direction groups   108
Sum equation   124
Supporting map   332, 346
Survey marker   70–77
  symbols for presentation   195, 196
  (See also Bench mark)

Tacheometers
  electronic   93, 199, 200, 209, 210
  optical   181–183
Targeting of points, photogrammetric
    263
  target forms   264
  target size   265
Technical city surveying   9
Thematic maps   330, 351–353
Theodolites   93, 94, 189–191
  handling of   102, 103
Thermal expansion, of tapes   177
Topographical maps   295, 350
Traffic studies, photogrammetric
    325
Transit (See Theodolites)
Transverse Mercator coordinates   15
  conversion into geographical
    coordinates   24, 25
  translation into grid coordinates
    25
Traverse
  in control surveys   51–53, 56,
    67–69
  in detailed surveys   162, 166, 167
  instruments and signals   91–96
  misclosure   119–121
Traversing, with wall monumentation
    72–76

Triangulation   50–52
  vs. triangulateration   66–69
  (See also Aerial triangulation)
Trigonometric functions, calculation
    of   19
Trigonometric leveling   157–159

Underground utilities (See Utilities)
Units, angular   46, 47
Units of area   46
Units of length   45, 46
Universal Transverse Mercator
    coordinates   16, 25, 26
Updating of maps   354
U.S. Geological Survey   286
U.S. National Ocean Survey   283
Utilities   221–233
  branch materials on   225
  cadastre of   222, 226–232
  electronic locators of   225, 226
Utility surveys   221–233
  accuracy of   222
  methods for   222–226
  (See also Utilities)
UTM (See Univeral Transverse
    Mercator)

Vanicek, P.   151
Vapor pressure   82
Variance   57
Variance-covariance matrix   54,
    59–61, 130, 131, 135
Variance factor   128, 129
Vertical angle measurements   157,
    158
Vertical control   139–160
  accuracy of   140–142
  computations and adjustment of
    149–156
  design of   142–144
  monumentation   144–146

Vertical pointing accuracy,
    photogrammetric   300–302
Vertical stadia surveying   208, 209

Wall marker   54, 72–76
Weighted mean   108
Weight matrix   61

Weights, unequal   109
Witness marker   69
Wollaston prism   187, 188

Zero correction   81, 83, 84
Ziemann, H.   283